Here are your

# 2000 SCIENCE YEAR
# Cross-Reference Tabs

## For insertion in your WORLD BOOK

Each year, SCIENCE YEAR, THE WORLD BOOK ANNUAL SCIENCE SUPPLEMENT, adds a valuable dimension to your WORLD BOOK set. The Cross-Reference Tab System is designed especially to help you link SCIENCE YEAR's major articles to the related WORLD BOOK articles that they update.

**How to use these Tabs:**

First, remove this page from SCIENCE YEAR.

Begin with the first Tab, **Aging**. Take the A volume of your WORLD BOOK set and find the **Aging** article. Moisten the **Aging** Tab and affix it to that page.

Glue all the other Tabs to the corresponding WORLD BOOK articles.

# SCIENCE YEAR 2000

## The World Book Annual Science Supplement

A review of science and technology during the 1999 school year

**World Book, Inc.**
a Scott Fetzer company
Chicago

**www.worldbook.com**

# THE YEAR'S MAJOR SCIENCE STORIES

From new advances in cloning to the start of construction on the International Space Station, it was an eventful year in science and technology. On these two pages are the stories the editors chose as the most memorable or important of the year, along with details on where to find information about them in the book.

**Evidence for an everlasting—and accelerating—universe**
Observations reported by two groups of American astronomers in late 1998 indicate that the expansion of the universe will go on forever, and at an ever-increasing rate. That unexpected finding will have to be confirmed by further research. In the Special Section, see WHAT IS THE ULTIMATE FATE OF THE UNIVERSE?

**A feathered dinosaur?** ▲
Scientists reported in June 1998 that a creature named *Caudipteryx zoui*, a fossil of which was found earlier in China, had feathers and may have been a dinosaur ancestor of birds. The finding rekindled a debate about the origin of birds. In the Special Reports section, see FOSSILS, FEATHERS, AND THEORIES OF FLIGHT.

**Building the International Space Station**
The construction of the International Space Station began in December 1998 with the connection of two modules, one built in Russia (at left in photo), the other in the United States. In the Science News Update section see SPACE TECHNOLOGY. ▶

World Book, Inc.
525 W. Monroe
Chicago, IL 60661

ISBN: 0-7166-0550-3
ISSN: 0080-7621
Library of Congress Catalog Number: 65-21776
Printed in the United States of America.

**Off to explore an asteroid** ▶

In October 1998, the National Aeronautics and Space Administration launched an advanced space probe called Deep Space 1 toward an asteroid, 1992 KD. The probe was the first to be propelled by an engine that generates thrust with a beam of *ions* (electrically charged atoms). Deep Space 1 was to make a close pass by the asteroid in July 1999 and then perhaps visit two comets. In the Science News Update section, see SPACE TECHNOLOGY.

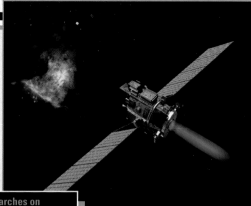

**An ancient burial in Mexico**

In October 1998, archaeologists working at the Pyramid of the Moon in the ancient city of Teotihuacan near Mexico City reported finding a tomb, dating from about A.D. 100, containing a skeleton and many valuable artifacts. The skeleton was of a person who was apparently sacrificed, perhaps as part of a royal burial or a building-dedication ritual. In the Science News Update section, see ARCHAEOLOGY.

**Cloning marches on**

Researchers at the University of Hawaii announced in July 1998 that they had created more than 50 *clones*—exact genetic duplicates—of adult mice from the mice's body cells. This achievement confirmed a cloning technique reported by Scottish scientists in 1996. In the Science News Update section, see GENETICS.

**Evidence that neutrinos have mass**

Scientists in Japan working at a huge underground facility designed to detect tiny, fast-moving particles called neutrinos, reported in June 1998 that the mysterious particles may have a very small amount of mass. That finding, if confirmed, could have an impact on many areas of science. In the Science News Update section, see PHYSICS. ▼

**The Millennium Bug**

As the year 2000 approached, computer experts were concerned about old programs that denote just the first two digits of a year, thus making it impossible for a computer to distinguish between dates in the 1900's and ones in the 2000's. Predictions varied widely about the consequences awaiting companies and countries that would not be able to fix the problem before the year 2000 arrived. In the Science News Update section, see COMPUTERS AND ELECTRONICS (Close-Up).

**Human stem cells isolated**

Researchers at two U.S. institutions reported in November 1998 that they had isolated human stem cells from embryos. Scientists in Baltimore later announced that they had extracted stem cells from adult bone marrow. The cells, which can grow into any type of tissue, could have important medical applications. In the Special Reports section, see HUMAN TISSUE ENGINEERING—FROM SCIENCE FICTION TO MEDICAL FACT.

# CONTENTS

Page 38

Page 132

# SCIENCE NEWS UPDATE .................. 176

Twenty-eight articles, arranged alphabetically, report on the year's most important developments in all major areas of science and technology, from *Agriculture* to *Space Technology*. In addition, five Close-Up articles focus on especially noteworthy developments:

Page 189

# SCIENCE YOU CAN USE .................. 285

Five articles present various topics in science and technology as they apply to the consumer.

Page 286

# WORLD BOOK SUPPLEMENT .......... 305

Seven new or revised articles from the 1999 edition of *The World Book Encyclopedia:* **Astronomy; Universe; Planet; Galaxy; Solar energy; Solar system,** and **Carbon.**

# INDEX .......... 337

A cumulative index of topics covered in the 2000, 1999, and 1998 editions of *Science Year.*

# CROSS-REFERENCE TABS

A tear-out page of cross-reference tabs for insertion in *The World Book Encyclopedia* appears before page 1.

# STAFF

# EDITORIAL ADVISORY BOARD

# CONTRIBUTORS

**Asker, James R.,** B.A.
Washington Bureau Chief,
*Aviation Week & Space Technology*
magazine.
[Special Report, *The Wireless World;
Space Technology*]

**Baars, Bernard J.,** Ph.D.
Professor,
Wright Institute.
[Special Section, *What is the Nature
of Consciousness?*]

**Black, Harvey,** B.S., M.S., Ph.D.
Free-Lance Writer.
[Science You Can Use, *How to
Protect Fine Woods in Your Home*]

**Bolen, Eric G.,** B.S., M.S., Ph.D.
Professor,
Department of Biological Sciences,
University of North Carolina at
Wilmington.
[*Conservation*]

**Brett, Carlton E.,** M.S., Ph.D.
Professor,
Department of Geology,
University of Cincinnati.
[*Fossil Studies*]

**Brody, Herb,** B.S.
Senior Editor,
*Technology Review.*
[Science You Can Use, *Biometric
Identification: Is That Really You?*]

**Cain, Steven A.,** B.A.
Communication Specialist,
Purdue University School of
Agriculture.
[*Agriculture*]

**Chiras, Dan,** B.A., Ph.D.
Adjunct Professor,
Environmental Policy and
Management Program,
University of Denver.
[*Environmental Pollution*]

**Cruz-Uribe, Kathryn,** B.A., M.A.,
Ph.D.
Professor of Anthropology,
Northern Arizona University.
[*Anthropology; Anthropology* (Close-
Up)]

**Ferguson, Edwin L.,** Ph.D.
Associate Professor,
Department of Molecular Genetics
and Cell Biology,
University of Chicago.
[Special Section, *How Does an
Organism Develop from a Single
Cell?*]

**Ferrell, Keith**
Free-Lance Writer.
[*Computers and Electronics;* Science
You Can Use, *Digital Cameras Gain in
Popularity and Utility*]

**Finkelstein, David B.,** Ph.D.
Science Writer.
[Special Section, *Why Do We Age
and Die?*]

**Graff, Gordon,** B.S., M.S., Ph.D.
Free-Lance Science Writer.
[*Chemistry*]

**Hay, William W.,** B.S., M.S., Ph.D.
Professor of Paleooceanology,
Geomar, Christian-Albrechts
University,
Kiel, Germany.
[*Geology*]

**Haymer, David S.,** M.S., Ph.D.
Professor,
Department of Genetics and
Molecular Biology,
University of Hawaii.
[*Genetics; Genetics* (Close-Up)]

**Hester, Thomas R.,** B.A., Ph.D.
Professor of Anthropology and
Director,
Texas Archeological Research
Laboratory,
University of Texas at Austin.
[*Archaeology*]

**Horowitz, Gary T.,** Ph.D.
Professor of Physics,
University of California at Santa
Barbara.
[Special Section, *What is the
Fundamental Nature of Space?*]

**Johnson, Christina S.,** B.A., M.S.
Science Writer, *North County Times*
[*Oceanography; Oceanography*
(Close-Up)]

**Kirshner, Robert P.,** A.B., Ph.D.
Professor of Astronomy,
Harvard-Smithsonian Center for
Astrophysics.
[Special Section, *What is the Ultimate
Fate of the Universe?*]

**Koka, Rahul,** A.B.
Research Technician,
Laboratory for Tissue Engineering and
Organ Fabrication,
Massachusetts General Hospital,
Boston.
[Special Report, *Tissue Engineering–
From Science Fiction to Medical Fact*]

**Kolberg, Rebecca.,** B.A., M.S.
Editorial Supervisor,
Health Week.
[Science You Can Use, *Perfumes and
Colognes: Romance in a Bottle*]

**Kowal, Deborah,** M.A.
Adjunct Assistant Professor,
Emory University Rollins School of
Public Health.
[*Public Health*]

**Limburg, Peter R.,** B.A., M.A.
Free-Lance Writer
[Special Report, *Deep into the Past* ]

**Lunine, Jonathan I.,** B.S., M.S.,
Ph.D.
Professor of Planetary Science,
University of Arizona Lunar and
Planetary Laboratory.
[*Astronomy*]

**Maran, Stephen P.,** B.S., M.A., Ph.D.
Press Officer,
American Astronomical Society and
Editor,
*The Astronomy and Astrophysics
Encyclopedia.*
[Special Report, *How the Moon Was
Born*]

**March, Robert H.,** A.B., M.S., Ph.D.
Professor of Physics and Liberal
Studies,
University of Wisconsin at Madison.
[*Physics*]

**Marschall, Laurence A.,** B.S., Ph.D.
Professor of Physics,
Gettysburg College.
[*Books About Science*]

**Martin, Larry D.,** B.S., M.S., Ph.D.
Professor of Ecology and Evolutionary
Biology and Curator of Vertebrate
Paleontology,
Natural History Museum,
University of Kansas.
[Special Report, *Fossils, Feathers, and
Theories of Flight* ]

**Maugh, Thomas H., II,** Ph.D.
Science Writer,
*Los Angeles Times.*
[*Biology; Biology* (Close-Up)]

**McKenna, James T.**
Transport and Safety Editor,
*Aviation Week & Space Technology*
magazine.
[Special Report, *Turbulence: Hidden
Threat in the Skies*]

**Moser-Veillon, Phylis B.,** B.S., M.S.,
Ph.D.
Professor,
Department of Nutrition and Food
Science,
University of Maryland.
[*Nutrition*]

**Murphy, Michael J.,** M.D., M.P.H.
Assistant Psychiatrist,
McLean Hospital, and
Instructor,
Harvard Medical School
[*Psychology*]

**Reed, Michael,** B.S.
Free-Lance Writer.
[Science You Can Use, *Aquariums,
Living Jewels of Light and Motion*]

**Riley, Thomas N.,** B.S., Ph.D.
Professor,
School of Pharmacy,
Auburn University.
[*Drugs*]

**Roberts, Kathryn R,** B.A., M.A.,
Ph.D.
Vice President of Programs at the
Minneapolis Foundation in Minnesota
and a former director of the
Minnesota Zoo.
[Special Report, *What's New at the
Zoo?*]

**Sforza, Pasquale M.,** B.Ae.E., M.S.,
Ph.D.
Program Director,
Graduate Engineering and Research
Center,
University of Florida.
[*Energy*]

**Snow, John T.,** B.S.E.E., M.S.E.E.,
Ph.D.
Dean,
College of Geosciences, and
Professor of Meteorology,
University of Oklahoma.
[*Atmospheric Sciences*]

**Snow, Theodore P.,** B.A., M.S.,
Ph.D.
Professor of Astrophysics and
Director,
Center for Astrophysics and Space
Astronomy,
University of Colorado at Boulder.
[*Astronomy*]

**Tamarin, Robert H.,** B.S., Ph.D.
Dean of Sciences,
University of Massachusetts.
[*Ecology*]

**Teich, Albert H.,** B.S., Ph.D.
Director,
Science and Policy Programs,
American Association for the
Advancement of Science.
[*Science and Society*]

**Trefil, James,** B.A., Ph.D.
Clarence J. Robinson professor of
physics,
George Mason University.
[Special Section, *A Thousand Years of
Discovery*]

**Trubo, Richard,** B.A., M.A.
Free-Lance Writer.
[*Medical Research*]

**Vacanti, Joseph P.,** M.D.
John Homans Professor of Surgery,
Harvard Medical School.
[Special Report, *Tissue Engineering–
From Science Fiction to Medical Fact*]

**Wolberg, Donald L.,** B.A., Ph.D.
Chief Executive Officer and President
of Natural History Development
Company, Inc.
[Special Section, *What Has Caused
Mass Extinctions?*]

**Wright, Andrew G.,**.B.A.
Associate Editor,
*Engineering News-Record.*
[Engineering]

Feature articles take an in-depth look at significant and timely subjects in science and technology.

••••••••••••••••••••••••••••••••••••••••••••••••••••

Page 84

Page 12

Page 98

Sophisticated technology is enabling archaeologists to conduct research at deep-water sites that had long been out of reach to them.

# Deep into the Past

By Peter R. Limburg

ruising the waters of the Mediterranean Sea in September 1998, a team of investigators from Odyssey Marine Exploration in Tampa, Florida, realized from their sonar readings that there was something worth checking out on the sea floor. The group had been searching for the remains of a British warship from the 1600's that was loaded with valuable coins, but it was about to discover something that most archaeologists would consider even more valuable. Seated in the research ship's control room, a pilot guided a *remotely operated vehicle* (ROV), an unmanned robotic minisubmarine, closer to the mysterious object. As the ROV approached the sea floor at a depth of more than 900 meters (3,000 feet), its video camera revealed the hull of an ancient ship lying half buried in sediments. A closer look showed some of the ship's cargo—dozens of *amphorae*, clay vessels that served as all-purpose containers. When archaeologists studied the videotapes later, they determined from the styles of the amphorae that they were about 2,500 years old.

Based on their preliminary investigation, the archaeologists concluded that the vessel was built by the Phoenicians, a seafaring people who

lived on the coast of what today is Israel, Lebanon, and Syria. By about 1,000 B.C., the Phoenicians had become famous as traders throughout the Mediterranean region and beyond. In spite of their renown as sailors, however, archaeologists had found little evidence of their ships. The discovery of the ship, which the exploration team named the *Melkarth* after a Phoenician god, was welcome news for archaeologists.

The *Melkarth* was not the first ancient ship to be discovered in deep water. In 1988, U.S. oceanographer Robert D. Ballard discovered an ancient Roman ship in the Mediterranean at a depth of about 800 meters (2,600 feet). During further investigations in the same region from 1988 to 1997, Ballard and his colleagues at the Institute for Exploration (IFE) in Mystic, Connecticut, discovered four more Roman shipwrecks, as well as a ship with Islamic artifacts dating from the 1700's or 1800's and two sailing vessels from the 1800's.

Archaeologists can learn much from sunken ships like these. Every ship on the sea floor is a snapshot of the past, a glimpse of life as it was the moment the vessel went down. Researchers can study *artifacts* (ob-

Powerful lights on a remotely operated vehicle illuminate the wreckage of an ancient Roman ship resting at a depth of about 800 meters (2,600 feet) on the floor of the Mediterranean Sea.

## Terms and concepts

**Amphora:** A clay jar that was used for storing a variety of substances in ancient times.

**Artifact:** An object made by human hands.

**Magnetometer:** A magnetic device used to detect iron objects.

**Remotely operated vehicle (ROV):** An unmanned, robotic minisubmarine.

**Sonar:** A device that uses sound waves to locate underwater objects and create images of them.

**Sub-bottom profiler:** A sonar device that penetrates the sea floor to locate hidden objects.

jects made by human hands) for clues to what people considered the basic necessities for life on board a ship. They can retrace ancient trade routes by examining the ship's cargo to determine who traded what with whom. They can also learn how ships were constructed and what kinds of wood were used by shipbuilders of different regions. And researchers can piece together a more complete picture of seafaring skills and technology by combining the findings of underwater research with the depictions of maritime activity in ancient works of art.

But underwater archaeology is concerned with more than just shipwrecks. For example, many former coastal settlements are also hidden beneath the sea, and modern dams and reservoirs have submerged the remains of many past cultures.

Archaeologists began to explore underwater sites in the 1950's, but they were restricted to shallow coastal water, lakes, and rivers—areas that could be reached by human divers. Various conditions determine the depth and duration of a dive with scuba gear, but, in general, underwater archaeological research with human divers has been limited to depths of no more than 50 meters (160 feet). Since the 1980's, however, advanced deepwater technology—such as ROV's, sophisticated research submarines, new kinds of sonar devices, and video cameras—has opened vast new underwater regions to archaeology. As the quality of deepwater investigative tools improved during the 1990's, archaeologists expanded their explorations and brought research on shipwrecks, historical sea battles, and submerged settlements into the mainstream of archaeological studies.

### Archaeology in the strange world of the deep

Deepwater archaeology has several advantages over explorations at lesser depths. Many shallow shipwrecks, like the tombs of ancient kings on land, have been plundered by treasure hunters. In some areas, fishing boats hauling large nets behind them have disrupted sites of archaeological interest. Natural processes, too, can ruin shallow sites: Storms and currents often break up ships' hulls, and sediments deposited by currents can quickly bury both a ship and its cargo. In deep water, on the other hand, human influences are usually absent, there is little current, and sedimentation is light. Moreover, the near-freezing temperatures and low concentration of oxygen in the deep retard corrosion and limit the presence of organisms, such as worms and bacteria, that contribute to the deterioration of artifacts in shallow areas.

In spite of these advantages, deepwater sites present difficult challenges to researchers. One is the sheer pressure of the water. For every 10 meters (33 feet) of depth, the pressure increases by 1.034 kilograms per square centimeter (14.7 pounds per square inch). At the depth of the *Melkarth*, the pressure of the water is 90 kilograms per square centimeter (1,300 pounds per square inch). At 3,800 meters (12,500 feet), the approximate depth at which the luxury liner *Titanic* rests, the pressure is a mind-boggling 390 kilograms per square centimeter (5,600 pounds per square inch).

**The author:**
Peter R. Limburg is a free-lance writer and the author of many science books for children and young adults.

**50 meters (160 feet)**
• Depth limit of diver with normal scuba gear for about 20 minutes.

**90 meters (300 feet)**
• *Hamilton* and *Scourge,* U.S. warships lost in Lake Ontario in 1813.
• Only a dim, blue light penetrates clear water.

**300 meters (1,000 feet)**
• Depth limit of NEWTSUIT.
• Water pressure is about 30 kilograms per square centimeter (450 pounds per square inch).

**600 meters (2,000 feet)**
• Human eye sees pitch black.

**900 meters (3,000 feet)**
• *Melkarth,* a Phoenician ship lost more than 2,500 years ago in the Mediterranean Sea.
• Water pressure is 90 kilograms per square centimeter (1,300 pounds per square inch).

**3,800 meters (12,500 feet)**
• *Titanic,* British passenger ship that sank in the North Atlantic Ocean in 1912.
• Water pressure is 390 kilograms per square centimeter (5,600 pounds per square inch).

**5,200 meters (17,000 feet)**
• *Yorktown,* U.S. aircraft carrier sunk by Japanese war planes in the South Pacific Ocean in 1942.
• Water pressure is 540 kilograms per square centimeter (7,600 pounds per square inch).

# The challenges of deepwater research

The two greatest challenges to archaeological research in deep water are the tremendous water pressure and darkness. As scientists designed equipment that could meet these challenges, they discovered many shipwrecks and historical sites that had previously been inaccessible.

A diver wearing a NEWTSUIT investigates the *Edmund Fitzgerald,* an iron-ore carrier that sank in Lake Superior in 1975. The NEWTSUIT is one of several armored, pressure-controlled suits that help scientists work safely at depths of as much as 300 meters (1,000 feet). Below that depth, manned or robotic submarines must be used.

These extreme pressures, of course, make such wrecks as *Melkarth* inaccessible to human divers. Even with an armored and pressure-controlled suit called an atmospheric diving system, a diver can descend no more than 300 meters (1,000 feet). But the severe water pressure can also damage ROV's and other mechanical equipment. If seals on underwater equipment break, seawater floods and damages electric motors, computer circuitry, and cameras.

The other major challenge of deepwater work is darkness. At a depth of about 90 meters (300 feet), only a dim, blue fraction of sunlight penetrates even the clearest water. If the water is full of sediment, visibility is cut off even sooner. At 600 meters (2,000 feet), water is pitch black, and high-intensity lights must be used for visibility and photography.

One of the most important tools for penetrating this darkness is sonar, short for *so*und *n*avigation *a*nd *r*anging. This device uses sound waves rather than light waves to "see" objects. The *Titanic* disaster of 1912 in the North Atlantic Ocean prompted the development of the technology so that ships could locate and navigate around icebergs at night or in fog. During World War I (1914-1918), British and French ships began using sonar to locate German submarines. Since then, scientists have made sonar a highly effective research tool.

## Searching for underwater sites

An underwater investigation often begins with scientists towing an underwater sonar device back and forth over a wide area, *right*. Sonar, or *sound navigation and ranging*, emits pulses of sound that travel through the water. The sound waves bounce off objects and return as echoes to a receiver. The elapsed time between the transmission of a pulse and the reception of its echo is translated into a measurement of distance.

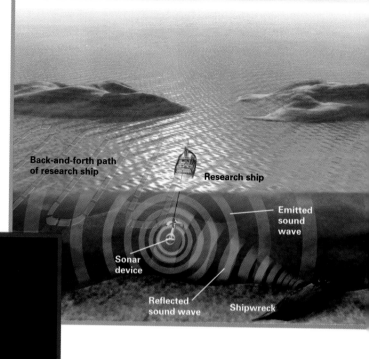

Back-and-forth path of research ship

Research ship

Emitted sound wave

Sonar device

Reflected sound wave

Shipwreck

A series of many distance measurements can be converted into a picture of an object on the sea floor. A sonar image of the *Scourge,* a U.S. warship that sank in Lake Ontario in 1813, shows the vessel's contours. The colors of the shipwreck are "false colors," computer-generated hues that make the sonar image easy to study.

## Seeing in the dark

Sonar works by sending out pulses of sound that travel through the water. When the sound waves hit an obstacle, they bounce back. The reflected sound is picked up by a receiver, and the sound energy is converted into an electrical signal that is translated into a measurement of distance. In water, sound travels about 1.6 kilometers (1 mile) per second. Thus, a time gap of two seconds from the sending of a pulse of sound until it is received back at the source means that the object reflecting the sound is 1.6 kilometers away—the sound wave took one second to reach the object and one second to return. A continuous series of distance measurements can be displayed as an image on paper or a computer screen. The result is a contour map of the sea floor. An *anomaly* (unusual feature) on the contour map could be a rocky outcrop, a collection of fuel drums tossed overboard from a freighter, or a shipwreck.

Researchers can manipulate sonar signals for different purposes. The various frequencies of sound waves emitted by sonar are measured in *kilohertz* (thousands of cycles per second). Most people can hear frequencies ranging from 0.02 to 20 kilohertz. A standard frequency setting for a sonar device is 30 kilohertz. At this frequency, sound waves can travel up to 2,500 meters (8,200 feet), but the resulting image is fuzzy. When a promising feature is sighted, the researchers will investigate the anomaly with a higher-frequency sound wave. For example, 100-kilohertz sound waves travel only about 400 meters (1,300 feet), but they produce a far clearer picture, making it possible to distinguish between a sunken ship and an outcrop of rock.

Researchers can apply sonar technology in a variety of ways. Side-scan sonar, for example, is a small underwater device that is towed by a research vessel and emits relatively low frequencies of sound across a wide path. This type of sonar allows scientists to conduct a general underwater survey. Another variation of sonar technology is the sub-bottom profiler, which uses sound waves at frequencies between 5 and 20 kilohertz to probe beneath the sea floor. The sound waves penetrate the bottom sediments, creating images of narrow vertical cross sections of sediment, like thin slices of a layer cake. This technology is particularly useful for identifying buried objects in underwater archaeological sites. Sonar devices also act as the "eyes" of ROV's to help a pilot in a surface research vessel steer an underwater vehicle.

Researchers can also explore the darkness of the deep sea or beneath the sea floor with a device called a magnetometer. This sensitive device measures the magnetic field of the Earth and can detect slight distortions of the field caused by the presence of iron. Consequently, the magnetometer can help locate sunken artifacts such as anchors.

## Underwater vehicles

Once an object of interest has been identified, researchers must go in for a closer look. For that, they use vehicles called submersibles. These fall into two general categories: manned and unmanned. Some manned submersibles are relatively small vehicles that hold just one or two people, along with cameras, lights, and other research equipment. Others are large, sophisticated research submarines.

One such vessel is the United States Navy's nuclear-powered research submarine NR-1. This sub measures about 46 meters (150 feet) in length and 3.7 meters (12 feet) in diameter and carries a crew of five plus two scientists. NR-1 can operate safely to a depth of 725 meters (2,375 feet) and remain submerged indefinitely, except to take on fresh supplies. The sub has three viewports that allow researchers to look at sites directly below the craft. Its equipment includes very powerful sonar devices, video and still cameras, and a manipulator arm that can be fitted with a variety of tools for cutting into or picking up objects on the sea floor.

But the real stars of deepwater research are the unmanned submersibles. Some of these are towed survey vehicles equipped with sonar, cameras, video equipment, and lights. These towed vehicles also serve as a dock for the more sophisticated unmanned submersibles, the ROV's. With a pilot on the research ship steering the ROV, the scientists can maneuver a large array of investigative tools around an archaeological site. Some ROV's are also equipped with manipulator arms that can recover objects as fragile as a china teacup or as heavy as a 15-kilogram (33-pound) hunk of iron. All unmanned submersibles are connected to the research vessel with an *umbilical*, a cable that links the submerged equipment to the computers and video monitors in the ship's control room.

The entire range of technology—from the farthest-reaching sonar to the most sophisticated manipulator arm—must operate together as a coordinated whole. Any one of these devices has limited use without the others. Therefore, archaeologists must plan their research carefully to ensure that they make maximum use of the tools at their disposal.

**Satellite**

# Investigating and documenting an underwater site

After researchers discover a shipwreck or other object of interest, they use a variety of high-tech devices to investigate the site and make a record of everything they find.

## Research ship

The research ship is the control center for the entire investigation. The scientists use the Global Positioning System (GPS), a satellite network in space, to determine and record the exact position of the shipwreck.

## Elevator

An "elevator," rigged with flotation devices, is often used to carry artifacts, collected on the sea floor by a remotely operated vehicle (ROV), up to the surface.

## Umbilical

The umbilical is a cable that connects all the underwater equipment to the computers and monitoring devices in the control room of the research ship.

## Towed submersible

A towed submersible, often called a sled, carries many pieces of equipment, such as lights, sonar, and video and still cameras. These submersibles sometimes serve as "garages" for ROV's.

### Debris field of a shipwreck

## Transponder

Transponders are sonar devices placed around an underwater site. Each transponder receives and sends back a particular frequency of sound wave. The different signals produced by the transponders enable researchers to always know where the ROV is in relation to a site and to navigate it through the site with precision.

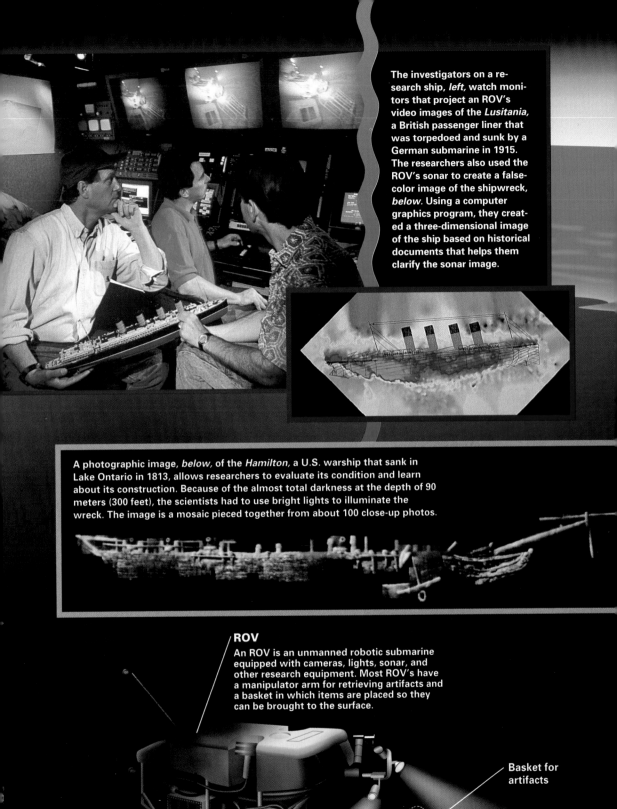

The investigators on a research ship, *left,* watch monitors that project an ROV's video images of the *Lusitania,* a British passenger liner that was torpedoed and sunk by a German submarine in 1915. The researchers also used the ROV's sonar to create a false-color image of the shipwreck, *below.* Using a computer graphics program, they created a three-dimensional image of the ship based on historical documents that helps them clarify the sonar image.

A photographic image, *below,* of the *Hamilton,* a U.S. warship that sank in Lake Ontario in 1813, allows researchers to evaluate its condition and learn about its construction. Because of the almost total darkness at the depth of 90 meters (300 feet), the scientists had to use bright lights to illuminate the wreck. The image is a mosaic pieced together from about 100 close-up photos.

## ROV

An ROV is an unmanned robotic submarine equipped with cameras, lights, sonar, and other research equipment. Most ROV's have a manipulator arm for retrieving artifacts and a basket in which items are placed so they can be brought to the surface.

Basket for artifacts

Manipulator arm

Shipwreck

## Conducting a deepwater investigation

When archaeologists want to locate a specific shipwreck, they begin by researching old naval records, historical accounts, and other sources, such as maps and charts. If researchers are interested in finding shipwrecks for which there are no written records, they check navigation charts for dangerous shoals and reefs where shipwrecks were likely to occur. Even when scientists are not looking for a specific site and simply want to find out what is located on a particular part of the sea floor, they will make highly detailed plans for their expedition.

The investigation of the selected area usually begins with the research ship towing side-scan sonar in a back-and-forth pattern known as "mowing the lawn" to create a low-resolution map of the sea floor. In the control room of the research vessel, video monitors display the contour map being generated by the reflected sonar signals. As the images appear, the scientists observe them carefully, watching for anomalies. If the scientists are looking for a ship, they keep a sharp lookout for bumps on the map that could signify debris. When a ship breaks up as it is sinking, cargo and pieces of the ship itself fall free of the hull and get carried away by currents. The heaviest objects sink closest to the wreck, and lighter ones settle farther away, forming a debris trail.

Once researchers have found a shipwreck—or some other site of interest—they send down an ROV and other submersibles loaded with equipment to conduct a preliminary investigation. Shorter-range sonar on an ROV can reveal more details about the wreck. Powerful floodlights illuminate parts of the wreck, allowing the researchers to take photos that can be processed later in the ship's darkroom or to view the scene with digital cameras or video recorders that immediately transmit electronic images to control-room monitors.

## Documenting and excavating a deepwater site

After completing the preliminary survey, the archaeologists begin documenting the site. This process usually begins with the use of an ROV to place transponders on the sea floor around the site. A transponder is an electronic device that receives sonar signals from the research ship and sends them back. Each transponder is designed to work at a different frequency so that the topside crew can identify it and use it to calculate the research ship's position in relation to the wreck. The transponders also act as electronic guide posts that help the pilot of an ROV maneuver the vehicle around the site.

The scientists record the exact position of the shipwreck on a navigational map using the research ship's Global Positioning System (GPS) equipment. The GPS is a collection of satellites, owned by the U.S. Department of Defense, that orbit the Earth at an altitude of about 17,500 kilometers (10,900 miles). The computerized GPS equipment on the research ship receives signals from several of the satellites and converts the information into a measurement of distance from the satellites. With that information, the research ship's GPS equipment determines the vessel's exact location on the globe. The calculation is important because it helps the ship's navigator keep the vessel directly over the shipwreck or return to the same location at a later time. Recording the precise location of a shipwreck can also aid future investigations. For example, by knowing the location of several shipwrecks in a particular area, scientists can "connect the dots" to discover possible trade routes in times past.

After researchers have surveyed a site and recorded its location, they begin the painstaking process of documenting it in detail. The archaeologists study close-up video and still images of the site to determine the condition of the wreck and its cargo and in some cases to identify the ship. For example, a particular feature, such as the figurehead on a ship's bow, may provide a vessel's identity, while the presence of cannons could possibly identify it as a warship.

The researchers also create a photographic image of the entire wreck. Because a site is usually too large to be captured in a single photograph, still cameras on an ROV shoot dozens of overlapping photographs of the site. Technicians later piece the photos together into a single image known as a mosaic.

To supplement the visual images, researchers might also use magnetometers or sub-bottom profilers to locate the parts of the ship that are hidden beneath the sea floor. They will then have extensive knowledge of the site and everything it contains.

Using the entire collection of visual images and the data from their instruments, the researchers create a detailed map of the wreck site and the location of every artifact. If an item is later retrieved from the sea floor, the scientists know exactly where it came from—a clue that could be important in understanding how a ship's cargo was stored, how the vessel was constructed, or why it sank.

After archaeologists have carefully documented an underwater site, they usually select certain artifacts to bring to the surface. An ROV usually has a retrieval basket, enabling it to carry light loads up to the surface ship. With heavier artifacts, the researchers may use the ROV to place the objects in an *elevator*, a sturdy metal basket with a flotation device attached to it. The elevator is carried down to the bottom with a heavy iron weight. When the basket is full, the ROV pilot sends out a sonar signal that releases the weight, and the elevator floats to the surface.

## Saving artifacts that have soaked for centuries

As soon as an object is brought to the surface, it gets special treatment. Nearly all artifacts that have spent a long time underwater have been damaged by the ravages of time and nature. Even though artifacts in deepwater sites usually deteriorate more slowly than those found in shallow sites, corrosion and other forms of damage still occur. Researchers must rely on the science of conservation, a variety of methods for cleaning and stabilizing artifacts to prevent objects from deteriorating further. Archaeologist George F. Bass of Texas A&M University in College Station, one of the fathers of underwater archaeology, has written that for every month of work at an underwater site, there are two years of conser-

## Glimpses of the past

Archaeologists have revealed a glimpse of past civilizations and maritime history through their discoveries of well-preserved shipwrecks on the floor of oceans, seas, and lakes.

Clay vessels, *above,* from the wreckage of a Phoenician ship in the Mediterranean Sea offer clues about ancient traders. The figurehead of the *Scourge,* a U.S. battleship, *right,* rises from the floor of Lake Ontario. The ship, which sank in 1813, has provided researchers with details about shipbuilding practices.

Researchers examine sections of marble columns, *above,* which they retrieved from the wreck of an ancient Roman ship in the Mediterranean Sea. Archaeologists believe that the columns were being shipped from a quarry to a building site when the ship sank.

vation. Without such measures, many artifacts retrieved from the water would quickly disintegrate. Because laboratory space on a research vessel is very limited, artifacts are usually given a bath in fresh water (if they were recovered from salt water) and then carefully wrapped in sheets of plastic to keep them from drying out until they can be treated on shore.

The primary challenge in conserving artifacts from underwater sites—both freshwater and saltwater—is counteracting water damage. Generally speaking, the more porous a substance is, the more vulnerable it is to damage. Wood, for example, absorbs water readily, becoming soft and spongy. When a waterlogged wooden object is exposed to air, the water evaporates from it and the wood shrinks and warps. To prevent that from happening, conservators must use a chemical treatment that stabilizes the remaining fibers of the wood while it is drying.

Even metals can be damaged by long exposure to water. All metals except gold react with the oxygen dissolved in water. The iron of a cannonball, for instance, combines with oxygen to form rust, a form of corrosion, on the surface of the ball. A similar reaction with dissolved oxygen occurs with objects such as copper and silver coins.

The salty water of oceans and seas creates additional problems for conservators. When metals are exposed to salt water, chloride ions—chlorine atoms that have gained electrons and thus acquired an electric charge—react with the metals and cause further corrosion. Over long periods, a metal object such as an anchor can corrode so much that it becomes spongy and crumbles to the touch. Earthenware pottery—the softest, cheapest, and most common kind of pottery—also suffers from immersion in salt water, becoming impregnated with salts. After a piece of pottery from an underwater site has dried, the salts expand and contract as the humidity in the atmosphere changes. These slight changes

create strains that can make the surface of the pottery flake off or cause the entire piece to break apart. Archaeological conservators, therefore, use various treatments on artifacts to remove or neutralize corrosive salt and strengthen the materials.

## New light on the past

These conservation efforts and the detailed research on artifacts recovered from old shipwrecks have begun to shed new light on past civilizations. For example, Ballard's discovery of the five ancient Roman ships in the Mediterranean revealed previously unknown details about Roman trade and seafaring.

The expedition's archaeologist, Anna Marguerite McCann of Boston University, coordinated the research on about 150 artifacts, including a Roman lamp, a millstone for grinding grain, and numerous amphorae. In artifacts from the 1988 discovery, the researchers found a copper coin bearing the image of Constantius II, the Roman emperor from A.D. 337 to 361. One of the ships discovered in 1997 carried sections of marble blocks and columns intended for the construction of buildings.

By comparing the amphorae to others found over the years at various archaeological digs, the researchers could determine that the jars had originated in both the eastern and western regions of the Mediterranean. Also, because archaeologists have learned what kinds of products were usually stored in each of the various styles of amphorae, McCann could conclude that the ships were probably transporting olive oil, wine, and *garum,* a popular sauce made from fermented fish. Some amphorae were stained blue on the inside and therefore may have held a pigment that was used for dying fabrics and formulating cosmetics.

The shipwrecks also provided new details for historians and archaeologists. Some amphorae contained tar, which, according to McCann, was the earliest known example of a substance used for the daily maintenance of a ship and its rigging. One wreck included a *bilge pump,* a device for pumping water out of the bottom of a boat, which provided an alternative to bailing out unwanted water by hand. The pump revealed a level of seafaring technology that was more sophisticated than modern experts had thought existed in the ancient world. In addition, the location of the wrecks far out at sea revealed the location of a trade route directly across the Mediterranean, which archaeologists and historians had long thought must have existed between Rome and the cities of North Africa.

Deepwater technology is not limited to searching for sunken relics of ancient maritime civilizations. Some work in the United States, for example, has focused on shipwrecks that are historically important in U.S. naval and warfaring history.

One of these is the U.S.S. *Monitor,* an ironclad gunboat used by the Union Army during the Civil War (1861-1865) that sank in a storm off Cape Hatteras, North Carolina, in 1862. A research team from Duke University in Durham, North Carolina, first discovered the wreckage of the *Monitor* in 1973. At the time, historians hoped that the famous warship could be brought ashore as a monument, but the ship was too badly corroded to be lifted from its resting place at a depth of about 70 meters (230 feet). In 1998, the U.S. National Oceanic and Atmospheric Administration and the U.S. Navy launched a program to document the site with video images, still photography, and maps of the wreckage and debris field. The researchers hoped to determine which portions could be stabilized, though left in place, and which items could be retrieved. They also wanted to learn as much as possible about the *Monitor* from its remains before the wreck became too corroded to be investigated.

Archaeologists have also examined two other historically significant American warships, the *Hamilton* and the *Scourge,* which sank in the deep waters of Lake Ontario during the War of 1812 (1812-1814). The U.S. government had purchased these two ships, which were civilian merchant schooners, and converted them into warships by mounting cannons on their decks. Unfortunately, the cannons made the ships top-heavy, and they sank during a storm in 1813.

Researchers from the Royal Ontario Museum in Canada first discovered the *Hamilton* and *Scourge* in 1973 with the aid of a magnetometer. The two vessels were resting about 400 meters (1,300 feet) apart at a depth of about 90 meters (300 feet). As underwater technology improved, researchers returned to the site several times. In 1990, Ballard, then at the Woods Hole Oceanographic Institution in Woods Hole, Massachusetts, used the ROV *Jason* to videotape and photograph the wrecks. Because the lake water contains no corrosive salts, Ballard found the ships to be remarkably well preserved. His investigation and ongoing research by other scientists have provided a valuable record of shipbuilding and maritime customs in the early 1800's.

## Modern shipwrecks and other sites

Some celebrated deepwater shipwrecks, such as the hulks of the *Titanic* and the German battleship *Bismarck,* sunk by the British in 1941, are not old enough to qualify as archaeological sites, but they nonetheless have great historical interest. Other famous vessels from the 1900's that have been investigated with modern underwater technology include the British passenger liner *Lusitania,* which was torpedoed by a German submarine off the Irish coast in World War I and went down in 88 meters (290 feet) of water; a Japanese World War II submarine, which sank to a depth of 5,500 meters (18,000 feet) in the Pacific Ocean in 1944; and the *Edmund Fitzgerald,* an iron-ore carrier that sank in a violent storm on Lake Superior in 1975 in 163 meters (535 feet) of water. In May 1998, Ballard and researchers from the National Geographic Society in

# Archaeology versus treasure hunting

Recovering objects from shipwrecks, or salvaging, dates back at least 2,000 years, when Roman divers retrieved cargo from shallow coastal wrecks. Salvaging began as a service to ship owners, but it also became a treasure-hunting venture.

Underwater archaeology has a much shorter history, dating back to the 1950's. Since then, archaeologists have claimed sunken vessels as cultural resources—sites that should be studied and preserved. But their efforts have run up against the complicated laws, financial interests, and adventurous spirit of salvaging and treasure hunting.

Many countries have laws that establish the rights and responsibilities of *salvors* (people who salvage ships or their cargo), and some of these laws were unified under two international treaties in 1910 and 1989. Generally, if a sunken ship is claimed by its owner, the salvor does not own the recovered property but must be compensated for retrieving it. If the ship is ancient or has been abandoned, the "law of finds" states that the person who discovers the ship can claim ownership. Consequently, shipwrecks, such as the numerous silver- and gold-laden Spanish ships that sank in the Caribbean Sea from the 1500's through the 1700's, have lured treasure hunters in search of lost riches.

In the United States, the Abandoned Shipwreck Act of 1987 provides some protection of historical sites in U.S. territorial waters. Under this law, the federal government claims ownership of any abandoned wreck in inland bodies of water and within 4.8 kilometers (3 miles) of U.S. coastlines. The federal government gives states the responsibility for wrecks off their shores, and the National Park Service establishes management guidelines. Other U.S. laws allow the government to create marine sanctuaries to protect underwater sites and to register wrecks as national historical sites.

The laws do allow for limited salvaging of some historical shipwrecks in inland and coastal waters, but frequently these operations are not carefully regulated. For example, many archaeologists have argued that salvors often retrieve artifacts from sunken ships without documenting where in the wreck the objects were found. Consequently, facts about a particular artifact's relationship to other objects on the ship and to the ship itself are lost forever. Many salvage companies have a staff archaeologist, but most scientists at universities and research institutions do not believe this arrangement guarantees "good archaeology," especially because many of the items are sold rather than studied.

Salvaging has also come under fire because many people see it as the disturbance of underwater graves. Some people, for example, see the sale of items from the *Titanic,* which sank in 1912 with the loss of more than 1,500 lives, as unethical.

In 1999, a delegation at the United Nations Educational, Scientific and Cultural Organization (UNESCO) was working on a treaty that addressed many of these issues. UNESCO delegates expected that the plan would help protect underwater sites and also provide regulations for managing historical resources responsibly and ethically. [PRL]

A pocket watch was one of numerous articles retrieved from the *Titanic,* which sank in 1912. Many people believe that the retrieval and sale of such items is a dishonor to those who died in the tragedy.

Washington, D.C., located and photographed the U.S. aircraft carrier *Yorktown,* resting at a depth of nearly 5,100 meters (16,700 feet) in the South Pacific Ocean. Japanese warplanes sank the *Yorktown* during the fierce fighting of the Battle of Midway in June 1942.

But deepwater archaeology is concerned with more than just shipwrecks. Underwater search techniques have, for instance, been used to explore natural sinkholes in southern Mexico known as cenotes (*se-No-tays*), which were important sources of drinking water and sacred sites for the Maya. Archaeologists have used deepwater technology to retrieve ritual objects lost for centuries in the cenotes. They have also employed

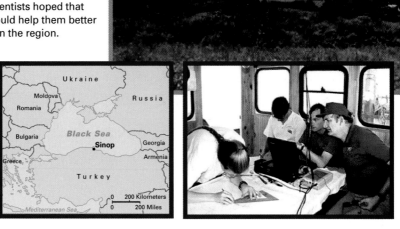

## In-depth study of the sea and land

A research team led by archaeologists at the University of Pennsylvania in Philadelphia in 1999 was in the first stages of a comprehensive study of both the land and sea around the ancient port of Sinop on the Black Sea, *right.* The scientists hoped that the Black Sea Trade Project would help them better understand centuries of trade in the region.

Sinop, which is located on the northern coast of Turkey, was a center of trade for the Greek and Roman empires and other ancient civilizations. Researchers on the Black Sea Trade Project, *far right,* examine the results of underwater sonar readings on a computer monitor and record their findings on a map of the sea floor.

it in the Southwestern United States, where giant dams have flooded canyons, burying prehistoric paintings and Native American artifacts in reservoirs that are more than 100 meters (330 feet) deep. In 1998, a joint expedition from the University of Connecticut in Storrs and IFE surveyed the coastal waters of southern New England—the first stage in a long-term project to find prehistoric campsites from the last Ice Age, which ended about 11,500 years ago. Because sea levels were about 90 meters lower during the height of the Ice Age than they are today, many areas that are now under water were once dry land. Researchers hope to find artifacts left by peoples who may once have lived in those areas.

## The future of deepwater archaeology

One of the most comprehensive underwater archaeological projects in the late 1990's was being conducted in the Black Sea. Under the direction of archaeologist Fredrik T. Hiebert of the University of Pennsylvania in Philadelphia, researchers in the Black Sea Trade Project were making a study of the land and sea near the ancient port of Sinop on the northern coast of Turkey. The scientists hoped that the archaeological data from both land and sea would yield a complete picture of ancient trade, manufacturing, farming, and daily life around the Black Sea.

The research was noteworthy because it was the first known attempt to apply some of the basic principles of land-based archaeology to the underwater domain. When archaeologists study a land site, they first conduct an exhaustive survey of the area by covering every bit of it on foot and recording all they find. The underwater equivalent of this process is the use of side-scan sonar, which was employed in 1998 and 1999 to survey the bottom of a selected area of the Black Sea. But the archaeologists were not simply trying to find a shipwreck or a lost settlement. Rather, their objective was to map a large region of the sea floor. In this approach, finding nothing could be as significant as finding a shipwreck or the remnants of a town because, for example, an area with no sunken ships is not likely to have been on a trade route.

In 1998, engineer David Mindell of the Massachusetts Institute of Technology in Cambridge coordinated the underwater survey near Sinop, investigating the sea floor to a depth of about 80 meters (260 feet). The survey revealed 10 shipwrecks and the remains of what appeared to be five buildings. During the 1999 season, the researchers planned to survey areas beyond the coast at depths of as much as 500 meters (1,640 feet). They did not expect to ever have a complete picture of the sea floor, but they hoped that the preliminary sonar scans would give them clues about other areas of the sea that merit investigation. The scientists planned to select some of the already identified sites for detailed study in the future.

Mindell believes that the rapidly advancing field of deepwater archaeology has developed an important relationship with the science of oceanography. Archaeology gives oceanographic engineers an incentive to design and test better robotic equipment, and the new equipment stimulates archaeologists to pursue more ambitious investigations. Scientists from the fields of geology and computer science are also making important contributions. The results of these combined efforts are promising. "I think the deep sea holds more history than all the museums of the world," says Ballard. Thanks to deepwater archaeologists and their scientific colleagues, much of that history will soon be coming to light.

## For additional information:

### Books and periodicals

Delgado, James P., Ed. *Encyclopedia of Underwater and Maritime Archaeology.* Yale University Press, 1998.

Blot, Jean-Yves. *Underwater Archaeology: Exploring the World Beneath the Sea.* Harry N. Abrams, Inc., 1996.

### Web sites

Nautical Archaeology at Texas A&M University Web site— http://nautarch.tamu.edu/

The U.S. National Park Service Submerged Cultural Resources Unit Web site—http://www.nps.gov/scru/home.htm

■···■···■

# Questions for thought and discussion

1. What does underwater archaeology contribute to our understanding of history and ancient civilizations?
2. What challenges do archaeologists encounter in underwater research?
3. Why do most archaeologists object to the work of salvagers and treasure hunters? What is your opinion of excavating historically important shipwrecks to make money?
4. If you were a state official responsible for managing the investigations and excavations of underwater sites, what rules would you establish?

# How
# the Moon
# Was Born

By the late 1990's, the most popular
theory for the formation of the moon
was that it was born when a giant object
struck the Earth with cataclysmic force.

By Stephen P. Maran

**The author:**
Stephen P. Maran is the Press Officer of the American Astronomical Society in Washington, D.C., and editor of *The Astronomy and Astrophysics Encyclopedia.*

T he moon has been an object of wonder and the subject of art and poetry since people first kept written records or decorated their caves with paintings. Many ancient civilizations based their calendars on the cycles of the moon. Lunar tides in seas and harbors helped determine when ships could set off on journeys or float safely into port. Before the development of electric lights and gas lamps, the moon lit the way for travelers at night. People imagined they saw a "man in the moon" and other shapes in the lunar surface, speculated about what the moon was made of, and wondered if it was inhabited. Scientists, too, have been curious about the moon and have scrutinized its cratered landscape with their telescopes. But for a long time, few of them gave much thought to how the moon came to be.

By 1998, the attitudes of scientists had changed. Not only were more researchers interested in the origin of the moon, but nearly all had come to believe in one particular model of lunar origin—the giant-impact theory. According to this theory, the moon was formed when a body about the size of the planet Mars—or a few times larger—struck the young Earth with cataclysmic force. Much support for this theory has come from the study of rock and soil samples from the moon, brought to Earth by U.S. astronauts. In 1998, a satellite called Lunar Prospector, in orbit around the moon, supplied evidence consistent with the theory, as did computer simulations of the giant impact.

## Early theories

Scientists' interest in the origin of the moon in the mid- and late 1900's stood in sharp contrast to attitudes about the moon during the early days of astronomy. The Italian astronomer and physicist Galileo Galilei first gazed at the moon in 1609 with the newly invented telescope, but he offered no theories about the moon's history, nor did most of the other astronomers who came after him.

It was only in 1879 that the British mathematician George H. Darwin, son of the great naturalist Charles Darwin, proposed one of the first theories of lunar origin. According to Darwin's "fission" theory, the material that became the moon was spun off from the very young, fast-spinning Earth, which was still in a molten state. Darwin's theory had little influence on astronomers for almost 100 years, in large part because they continued to be uninterested in the moon's origin.

Scientists finally started to investigate the question in the 1950's. By 1952, a "capture" theory of the moon's origin, proposed by an Ameri-can chemist, Harold C. Urey, had become the prevailing view. Urey believed that the moon was a *primordial* (ancient) body that formed at the dawn of the solar system and was later captured by the Earth and pulled into its present orbit.

Urey's views had great influence on the planning of the National Aeronautics and Space Administration's (NASA) Apollo program, the series of manned flights to the moon that took place between 1969 and 1972. Many scientists believed that the moon rock samples that Apollo astronauts were to bring back would reveal the minerals and compounds—incorporated into the primordial, virtually unchanging moon—that were present when the solar system formed. They also expected the samples to help prove or disprove the capture theory. If the moon formed far away from the Earth, they reasoned, it might have a different chemical composition than the Earth.

But long before the lunar samples were gathered, some scientists had doubts about the capture theory. If the moon had formed far away in the solar system, mathematical calculations showed, it would have had to follow a long, looping orbit to reach the Earth, which it would have zipped past at high speed. It would not be possible, the scientists calculated, for a moon captured from such an orbit to end up in its present position, with the Earth and moon spinning at their current rates, once every 24 hours and once every 28 Earth days, respectively.

Some scientists proposed in the early 1960's that the moon formed at roughly the same time and at the same distance from the sun as the Earth, though not alongside it. The infant moon would have been orbiting the sun at nearly the same speed as the Earth and would sometimes move slowly past it. The moon would then have been easy to capture. If this scenario was true, the moon should be made of the same materials as the Earth.

## Scientists meet to debate lunar origins

By 1964, the question of how the moon originated had generated sufficient interest to merit a conference devoted entirely to that subject. At that meeting in New York City, much attention was paid to the fission theory proposed by George Darwin. Darwin's suggestion that the moon was created from molten material spun off by the Earth seemed plausible for a time when our planet rotated much faster than it does now, possibly once every 4 hours instead of once every 24 hours. By the time of the conference, however, calculations had shown that even such a high rotational speed would not have been sufficient to form the moon. Friction in the molten rock would not have permitted a bulge of matter to rise high enough to be flung free of Earth's gravity.

In light of that finding, proponents of the fission theory at the conference suggested another possibility. The material that formed the moon, they said, might have been spun off when the Earth was rotating not once every 4 hours but once every 2.1 hours. If a molten Earth had ever rotated that fast, they pointed out, its mechanical strength and gravity would simply not have been able to hold the planet together. The Earth would have hurled great amounts of matter into space in a process called rotational instability.

Proponents of the modern fission theory calculated exactly how such a process could take place. The birth of the moon would not have occurred as a catastrophic outburst, such as an explosion, but as a progressive change in Earth's shape. First, the Earth's rapid rotation would have flattened the planet at the poles, deforming it into an *oblate* shape—wider at the equator than across the poles. Next, the spinning, flattened Earth would have produced a neck of material from a point on its equatorial bulge. The neck would have flown off into space as a large blob trailed by smaller blobs that would soon have fused into one. But a major flaw in the fission theory, other scientists noted, was that in order for fission to occur, the Earth would have been spinning so fast that it would still be turning much more rapidly than it is today.

Another theory for the origin of the moon discussed at the 1964 conference was that the moon formed together with the Earth at the same time and in the same place. Just as the Earth was believed to have been created by *accretion* (the accumulation of solid particles and larger objects) about 4.6 billion years ago when the solar system was taking shape, so, according to this theory, was the moon.

But the "coaccretion" theory presented problems as well. Astronomers had calculated the moon's density by using a formula based on the moon's mass, as determined by its distance from the Earth and its orbital velocity, divided by its volume. The moon's density, calculated to be 3.3 times the density of liquid water, is much less than the density of the Earth, which is 5.5 times the density of water. If the Earth and the moon were formed from the same cloud of matter, at the same distance from the sun, how could their composition be so different?

In addition, other doubts plagued scientists as well. If the moon formed near Earth by the same processes that formed the other planets, why don't the other planets of the inner solar system have large satellites like the moon? Mercury and Venus have no moons, and Mars has only two tiny satellites that may be captured asteroids.

For such reasons, some scientists by the time of the 1964 conference were already doubtful about the validity of the coaccretion theory as well as the earlier theory that the moon formed at about the same distance from the sun as the Earth. Instead, most participants at the meeting favored some version of Urey's theory in which the moon was formed elsewhere and then captured by the Earth, or some version of the fission theory. After all, they noted, because fission would take place in the upper layers of the Earth, which are much less dense than its heavy iron core, fission could account for the moon's low density and apparent lack of a significant iron core. Though fission caused by Earth's rapid rotation had been ruled out, the scientists speculated that a portion of the Earth might still have been flung off to form the moon in some other way.

### A revised capture scenario

In 1972, Ernst J. Opik, an Estonian-born astrophysicist who worked in Northern Ireland, proposed a totally different theory. Opik suggested that a primordial object may have streaked past the Earth in the early solar system and come within *Roche's limit,* a boundary located about 18,500 kilometers (11,500 miles) from the center of the Earth. Inside Roche's limit, the Earth's gravity can pull a weak body apart. (Every planet has its own Roche's limit, which varies according to the planet's mass.) Some debris from the disintegrating body may have gone into

## Early theories of moon formation

British mathematician George H. Darwin, son of the naturalist Charles Darwin, proposed the "fission" theory in 1879. According to Darwin, the young, molten Earth spun much faster than it does now. A piece of the Earth was flung off and formed the moon.

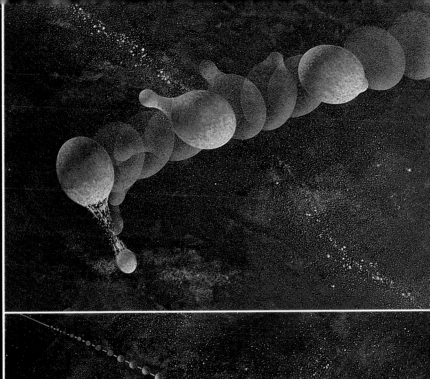

American chemist Harold C. Urey introduced the "capture" theory of lunar origin in 1952. Urey believed that the moon formed in a separate part of the solar system from Earth. As the moon traveled past the Earth, it was captured by the Earth's gravity and pulled into its present orbit.

Some scientists of the 1960's suggested that the moon was formed at the same time and in the same place as the Earth. According to the "coaccretion" theory, solid particles of interplanetary dust collided and stuck together. Gravity pulled the clumps into small bodies, which in turn collided to form asteroids, planets, and the moon.

orbit around the Earth and then coalesced to form the moon.

Scientists agreed that Opik's "disintegrative capture" theory convincingly explained how the moon could have been made from different material than the Earth—material from a distant part of the solar system. And the theory provided an answer to the question of how a large object—or at least part of it—could have been pulled into orbit around the Earth, a test that the previous capture theory had failed. Nonetheless, by 1985, researchers had made detailed calculations showing that a fast-moving object would not have spent enough time within Roche's limit to be pulled apart and leave debris around the Earth before it flew on. Thus, until better calculations might somehow yield contrary results, the disintegrative capture theory was shelved.

## New evidence—and a new theory

Even while scientists were evaluating Opik's theory, the first physical evidence from the moon had become available. From 1969 to 1972, Apollo astronauts landed on the moon six times, collecting more than 2,000 rock and soil samples. Researchers who analyzed the samples found that the specimens shared a unique characteristic: The relative amounts of oxygen *isotopes* (atoms of the same element that differ in atomic weight) in the moon rocks proved to be almost identical to those found in the Earth's mantle, a thick layer of hot rock between the Earth's outer core and its crust. The relative amounts of these isotopes are believed to have varied in different parts of the solar system, so the oxygen-isotope ratios strongly implied that the moon and Earth's mantle had a common origin.

The moon rocks were also extremely dry, whereas a great many rock types and minerals on Earth contain water. Compared with Earth rocks, there were very few other substances in moon rocks that, like water, are *volatile* (vaporize easily).

Finally, the dating of *radioactive isotopes* in the lunar samples told the scientists how long it had been since minerals in the samples had solidified from a molten state. (A radioactive isotope is one that decays spontaneously to become another substance. For example, an isotope of hafnium decays to form an isotope of tungsten.) By comparing the relative number of such isotopes, scientists can tell the age of the sample. The dating process showed that all or most of the moon had been molten at one time, solidifying by 4.4 billion years ago. This finding, in particular, convinced most scientists that the moon could not be the primordial body that Urey thought had been captured by the Earth.

Thus, studies of the moon rocks clearly showed that the moon could not be an intact, or nearly intact, remnant from the earliest days of the solar system. The evidence suggested that the moon was born from the Earth's mantle by a process that released enormous heat. The heat would have driven off the volatiles and vaporized the rock.

Influenced by the information from the moon rocks, in the mid-1970's two teams of planetary scientists separately proposed what is now called the "giant-impact" theory of the origin of the moon. The groups were led by A. G. W. "Al" Cameron and William R. Ward at Harvard University in Cambridge, Massachusetts, and William K. Hartmann and Donald R. Davis at the Planetary Science Institute in Tucson, Arizona. The researchers suggested that a huge Mars-sized object struck the Earth, blasting away enough of the planet's crust and mantle to create the moon from its debris. According to the theory, the object struck the Earth a glancing blow, rather than hitting it head-on. The blow set the Earth spinning faster than before. This may explain why the *angular momentum* of the Earth-moon system (a measure of the rate at which both bodies are rotating and orbiting) is unusually high, compared with other planets and moons in the solar system.

At first, many scientists were reluctant to accept the giant-impact theory, because the theory was based on a single event—a catastrophic collision—that most researchers believed had occurred very rarely in the solar system. However, the more that lunar and planetary scientists thought about the theory, the better they liked it. Surveys of lunar geography made with both Earth-based telescopes and telescopes and

**Evidence from the moon**

United States astronaut David R. Scott uses a lunar drill to collect moon rock and soil samples during the 1971 Apollo 15 mission. The chemical composition of such samples, collected during the Apollo missions of 1969 to 1972, suggested to scientists that the moon was born from the Earth's mantle by a process that released tremendous heat. One of the rock specimens is displayed in a metal holder (inset).

cameras on spacecraft showed that strikes on the moon by asteroids and *meteoroids* (metal or rocky objects smaller than asteroids) after it had formed and solidified had produced many huge craters, called impact basins. Why couldn't the impact of an even more massive object on the Earth have made the moon itself?

At the same time, new findings about the formation of the solar system suggested that large numbers of massive objects, called protoplanets or planetesimals, had been orbiting the sun at the time the Earth formed. Gradually, scientists came to believe, these objects collided with one another and were broken up, or massed together to make the planets, or were flung out of the solar system.

At the second major conference on the origin of the moon, held in Kona, Hawaii, in 1984, the giant-impact theory was the center of attention. It became the prevailing view of most astronomers.

As more researchers tested the giant-impact theory, they had to contend with the fact that many of the early calculations of the theory were relatively crude. Although the calculations showed that an enormous collision could have blown enough matter out of the Earth (and from the vaporized outer layers of the impacting object itself) to make the moon, they did not trace the events in detail. So while the theory was very promising, astronomers did not know how realistic it was.

## The help of computer simulations

During the 1990's, a new generation of planetary scientists, including Robin M. Canup, now at the Southwest Research Institute in Boulder, Colorado, developed the theory further. They made improved calculations with better computers and more advanced programs that more realistically simulated cosmic collisions and their consequences. By 1996, Canup and her associate Larry W. Esposito of the University of Colorado in Boulder had demonstrated that the giant-impact theory required a much more massive impacting object than previously suspected. They found that the object had to be at least twice as massive as Mars, or even larger, for the debris from the collision to coalesce into the single moon that we have today. Otherwise, the calculations indicated, there would be several moons circling the Earth.

Though Canup and Esposito's calculations were more sophisticated than those of the past, they still were not precise enough. The team used a simplifying assumption, called gas dynamic theory, that represented the processes in the formation of the moon after the giant impact on Earth as though the debris were all gas. In reality, even though the material hurled up by the collision was so hot that it took the form of vapor, it soon cooled and condensed. So the actual conditions must have been much more complicated than those represented by the gas dynamic calculations, with clumps of rock cooling, solidifying, and colliding. In fact, the various proponents of the giant-impact theory had shown only that an enormous collision could have produced the necessary conditions to form the moon. But they had not proven that such an event had actually occurred.

In 1997, the Japanese geophysicist Shigeru Ida at the Tokyo Institute of Technology, together with Canup and Canup's University of Colorado colleague Glen R. Stewart, presented a more advanced simulation of the impact. They used a mathematical technique called the N-body method. This model recreated the processes inside the impact debris cloud as a series of interactions between a large number ("N") of individual clumps or objects that collided with one another and were affected by one another's gravitational forces. Thus, the N-body method represented real processes in space much better than the gas dynamic technique or other methods did. The Ida team found that the debris from a giant impact with the Earth would indeed have formed a large moon and that this process might have taken as little as one year. Refined calculations by the team and others showed that the most likely mass for the body that collided with Earth was three times that of Mars.

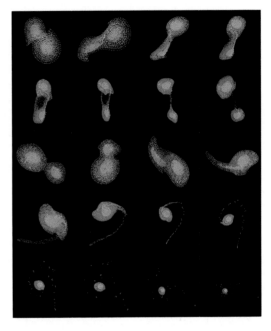

## Further support for the giant-impact theory

Besides conducting computer simulations, researchers pursued other ways of testing the giant-impact theory. During 1998, Lunar Prospector, NASA's first moon mission since Apollo, was launched to gather a variety of data about the moon. As Prospector orbited just 100 kilometers (60 miles) above the moon's surface, researchers used Earth-based radio telescopes to monitor the moon's gravitational pull on the spacecraft. By mapping the results, the scientists were able to get a far more accurate picture of how the moon's mass is distributed than they had had before. The researchers calculated that the moon has a core, most likely composed of iron, with a radius of 220 kilometers to 450 kilometers (135 miles to 280 miles). This is well within the range expected if the moon had formed as the result of a devastating impact on Earth and thus is consistent with the giant-impact theory.

In December 1998, at an origin-of-the-moon conference in Monterey, California, Canup and planetary scientists Craig Aignor of the University of Colorado at Boulder and Harold Levison of the Southwest Research Institute announced further calculations in support of the giant-impact theory. They used a computer method called "symplectic integration," which allowed them to simulate much greater lengths of time and thousands more sun-orbiting objects than ever before possible. Canup's group found that impacts on Earth, such as the one that may have formed the moon, were almost certainly common during the first 100 million years after the birth of the solar system. This conclusion was among the strongest indications available to astronomers that the giant-impact theory was not based on an unusual

**Computer modeling**
Computer models have helped to test theories of lunar origin. In the mid-1990's, planetary scientist Alastair Cameron at Harvard University simulated many ways that a large object could have hit the Earth. In each one, the object was destroyed, and enough material was flung into orbit around the Earth to form the moon. One simulation, *above*, depicts the Earth being struck twice by such a body. After the second strike, most of the object's core (shown in blue) merges with the Earth's. The remnants of the object form a tail that gradually becomes a diffuse cloud of debris. The debris will later coalesce to form the moon.

## The "giant-impact" theory

By the late 1990's, most planetary scientists believed that the moon was formed when a giant Mars-sized object struck the Earth a glancing blow, *left*. Part of the object's iron core, they believe, sank to the Earth's center. At the same time, the force of the blow vaporized a good part of the Earth's mantle, as well as that of the object, *right*.

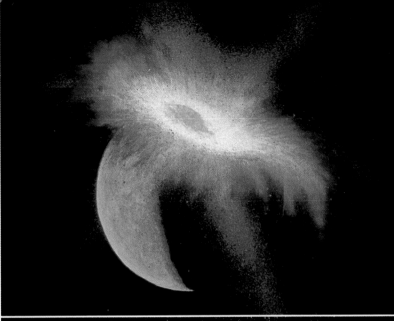

The collision threw a great, expanding cloud of fiery vapor composed of gasified rock to a height of about 22,500 kilometers (14,000 miles), *right*. Much of the vapor settled into a ring of cooling debris—like the ash particles that condense from a volcanic eruption—in an orbit above the Earth.

Over the course of about one year, the debris condensed into countless solid particles that caught the rays of the sun and formed a bright ring around the Earth. The particles slowly clumped together, forming tiny rocks, then bigger and bigger ones, until finally the moon began to form, *right*. Eventually, all the material in the ring was incorporated into the moon.

accident of nature, a finding that made the theory even more credible.

Nevertheless, the theory faced the same stumbling block that early theories did. The impact of a body three times the mass of Mars would have set the Earth spinning so rapidly that, even today, it would be turning faster than it is. Cameron and others proposed variations of the theory to get around this problem. One possibility they explored was that another large object might have struck the Earth from the opposite direction after the moon was formed, slowing the planet's spin.

Despite the one remaining difficulty, astronomers in 1999 considered the giant-impact theory the best explanation for the origin of the moon. They think they can now sketch fairly accurately how the moon was born and how it developed over time into the body that today illuminates our night sky.

## How scientists think it happened

Here's the scenario: A little more than 4.5 billion years ago, a young, hot Earth, constantly bombarded by thousands of asteroid-sized objects, had grown to almost its present size. Most of the iron it contained had sunk toward the center, forming a huge iron core that was much denser than the rest of the planet. Surrounding the molten core was a slowly hardening mantle of lighter rock.

Suddenly, a round, fast-moving, fully formed planet the size of Mars or perhaps larger, with its own iron core and rocky mantle, loomed from space. Traveling at a speed of 40,000 kilometers (25,000 miles) per hour, it struck the Earth a glancing blow. The object's *kinetic energy* (energy of motion) was instantly converted into heat that vaporized much of the object's mantle along with a good part of the Earth's. The collision produced a great, expanding cloud of fiery vapor composed of gasified rock. Thrown to a height of perhaps 22,500 kilometers (14,000 miles), much of the vapor formed a diffuse cloud that orbited the Earth. At the same time, most of the iron core of the body that struck Earth looped around the planet and struck again, this time penetrating and merging with the Earth.

Over the course of about a year, the debris in the cloud condensed into solid particles and formed a ring around the Earth. The particles slowly clumped together, forming tiny rocks, then bigger and bigger ones. For a time there were thousands of these "moonlets" orbiting Earth. But over a period of less than 100 years, the larger moonlets swept up the smaller ones, until they all merged into one large body. The ring was gone, and in its place there was the infant moon. At that point, the Earth-moon system resembled a double planet. The moon circled Earth rapidly at a small fraction of its present distance, and Earth spun rapidly, thanks to the blow it had suffered in the impact.

The newborn moon was at first covered by *magma* (molten rock). This feature, which geologists call the magma ocean, was at least a few hundred kilometers deep. The magma ocean was created by heat from the many large final impacts of moonlets on the by-then largely formed moon. Liquid iron sank to the moon's center, and electrical currents in the molten core generated a magnetic field. As the magma ocean cooled, about 4 billion years ago, it solidified. Heavier minerals sank, while lighter ones rose to form a crust.

Even as the lunar surface hardened, it was being peppered by meteoroids, asteroids, and comets. The largest of these objects produced huge basins up to 2,500 kilometers (1,600 miles) across. Later, heat released by radioactivity deep inside the moon caused magma to well up from the interior, partially filling and leveling many of the basins.

Bodies large enough to carve out basins stopped striking the moon by about 3.2 billion years ago, but smaller objects continued to hit the surface, forming many craters. At the centers of some craters the rebound of surface material after the impact created a mountain. The effects of impacts also produced mountain chains at the boundaries of many basins and craters. No lunar mountains were formed by the folding and upthrust of surface layers, as occurs to form many mountains on Earth. Nor did any large volcanic mountains develop on the moon.

The countless impacts of meteoroids slowly fragmented the lunar surface. This created a *regolith* (surface layer) of broken rocks and soil particles as deep as 15 meters (50 feet) in some areas.

Over 4.4 billion years, lunar tides caused the Earth to slow to its present spin rate of once every 24 hours and the moon to move gradually away from the Earth to its present distance of about 385,000 kilometers (240,000 miles). Even now, the process continues, though more slowly, as the moon recedes from the Earth at 3.75 centimeters (1.5 inches) per year, or about 3.6 meters (4 yards) per century.

The moon has changed little for several billion years. Newer craters have formed atop older ones and the regolith has gradually deepened, but there are no more magma flows—most scientists think the moon long ago became totally cold and solid—and no more huge impacts.

Scientists believe this is probably how the moon was created and how it developed, but even now they aren't certain that they have learned the true history of the moon. The giant-impact theory best fits the evidence we have, but as scientists continue their research, other theories may yet replace it. However, one thing will probably never change— the sense of wonder that people feel when they gaze up at the moon.

## For additional information:

### Books and periodicals

Jayawardhana, Ray. "Deconstructing the Moon," *Astronomy,* September 1998, pp. 40-45.
Spudis, Paul D. *The Once and Future Moon.* Smithsonian Institution Press, 1996.
Taylor, Stuart Ross. "The Moon" in *Encyclopedia of the Solar System.* Academic Press, 1999.

### Web sites

National Aeronautics and Space Administration Web pages— http://sse.jpl.nasa.gov/features/planets/moon/moon.html
Lunar and Planetary Laboratory, University of Arizona Web pages— http://seds.lpl.arizona.edu/billa/tnp/luna.html

# Turbulence:
## Hidden Threat in the Skies

By James T. McKenna

Aviation researchers are turning to technology in an effort to help planes avoid air turbulence—a serious problem that has plagued airliners for decades.

High over the Pacific Ocean, about two-and-a-half hours southeast of Tokyo on the evening of Dec. 28, 1997, most of the 374 passengers on United Airlines Flight 826 had just finished their dinner. As the passengers tried to get comfortable for the rest of the nine-hour trip to Honolulu, Hawaii, flight attendants worked along the aisles of the Boeing 747, collecting meal trays. Suddenly and without warning, the jumbo jet lurched sideways, heaved sharply upward, then plunged toward the ocean. Anything that was not strapped down—including flight attendants and passengers not wearing their seat belts—was immediately thrown about the cabin. For several seconds, the plane was filled with tumbling debris and the screams of terrified passengers.

After the incident, the pilot advised ground controllers that the plane was heading back to Tokyo. A flight attendant and 13 passengers were seriously hurt, and 83 others had minor injuries. One passenger, a Japanese woman, later died in the hospital of severe head injuries. The 26-year-old plane was so badly damaged that United retired it from service six months ahead of schedule.

While they are rarely fatal, encounters with turbulence damage more aircraft and injure more passengers and flight personnel than any other aviation mishap except crashes. According to statistics kept by the Federal Aviation Administration (FAA), turbulence incidents involving U.S. airlines—including the Flight 826 mishap—were blamed for 3 deaths, 76 serious injuries, and nearly 1,000 minor injuries from 1981 to 1998. Each year, turbulence costs airlines hundreds of millions of dollars in payments for injuries, damage to aircraft, and revenue lost while planes are out of service for repairs. As widespread as the problem is, however, the only defense against turbulence, even after more than four decades of commercial jet travel, is to avoid it. But steering clear of turbulent air can be more difficult than it might seem. The most dangerous kind of turbulence—because it is the hardest to detect—occurs in clear air, where pilots have no indication of disturbed air in the aircraft's path until they fly into it.

Things are beginning to change, however. Growing concern about the harmful effects of turbulence on commercial aircraft since the 1980's, combined with a number of promising technological advances, has led to turbulence detection and avoidance being regarded as a problem that can be solved. Airlines, government agencies, and private companies have begun cooperating in international efforts aimed at finding ways to detect and avoid this chronic threat to safe air travel.

## What is air turbulence?

To understand why turbulence poses such a persistent problem to commercial air travel, it is important first to understand what turbulence is and what causes it. In the atmosphere, currents of air flow in all directions and at a variety of speeds. Turbulence is, essentially, any sort of erratic flow of air produced when an otherwise smooth-flowing air current becomes disrupted. A particular form of turbulence, called wind shear, also commonly occurs as the result of changes in the direction and

## Terms and concepts

**Doppler radar:** An advanced radar system that measures the radar echo from a distant object for changes in frequency to determine the speed of the object and the direction it is moving in relation to the radar transmitter.

**Downburst:** A powerful current of cold air plunging to the ground.

**Jet stream:** An enormous current of high-altitude air that flows eastward at more than 100 kilometers (60 miles) per hour.

**Lidar:** A system similar to Doppler radar that detects turbulence using a laser beam to measure the speed and direction of particles in the air.

**Microburst:** A narrow and brief but extremely powerful type of downburst.

**NEXRAD** (Next Generation Weather Radar): A nationwide network of Doppler weather radars operated by the National Weather Service.

**Turbulence:** Sudden changes in the direction and speed of air currents over a short distance.

**Wind shear:** An extremely focused type of turbulence that occurs at the boundary between convection currents that are moving in different directions.

**The author:**
James T. McKenna is the transport and safety editor for *Aviation Week & Space Technology* magazine.

speed of adjacent air masses. A sometimes extremely dangerous occurrence, wind shear plagued commercial aviation until the 1980's, when research and development efforts all but eliminated it as a threat to airline safety. Severe wind shear, which is associated with stormy weather, occurs in small patches of the sky, at the boundary between convection currents that are moving in different directions. It can be powerful enough to snap a plane's wings. Abrupt shifts in wind currents can also result from other weather conditions and even terrain.

The most common type of atmospheric turbulence, called convective turbulence, develops with a rising mass of air. Energy from the sun is absorbed by the Earth's surface. The ground then gives up this energy in the form of infrared radiation, or heat. A mass of air near the surface absorbs this heat, which causes it to expand, forming a current of ascending air. As it rises, the air cools and the water vapor within it condenses, resulting in the formation of clouds. Eventually, the cooling air mass begins to drop, forming a descending current of cold air. This cycle of rising and falling air is known as convection, and the columns of air are called convection currents. These currents of air can ascend or descend at speeds up to 100 kilometers (60 miles) per hour, swirling the air between them into a highly turbulent flow. Strong convection currents are an essential ingredient in the development of a thunderstorm and can generate extremely powerful wind shear. For this reason, pilots are prohibited from flying into thunderstorms.

A thunderstorm juts up into the atmosphere like a boulder in a river, a situation that generates turbulence even beyond the immediate area of a storm. Winds are forced over and around the storm, and the disturbed air can roil for thousands of meters above or behind a thunderstorm. Pilots can often identify areas of convective turbulence ahead by looking for cumulus clouds—piled-up masses of white cloud. If the air in a particular region is too dry, however, there may not be enough water vapor present to form clouds, even though convection is at work.

As a rule, pilots avoid flying over the top of a thunderstorm. Flying below a storm, however, is sometimes unavoidable, as when an aircraft has to land. Cold air within a storm can plunge through the cloud base and continue all the way to the ground, producing an effect called a downburst. Occasionally, an extremely localized downburst—4 kilometers (2.5 miles) or less across—will form. This type of downburst, called a microburst, is narrow and brief but extremely strong. On Aug. 2, 1985, a Delta Air Lines Lockheed L-1011 bound for Dallas/Fort Worth International Airport attempted to land beneath a storm when a sudden microburst struck the plane. Because it was landing, the airliner was flying slowly and at a low altitude. The microburst slammed it into the ground short of the runway. Of the 166 people on the flight, only 30 survived.

## An invisible danger: clear-air turbulence

Even though pilots are taught to avoid turbulent air by looking for cumulus clouds, turbulence can strike even in the absence of clouds. This type of turbulence—especially dangerous because of its invisibility—is known as clear-air turbulence. It accounts for most turbulence-related injuries, mainly because pilots have no time to warn passengers and flight attendants to get strapped into their seats. Nearly 7 out of 10 turbulence incidents are the result of encounters with the clear-air variety.

Clear-air turbulence most often occurs as the result of a surface *cold front* (a large, moving mass of cold air) encountering warmer air in its path. Colder air is denser than warm air and tends to remain closer to the ground. Therefore, as the leading edge of a cold front advances, it forces its way under the warmer, lighter air in its path. This effect can create a trail of turbulent air thousands of meters above and 80 to 160 kilometers (50 to 100 miles) behind the leading edge of the cold front.

Encounters with clear-air turbulence most often occur along the borders of the jet streams, enormous currents of air thousands of miles long flowing eastward at more than 100 kilometers per hour. The major jet streams are usually found at least 10 kilometers (6 miles) above the surface. There are six jet streams that meander around the Earth. Three are in the Northern Hemisphere and three in the Southern Hemisphere— one each near the North and South poles, two along each side of the equator, and two running roughly across the center of each hemisphere.

Airlines prefer to fly their eastbound aircraft in a jet stream, because the high-speed eastward tailwind allows a plane to fly faster while burning less fuel. But this benefit has a downside. As a jet stream races along, eddies of air swirl from its boundaries into the slower air around it, creating turbulence. Therefore, using a jet stream in this way often means risking an encounter with clear-air turbulence as an airplane enters and leaves the stream.

Clear-air turbulence also can swirl near a deep upper trough, an elongated area of low air pressure at an altitude of 3,000 meters (10,000 feet) or more. Like water running down a drain, air flows counter-clockwise around troughs in the Northern Hemisphere and clockwise in the Southern Hemisphere. This circulation pattern can create turbulence along the trough's edges. The most turbulent air is often found upwind of the trough's base and along its centerline. Upper-trough turbulence can develop in small areas of the sky. They may last only 30 minutes or persist for an entire day. Such an elusive culprit is as difficult for *meteorologists* (weather forecasters) to predict and study as it is for pilots to avoid.

## Other causes of turbulence

Turbulent air can also be created by factors other than weather patterns. Features of the terrain like mountains and even tall buildings can create turbulence, disrupting smooth wind currents and causing them to burble in a turbulent flow. This phenomenon is called mechanical turbulence. Wind currents that become disrupted as they pass over mountainous areas can swirl chaotically for more than 800 kilometers (500 miles) beyond the peaks. Buildings disrupt wind currents on a smaller scale, creating turbulent eddies at lower altitudes, where they can pose a threat to airplanes that are taking off or landing.

Mechanical turbulence is sometimes a significant local problem. The

## Five major types of air turbulence

Atmospheric turbulence is caused by sudden changes in the direction and speed of air currents over a short distance. These changes can result from several factors, including weather conditions, geographic features, or even the aircraft themselves.

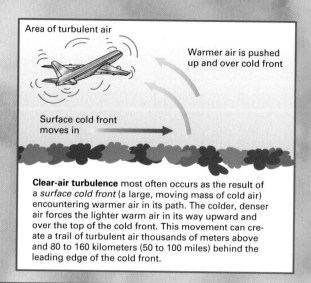

Area of turbulent air

Warmer air is pushed up and over cold front

Surface cold front moves in

**Clear-air turbulence** most often occurs as the result of a *surface cold front* (a large, moving mass of cold air) encountering warmer air in its path. The colder, denser air forces the lighter warm air in its way upward and over the top of the cold front. This movement can create a trail of turbulent air thousands of meters above and 80 to 160 kilometers (50 to 100 miles) behind the leading edge of the cold front.

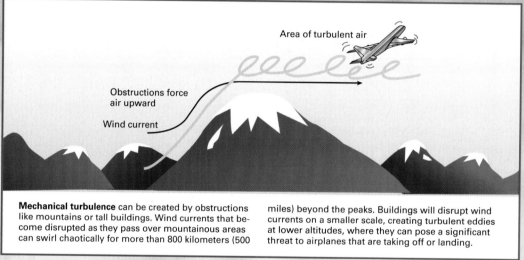

Area of turbulent air

Obstructions force air upward

Wind current

**Mechanical turbulence** can be created by obstructions like mountains or tall buildings. Wind currents that become disrupted as they pass over mountainous areas can swirl chaotically for more than 800 kilometers (500 miles) beyond the peaks. Buildings will disrupt wind currents on a smaller scale, creating turbulent eddies at lower altitudes, where they can pose a significant threat to airplanes that are taking off or landing.

international airport in Juneau, Alaska, for example, is surrounded by mountains that create turbulent wind conditions. Under certain weather conditions, aircraft taking off from Juneau International must make a sharp 180-degree turn during take-off to avoid turbulent air.

Mechanical turbulence can even be created by aircraft themselves. This phenomenon, called wake turbulence, is created as an aircraft's wings slice through the air. Some moving air along the top and bottom of the wings slips sideways toward the wing tips. When these two slips of air meet at the tip, they swirl off in invisible but powerful vortices, or spirals, that can trail off for miles behind and below the aircraft. The bigger the aircraft, the more powerful and persistent the wake turbulence it creates. An aircraft that inadvertently flies into the wake of another plane

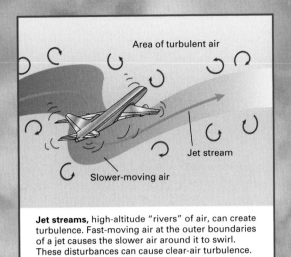

Area of turbulent air

Jet stream

Slower-moving air

**Jet streams,** high-altitude "rivers" of air, can create turbulence. Fast-moving air at the outer boundaries of a jet causes the slower air around it to swirl. These disturbances can cause clear-air turbulence.

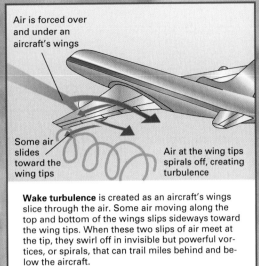

Air is forced over and under an aircraft's wings

Some air slides toward the wing tips

Air at the wing tips spirals off, creating turbulence

**Wake turbulence** is created as an aircraft's wings slice through the air. Some air moving along the top and bottom of the wings slips sideways toward the wing tips. When these two slips of air meet at the tip, they swirl off in invisible but powerful vortices, or spirals, that can trail miles behind and below the aircraft.

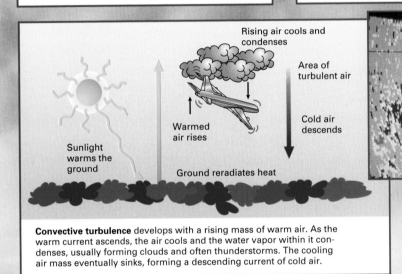

Rising air cools and condenses

Area of turbulent air

Cold air descends

Warmed air rises

Sunlight warms the ground

Ground reradiates heat

**Convective turbulence** develops with a rising mass of warm air. As the warm current ascends, the air cools and the water vapor within it condenses, usually forming clouds and often thunderstorms. The cooling air mass eventually sinks, forming a descending current of cold air.

By revealing small air masses that are moving toward the transmitter (blue) or away from it (red), Doppler radar can spot a microburst.

can be pummeled by these unseen swirls. In severe cases, the pilot can lose control of the aircraft.

## Turbulence becomes a significant threat

Turbulence has plagued commercial air travel since the beginning of the "Jet Age" in the late 1950's. Prior to that time, people did not worry much about turbulence because the propeller-driven airplanes of the prejet era rarely encountered it. For one thing, these planes cruised at an altitude of about 5,800 meters (19,000 feet), where turbulence is not a major threat, and flew at only about 325 kilometers (200 miles) per hour, which made recovery from wind-shear encounters easier. In addi-

## The effects of turbulence on passengers

A turbulence encounter can give airline passengers a bumpy ride. Sudden movements of the plane create forces on unrestrained passengers that can lead to injury. According to the Federal Aviation Administration, most people injured by turbulence are not wearing their seat belts during the incident.

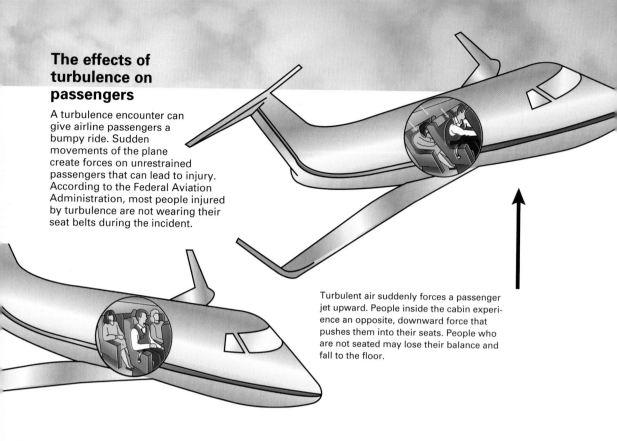

Turbulent air suddenly forces a passenger jet upward. People inside the cabin experience an opposite, downward force that pushes them into their seats. People who are not seated may lose their balance and fall to the floor.

tion, the lack of sophisticated navigation systems at that time prevented prop-driven aircraft from flying in stormy weather—conditions in which wind shear is likely to occur. Of course, wind-shear accidents probably did occur, but the less-sophisticated accident investigation techniques of that era made it difficult to identify wind shear as the cause.

By 1958, however, passenger jets had become commonplace, and commercial aircraft began flying higher and faster. Modern jets typically cruise at altitudes in the vicinity of 9,000 meters (30,000 feet), where turbulence can be a deadly threat. They also fly at speeds of more than 800 kilometers (500 miles) per hour, which greatly magnifies the force that turbulence can create on a plane and its occupants.

In 1958, a Soviet-built Tupolev Tu-104 jet airliner ran into severe turbulence while cruising at 11,000 meters (36,000 feet). The pilots lost control of the aircraft and it crashed, killing everyone on board. Since then, official records have logged more than 180 cases in which turbulence damaged an aircraft or injured its passengers. But that is only the tip of the iceberg. Safety experts agree that most turbulence encounters are never reported because they result in no damage or injuries.

Still, even the official figures are sobering. Turbulence accounts for more than 20 percent of airline *accidents*, an official record-keeping category that includes every incident in which an aircraft is substantially damaged or a person is seriously injured. And though modern airliners are built to withstand significant stresses in flight, turbulence severely damages at least one commercial jet per year.

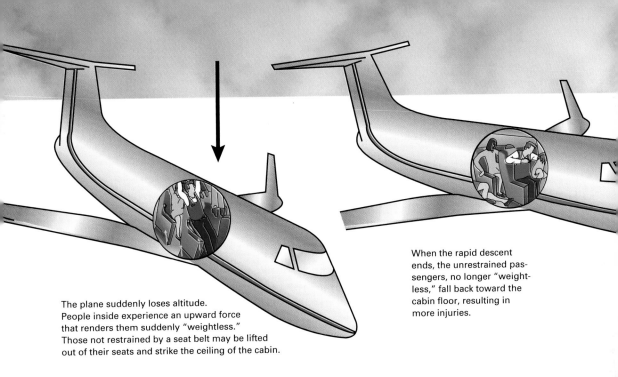

The plane suddenly loses altitude. People inside experience an upward force that renders them suddenly "weightless." Those not restrained by a seat belt may be lifted out of their seats and strike the ceiling of the cabin.

When the rapid descent ends, the unrestrained passengers, no longer "weightless," fall back toward the cabin floor, resulting in more injuries.

## Traditional approaches to dealing with turbulence

At present, the most effective way to deal with turbulence is to avoid it, and the best means of avoiding turbulence is pilot reports. Pilots that encounter turbulent air radio their altitude and position to air traffic controllers, who warn other aircraft of the trouble spot if their workload permits. However, this does little to spare most passengers and flight attendants from a bumpy and often dangerous ride, and it does nothing to help scientists study the phenomenon. Furthermore, these reports are sparse (not all pilots bother to radio them in) and subjective (what one pilot calls "light" turbulence another might consider "moderate").

Turbulence-avoidance efforts today also depend on thorough pre-flight planning. The computer-aided flight-planning tools used by most airlines provide information, such as suspected areas of wind shear, that can provide clues to locations of possible turbulence. Most airlines also employ flight dispatchers, specialists trained in meteorology and aircraft performance. In addition to developing a plan for a flight's course, altitude, speed, and fuel consumption, dispatchers monitor a flight's progress and advise the pilots by radio of possible problems along the way, including areas of potential turbulence.

Airlines have also tried to educate pilots and flight crews about the need to anticipate turbulence and prepare for its effects. In 1997, the FAA produced a turbulence training kit in partnership with the aircraft maker McDonnell Douglas and the Air Transport Association, a trade group for the largest North American airlines. Distributed free to airlines, the kit includes a primer on turbulence, outlines of model training programs, and briefings advising airline executives and employees to take the problem seriously.

Studies have found that poor communication is the root cause of most turbulence-related injuries and aircraft damage. For instance, Delta Air Lines pilot Robert Massey recalls an incident in which the Airbus Industrie A310 he was piloting across the Atlantic from Europe to New York City suddenly plunged 300 meters (1,000 feet) in clear air near Greenland. The plane was then rocked by rough air for more than 10 seconds. After the mishap, Massey learned that the airline's dispatchers had expected turbulence in the area but issued no warning about it. The dispatchers had decided to wait until a pilot actually reported a turbulence encounter before issuing a warning.

Safety experts say that better communications among flight crews and between the crew and passengers can also help prevent turbulence-related injuries. They urge flight crews to tell passengers to keep their seat belts fastened whenever they are seated. Although airlines are not required to report injuries caused by turbulence, an estimated 90 percent of such injuries happen to people who were not using their seat belts when the turbulence struck—even, in some case, when the aircraft's "Fasten Seat Belt" signs were on. Moreover, many passengers interpret a turned-off sign as a signal to unbuckle, unaware that a plane can be buffeted by turbulence before the pilots have a chance to relight the sign.

## Predicting where turbulence will strike

Experts agree that a greater use of seat belts, while helpful, can only do so much. In order to solve the problem of turbulence, they contend, a way must be found to detect and avoid it. With that goal in mind, some researchers are seeking to identify patterns in weather data that can be used to predict when and where turbulence might occur.

The huge amounts of of weather data routinely collected by federal agencies represent a tremendous resource for investigators studying turbulence. For several decades, meteorologists have used radar to monitor and help predict the weather. Radar (which stands for *r*adio *d*etection *a*nd *r*anging) detects approaching storms by bouncing radio signals off aerosols moving within currents of air. Aerosols are particles in the atmosphere, such as rain, snow, ice, or dust. The most significant advance in radar systems in recent years has been the development of a technology called Doppler radar, which employs a principle known as the Doppler shift in analyzing the reflected radar signal, or echo. An echo from a distant object that is in motion will have a slightly different frequency than the original signal, depending on the speed of the object and the direction it is moving in relation to the radar transmitter. This change in frequency is called the Doppler shift. In weather forecasting, computers connected to a Doppler radar system watch for Doppler shifts in air masses and use them to calculate the speed and direction at which the air is moving. By revealing areas of wind shear, Doppler radar can pinpoint areas of potentially turbulent airflow. This has made Doppler radar an extremely valuable tool in aviation safety.

A growing volume of Doppler radar data is being supplied by NEXRAD (Next Generation Weather Radar), a nationwide network of advanced weather radars operated by the National Weather Service (NWS). The NWS began installing NEXRAD in the 1980's to help predict severe weather. Completed in 1999, NEXRAD consists of more than 150 ground stations that collect weather data on more than 90 percent of the skies over the United States at altitudes of 3,000 meters and up. The system is already proving its worth. Upon receiving the time and location of a aircraft turbulence report, NWS researchers can compare NEXRAD records from that region for patterns that might reveal recognizable "signatures" of turbulence.

Another NWS advance, called the Rapid Update Cycle 2 (RUC-2), also promises to help meteorologists identify areas of potential turbulence. This forecasting tool is an advanced computer *model* (simulation) of the atmosphere above the United States that analyzes weather data more quickly than the previous model and generates forecasts more frequently. The old model put out nationwide forecasts (broken down by subregions) every three hours. RUC-2 generates more precise three-hour forecasts and distributes them hourly. RUC-2 provides forecasts for 25-square-mile (65-square-kilometer) sectors, compared with the 37-square-mile (96-square-kilometer) sectors used in the previous system, and divides atmospheric data into 40 different altitude layers instead of the previous 25. This greater resolution should allow government and airline meteorologists to be more precise in forecasting the times and locations of probable turbulence.

### Airborne turbulence detection systems

To help researchers learn more about turbulence, an airborne data collection system was being tested in 1999. It involved loading a special computer program into the onboard computers of more than 200 aircraft operated by United Airlines. The program, developed by scientists at the National Center for Atmospheric Research (NCAR) in Boulder, Colorado, calculates precisely how much turbulence a plane encounters in flight. The software was designed for use with the existing computers and sensors that make up a modern airliner's flight manage-

### Predicting areas of turbulence

Recent technological advances have helped *meteorologists* (weather forcecasters) monitor the weather for conditions likely to cause turbulence as they develop.

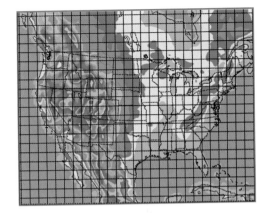

RUC-2, an new computer model, creates three-hour weather forecasts several times a day for smaller regions than ever before. Such frequent and detailed forecasts help airlines plan routes for their flights that avoid severe weather (purple and red areas).

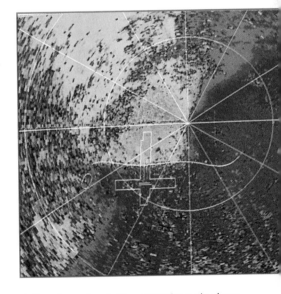

A Doppler radar station spots the early signs of a developing storm by detecting two large air masses (yellow and blue) that have collided and begun to swirl around each other. Early warnings of such conditions can help airliners avoid turbulence.

## Onboard turbulence warning systems

Several companies are developing onboard systems to enable individual aircraft to detect turbulent air in their path in time to evade it.

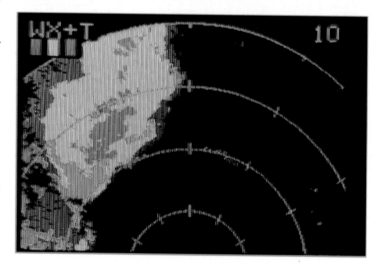

An onboard radar system being developed for use on commercial aircraft by the electronics firm Rockwell Collins enables pilots to see thunderstorms ahead of the aircraft as red patches, *top*. The system also incorporates Doppler radar. With the Doppler function activated, the system also reveals potentially dangerous turbulence ahead of the storms, shown as purple areas, *bottom*.

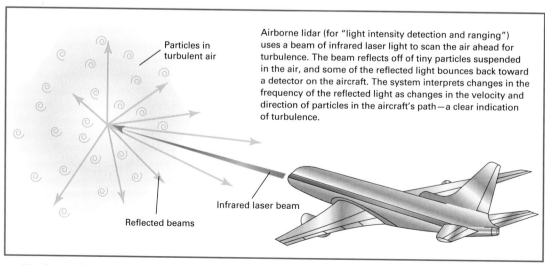

Particles in turbulent air

Airborne lidar (for "light intensity detection and ranging") uses a beam of infrared laser light to scan the air ahead for turbulence. The beam reflects off of tiny particles suspended in the air, and some of the reflected light bounces back toward a detector on the aircraft. The system interprets changes in the frequency of the reflected light as changes in the velocity and direction of particles in the aircraft's path—a clear indication of turbulence.

Infrared laser beam

Reflected beams

## Future technology

Future turbulence prediction and detection systems may include a unified display giving pilots and air traffic controllers weather information from a variety of sources. Researchers in 1999 were developing prototypes of such "Forecast/Nowcast" systems.

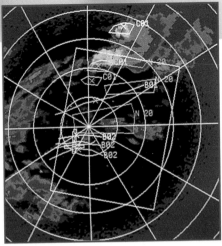

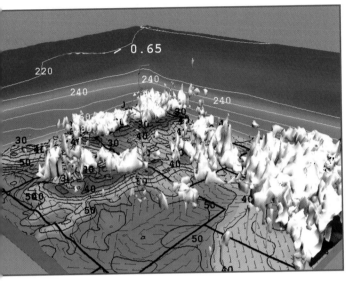

An integrated display, *left,* combines temperature and wind measurements, radar images, and satellite data that pilots can use to monitor conditions at their destination. Another prototype system, *above,* gathers and combines various types of weather data and uses a computer program to generate short-term predictions of weather conditions surrounding an airport.

ment system. The system makes routine measurements of the airplane's speed, attitude, and direction of flight, as well as of atmospheric forces being exerted on the aircraft. The NCAR program creates a detailed description of turbulent air recorded by an aircrafts's sensors by analyzing how the turbulence affects the aircraft's motion. It then transmits that information to stations on the ground through the plane's Aircraft Condition and Reporting System (ACARS), a data link that airlines use in everyday operations.

Even the best predictions can be inaccurate, however. More must still be learned about the nature of turbulence before it can be accurately forecast. Therefore, many researchers are more interested in fitting aircraft with equipment that detects turbulence ahead of a plane and warns the pilots. Engineers had yet to perfect such a device in 1999, but they were working on several different systems.

Given the success of ground-based Doppler radar in helping aircraft avoid wind shear, engineers have been trying to develop an onboard Doppler radar system to detect turbulence ahead of an aircraft. By 1999, two electronics firms, AlliedSignal Aerospace and Rockwell Collins, had developed an airborne system that could detect wind shear. However, these devices can only be used during landing approaches because the radar transmitter is located beneath the airplane's nose and points slightly downward. To make the system useful throughout a flight, engineers were working on a way to point the beam straight ahead of the aircraft while it is at cruising altitude.

Another manufacturer has been working with NASA and NCAR to develop an airborne turbulence detector that is similar in principle to Doppler radar but employs light waves. The system, called Airborne Coherent Lidar for Advanced In-flight Measurement (ACLAIM), was developed for NASA's Dryden Flight Research Center in California, by Coherent Technologies Incorporated, of Lafayette, Colorado. (*Lidar* stands for "light intensity detection and ranging.") The system transmits a laser beam that reflects off dust and other tiny particles in the air and uses the Doppler shift to measure the speed and direction of the particles. Any significant changes in the motion of particles in the skies ahead of the aircraft is a clear indication of turbulence.

A ground-based version of the ACLAIM system was ground-tested in mid-1997 and was able to detect turbulence in the path of airplanes that were coming in for a landing. However, the test unit employed a fixed beam. For use on aircraft, the beam must be able sweep across the sky ahead of the plane in search of turbulent conditions.

Another system under evaluation in 1999 used a laser beam to detect wake turbulence by measuring not the motion of particles in the air but the *acoustic* (sound) energy generated by their movement under different weather conditions. The system, dubbed SOCRATES (Sensor for Optically Characterizing Ring-Eddy Atmospheric Turbulence Emanating Sound), was developed by Flight Safety Technologies of New London, Connecticut. SOCRATES is based on the idea that turbulence causes density changes in air that can be detected as sound. A two-week test in May 1998 at New York City's Kennedy International Airport confirmed this theory, but further testing was needed.

So far, the best turbulence-detection systems under evaluation would give pilots reliable warnings no more than one minute before the plane hits rough air. For any such system to be useful, however, experts agree that it would need to provide a warning at least two minutes ahead of time. This is considered to be the minimum amount of time needed to give people time to sit down and strap in. Such a system must also be extremely reliable. Too many false alarms would cause passengers to begin ignoring the warnings, making the system almost useless.

## Integrated systems—tracking turbulence in the future

In the future, some experts predict, a variety of turbulence prediction and detection methods and technologies will be combined in a comprehensive detection and warning system. Researchers at NCAR are working toward the goal of equipping commercial airliners with an integrated system that would merge onboard sensor data, radar and satellite data, and short-term computer weather models in a unified cockpit display. Such a "Forecast/Nowcast" system would give pilots comprehensive data on the present situation as well as projected flight conditions up to an hour in the future. For the rare occasions when turbulent air cannot be avoided, these systems might also incorporate the capability to minimize the effects of turbulence by using the plane's automatic pilot to compensate for any sudden movements.

## If There's Turbulence, Even The Most Loving Arms Can't Hold Him.

You make sure your carry-on bag fits overhead. You tuck your purse under the seat in front of you. You secure your seat backs and tray tables in their upright position. Doesn't your *child* deserve the same protection? If there's unexpected turbulence during a flight, a **child safety seat** is the safest and most secure place for your little one. To learn more about safe air travel with children, call 1-800-FAA-SURE.

TURBULENCE happens.

U.S. Department
of Transportation
**Federal Aviation
Administration**

A major campaign sponsored by the Federal Aviation Administration (FAA), individual airlines, and other airline industry organizations has focused on educating passengers about the dangers of turbulence. Aviation experts agree that increased awareness of the danger can help reduce the number of turbulence-related injuries.

However, even after turbulence forecasting techniques are made more accurate and detection systems are refined, airline passengers will still be best served by taking one simple precaution—always keeping their seat belts fastened. A seat belt may be uncomfortable, but the alternative is to risk a possibly serious injury. Even in the best flying conditions, it pays to be cautious, because—as history and experience prove—you never know when turbulence will strike.

## For additional information:

### Books and periodicals

Cowen, Ron. "Clearing the Air About Turbulence," *Science News,* June 27, 1998, pp. 408-410.
"Turbulence" (special section). *Aviation Week & Space Technology,* July 27, 1998, pp. 40-80.

### Web sites

ACLAIM Project, NASA Dryden Flight Research Center
(www.dfrc.nasa.gov/PAO/PAIS/HTML/aclaim/aclaim.html)
Aviation Weather Development Laboratory (AWDL)
(www.rap.ucar.edu/projects/awdl/awdl.html)
Federal Aviation Administration Office of Public Affairs
(www.faa.gov/apa/turb/turbhome/Hometxt.htm)

By Larry D. Martin

# Fossils, Feathers, and Theories of Flight

**Many scientists claimed in 1999 that recently discovered fossils prove that birds evolved from dinosaurs, but others disagreed.**

The Liaoning province of northern China is a fossil hunter's paradise. During the 1990's, local residents—and later scientists from around the world—unearthed numerous well-preserved fossils from about 140 million to 120 million years ago. Many of these specimens became the primary focus in an ongoing effort to explain the origin and evolution of birds. For example, in 1995, the Chinese site introduced the world to *Confuciusornis sanctus,* the oldest known bird with a modern-looking, toothless bill. And in 1998, scientists announced a find that was even more exciting—a dinosaur with feathers. That specimen, named *Caudipteryx zoui,* promised to play a key role in one of the hottest controversies in modern science: Did birds evolve from dinosaurs or from an earlier common ancestor of both birds and dinosaurs?

Many *paleontologists* (scientists who study prehistoric life) declared that they had answered that question once and for all with the discovery of *Caudipteryx.* In fact, some researchers maintained that the new specimen proves that birds not only evolved from dinosaurs but were remnants of the dinosaur family. A few paleontologists drove their point home by suggesting that hummingbirds are the smallest dinosaurs and that certain fast-food restaurants were serving up buckets of "dinosaur wings." Other scientists, however, expressed skepticism, arguing that *Caudipteryx* had been misidentified. They claimed that the creature was not a feathered dinosaur but a flightless bird.

To determine which of those interpretations is correct—and to settle the debate in general—scientists must compare the anatomical features of many species, including ones known to be either dinosaurs or birds and those believed to be more distantly related. The researchers must also determine what the various similarities and differences between dinosaurs and birds tell us about the bird-dinosaur relationship. Finally, they must contend with the gaps in the fossil record that leave many questions unanswered. In 1999, scientists continued to address these issues, gather new evidence, and reevaluate the conclusions of their colleagues in their attempt to understand the origin of birds.

## Two theories of the origin of flight

All paleontologists and *ornithologists* (bird specialists) agree that birds and dinosaurs descended from a group of animals known as archosaurs, or "ruling reptiles," which include modern crocodiles. But these scientists have differing opinions on how dinosaurs and birds fit into the archosaur family tree.

Many paleontologists believe that birds evolved during the Late Jurassic Period (170 million to 138 million years ago) from a group of meat-eating dinosaurs called theropods. According to most researchers who see a direct relationship between dinosaurs and birds, the theropods were *endothermic,* or warm-blooded. This means that, like modern birds and human beings, they maintained a constant, relatively warm body temperature regardless of the temperature of their environment. In this view of theropods, feathers evolved for warmth, as a means of conserving body temperature. The ability to fly developed later as a re-

*Opposite page:*
A model based on the fossil remains of *Caudipteryx zoui,* a species discovered in China in 1998, depicts a dinosaur-like creature with feathers. Some researchers claim that the species was a feathered dinosaur, while other scientists believe it was a flightless bird.

## Terms and concepts

sult of *bipedal* (two-footed) running and jumping to catch prey. This scenario is often called the "ground-up" theory of bird evolution.

The alternative explanation of bird evolution is called the "tree-down" theory. In this scenario, birds arose from an ancestral line of small, lizardlike *arboreal* (tree-dwelling) creatures that predated the appearance of dinosaurs about 230 million years ago. These bird ancestors, according to the theory, were—like modern reptiles—*ectothermic*, or cold-blooded, with body temperatures changing with the temperature of the environment. Ornithologists and some paleontologists in the tree-down camp argue that the ability to fly evolved from jumping among branches and later gliding from tree to tree—landing on the ground was initially an unhappy accident. The theory holds that feathers evolved for flight and only later were used for insulating birds' bodies. Birds' bipedal posture would have evolved from living and perching in trees.

## A century-old debate

These two theories have been in contention for well over 100 years, since shortly after the discovery of what most scientists think was the first bird. This primitive crow-sized bird, called *Archaeopteryx*, lived about 140 million to 150 million years ago. The unearthing of an *Archaeopteryx* fossil in 1861 in Bavaria (part of present-day Germany) provided the first evidence of a transition between birds and reptiles. *Archaeopteryx* had feathers and wings much like those of modern birds, but it also had reptilian features such as teeth, clawed fingers, and a long tail.

Another fossil find in 1861 set the stage for the bird-dinosaur debate. This was the discovery of a nearly complete skeleton of *Compsognathus*, a theropod dinosaur only slightly larger than *Archaeopteryx* that also lived 150 million to 140 million years ago. The realization that most dinosaurs were bipedal suggested that dinosaurs may have been ancestors of birds, a relationship first popularized in 1868 by the English biologist Thomas H. Huxley. Most paleontologists agreed with Huxley's theory, though they differed in their specific choice of dinosaurian ancestors. They also began to debate the question of how flight originated.

In the 1920's, a Danish biologist, Gerhard Heilmann, proposed that the closest relatives of birds were arboreal archosaurs, called pseudosuchians or thecodonts. These creatures were identified from fragmentary fossils from the Early Triassic Period (240 million to 220 million years ago) found in South Africa.

Variations on the theories originated by Huxley and Heilmann have been the basis of the bird-dinosaur debate ever since. Geologist John H. Ostrom of Yale University in New Haven, Connecticut, resurrected Huxley's research in 1973 and has since championed the ground-up theory. The most recent overview of the tree-down theory—the argument I began to endorse in 1980—is the 1996 book *The Origin and Evolution of Birds* by ornithologist Alan Feduccia of the University of North Carolina (UNC) in Chapel Hill.

Numerous discoveries over the years have shifted the argument in favor of one view or the other. For example, tree-down theorists cited the

---

**Terms and concepts**

**Archaeopteryx:** The oldest known bird, which lived about 150 million to 140 million years ago.

**Archosaurs:** A group of animals that were the ancestors of some modern reptiles and from which birds and dinosaurs evolved.

**Caudipteryx zoui:** An animal, which lived more than 120 million years ago, that many scientists believe was a feathered dinosaur.

**Cladistics:** A method of analysis that compares similarities among several animals to construct a history of their evolution.

**Confuciusornis sanctus:** A bird from more than 120 million years ago with a modern toothless bill.

**Ground-up theory:** A theory of bird evolution holding that birds arose from ground-dwelling theropod dinosaurs.

**Protarchaeopteryx robusta:** An animal that lived more than 120 million years ago and that may have been a feathered dinosaur.

**Protofeathers:** Bristlelike structures that may have evolved into feathers.

**Sinosauropteryx prima:** A theropod dinosaur that lived more than 120 million years ago and that may have had protofeathers.

**Theropods:** Meat-eating dinosaurs that may have evolved into birds.

**Tree-down theory:** A theory of bird evolution holding that birds arose from small, tree-dwelling lizardlike animals.

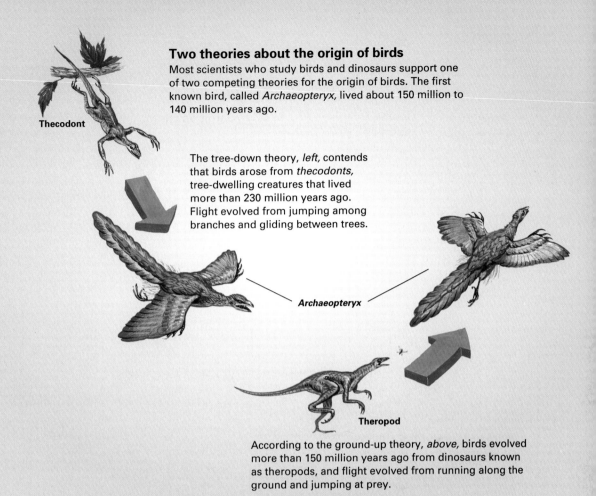

## Two theories about the origin of birds

Most scientists who study birds and dinosaurs support one of two competing theories for the origin of birds. The first known bird, called *Archaeopteryx,* lived about 150 million to 140 million years ago.

**Thecodont**

The tree-down theory, *left,* contends that birds arose from *thecodonts,* tree-dwelling creatures that lived more than 230 million years ago. Flight evolved from jumping among branches and gliding between trees.

*Archaeopteryx*

**Theropod**

According to the ground-up theory, *above,* birds evolved more than 150 million years ago from dinosaurs known as theropods, and flight evolved from running along the ground and jumping at prey.

absence in theropods of a wishbone—a bone common to all birds—as evidence that there was no link between dinosaurs and birds. When a theropod with a wishbone was first discovered in 1976, the ground-up theorists celebrated a victory. But the discovery of a wishbone in a thecodont that predated both birds and dinosaurs made the contested feature less relevant to the story, at least according to some scientists.

In 1998, many paleontologists contended that the discovery of *Caudipteryx* proved that dinosaurs were the true ancestors of birds. Other fossils from the Chinese site, as well as additional specimens from locations elsewhere in the world, also joined the growing body of evidence in favor of the theory. Nonetheless, there were still a great many scientists who did not accept these conclusions, and so the debate continued.

One issue central to the disagreement about bird ancestry is a method of analysis known as cladistics. In this technique, researchers evaluate numerous anatomical characteristics in the fossil remains of at least three different organisms that are believed to share some kind of ancestral history. They note similarities and differences in the pelvic bones, feet, skull, limbs, and whatever other parts of the body can be identified.

**The author:**
Larry D. Martin is professor of ecology and evolutionary biology and curator of vertebrate paleontology at the Natural History Museum and Biodiversity Research Center at the University of Kansas in Lawrence.

## Another story of flight

Before the story of dinosaurs or birds began—perhaps 240 million years ago—a group of reptilian creatures known as pterosaurs began to rule the skies. Some of these flying reptiles were as small as sparrows, but ones called pterodactyls had wing spans of up to 12 meters (39 feet).

Pterosaurs' wings were made of skin that stretched from the side of the body and along the arm to the end of a very long fourth finger. The animals had no feathers, but their bodies may have been covered with fur.

Although pterosaurs lived alongside both birds and dinosaurs until about 65 million years ago, they do not fit into the story of bird evolution. Pterosaurs belong to a much older branch of the reptilian family tree that left no modern ancestors.

After noting all the similarities and differences, the researchers generally find that some characteristics are shared by all of the animals. These common features are considered "primitive"—they are features that the organisms all inherited from an early common ancestor.

Further analysis of the similarities among the animals will almost always reveal that two of the organisms have more traits in common with each other than they do with the third organism. Such a finding indicates that they probably have a common ancestor that is less closely related to the third. The characteristics shared by the two organisms are called *derived features*—traits that are derived from the same source. The resulting information about which features are primitive and which are derived allows researchers to construct a *cladogram,* a kind of family tree that illustrates how the organisms are related to one another.

### Problems with comparing fossils

Not all scientists agree with the conclusions drawn from cladistics. A weakness of cladograms, many researchers say, is that they do not include the chronological order in which specimens appear in the fossil record. For instance, if birds are the descendants of dinosaurs, the specific dinosaurs that gave rise to birds should appear in the fossil record before the first real bird. In 1998, paleontologists at the National Geological Museum of China in Beijing and colleagues from the United States created a cladogram for *Caudipteryx.* Their analysis led them to conclude that the animal was a feathered dinosaur, a finding they cited as strong evidence that birds evolved from theropods. *Caudipteryx,* however, is dated at about 140 million to 120 million years old. *Archaeopteryx* is dated at 150 million to 140 million years old. Therefore, the age of the fossils suggests that the two species either were alive at the same time or that *Archaeopteryx* actually lived before *Caudipteryx.*

Proponents of cladistic analysis note that such chronological discrepancies do not necessarily rule out the conclusions of a cladogram. They argue that the similarities still indicate a significantly close relationship. For example, the features of *Archaeopteryx* and *Caudipteryx* might be similar enough to suggest that if they were not directly related, they at least shared a common ancestor that lived not too long before them and that looked very much like *Caudipteryx.* This argument, of course, rests on the assumption that fossils of such an ancestor will eventually be found.

Another problem with constructing a family tree based on shared physical features is a phenomenon known as convergence, by which two unrelated organisms independently develop similar characteristics because they have similar patterns of behavior. An example of convergence among modern organisms is the presence of a *dorsal* (back) fin on both sharks and dolphins. A shark is a fish, and a dolphin is a mammal—two groups of animals only distantly related to each other. Nonetheless, both sharks and dolphins evolved dorsal fins to help them navigate in the water. This shared feature does not help us understand how the two groups of animals are related and how far back in time one has to go to find a common ancestor. In fact, when scientists trace the evolutionary lineage of dolphins and sharks back in time, they find very few features in common between the animals' ancestral lines.

## Reaching different conclusions from the same evidence

In the debate among paleontologists and ornithologists, some researchers argue that similarities between birds and theropod dinosaurs are evidence of a close evolutionary relationship. Others argue that the similarities are the result of convergence. A good example of this disagreement is a 90-million- year-old dinosaur called *Unenlagia comahuensis,* discovered in Argentina in 1996. Standing about 1.3 meters (4 feet) tall and with proportionally short forelimbs, the dinosaur would not have been able to fly even if it had feathers, but Argentine researchers noted birdlike qualities in the skeleton. Paleontologists who support the ground-up theory of bird evolution were particularly interested in the shoulder joint, which they concluded faced outward. This feature would have allowed *Unenlagia* to hold its arms out to the side and move them up and down. Other theropod dinosaurs had a shoulder joint that faced down and backward so that they could not easily be brought out to the side and form a wing. The researchers maintained that, by holding out its arms, *Unenlagia* could stay balanced while running.

Birds also have shoulder joints facing outward that allow them to flap their wings. Many paleontologists contended that this similarity in the shoulder joints shows a close ancestral relationship between birds and dinosaurs. They argued that birds and *Unenlagia* shared a common theropod ancestor that developed this feature in order to run with more control and balance. This ancestral theropod's descendants then evolved along two separate paths. Animals in one ancestral line learned to make a flapping motion and eventually to fly. The other branch led to *Unenlagia,* which retained this particular feature but could not fly.

Researchers who do not support the ground-up theory argued that the anatomical similarity between birds and *Unenlagia* was simply the result of convergence. They noted that the similar feature could have evolved independently in two unrelated groups of animals that moved their arms in a similar fashion for different reasons—wing flapping versus the need to maintain balance. The similarities, therefore, did not necessarily suggest an ancestral link. They also noted that the appearance of *Archaeopteryx* in the fossil record at least 50 million years before *Unenlagia* supported the argument for convergence. During that amount of time, two animals that share a common ancestor would probably "grow apart" and appear less like each other rather than more so.

Another disagreement over the conclusions about *Unenlagia* was the condition of the specimen. The researchers did not have a complete skeleton—an occurrence that is not uncommon in the study of fossils. In this case, the scientists were working with pelvis, leg, forearm, and shoulder bones, and a few other fragmentary pieces. Researchers who disagreed with the conclusions noted that there was not enough evidence to determine how the shoulder joint worked and that the joint may very well be no different than ones seen in other dinosaur shoulders.

This problem highlights another general disagreement over cladistic analysis. Cladograms are based on the similarities identified among the various organisms, but this analysis only works if researchers are certain that similar features really are alike. Paleontologists often do not agree on such similarities because of incomplete specimens, the poor condition of fossils, or different approaches to comparing features.

The details of bird and dinosaur hands reveal this conflict well. (Although we may not think of birds as having hands, they have anatomical features in their wings that are the equivalent of a human hand.) A commonly cited similarity between the hands of theropod dinosaurs and birds is the presence of a wrist bone called the semilunate carpal. This bone, shaped like a half moon, allowed for wrist motion in theropods similar to that of birds. For a bird, this motion is necessary for tucking its wings up against its body when it is not flying. In the 1970's, Ostrom noted similar bones in *Archaeopteryx* and a theropod known as *Deinonychus,* which lived about 110 million to 100 million years ago. But scientists still do not agree on whether they are actually the same bone.

The ancient reptilian ancestors of birds and dinosaurs that lived during the Pennsylvanian Period (330 million to 290 million years ago) had

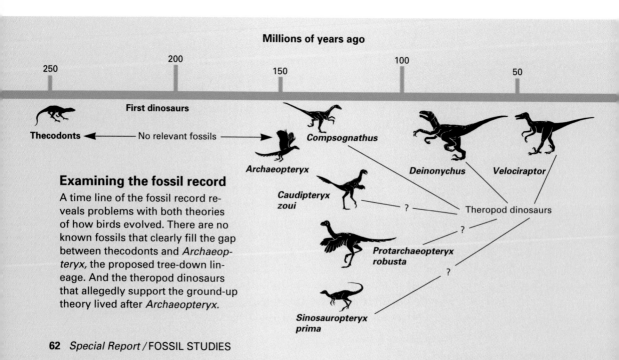

**Millions of years ago**

**Examining the fossil record**

A time line of the fossil record reveals problems with both theories of how birds evolved. There are no known fossils that clearly fill the gap between thecodonts and *Archaeopteryx,* the proposed tree-down lineage. And the theropod dinosaurs that allegedly support the ground-up theory lived after *Archaeopteryx.*

## The evidence for a dinosaur origin

Researchers base most of their conclusions about the origin of birds on the comparisons of fossils. Many scientists claim that their interpretation and reconstruction of the *Archaeopteryx* skeleton, *top,* reveal dozens of similarities to the theropod dinosaurs, such as one called *Compsognathus, right.* These proposed similarities are the basis for the theory that birds evolved from dinosaurs.

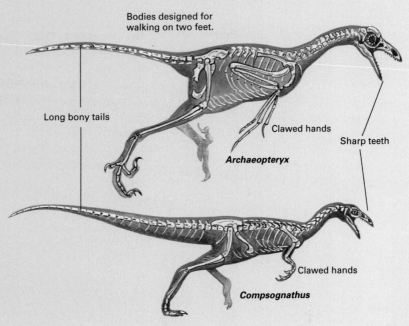

Bodies designed for walking on two feet.

Long bony tails

Clawed hands

Sharp teeth

*Archaeopteryx*

Clawed hands

*Compsognathus*

wrists with several bones. During millions of years of evolution, some of those bones were lost and some fused together to create "new" bones. Consequently, birds and theropods ended up with fewer wrist bones.

A similarity between the semilunate carpal of *Deinonychus* and *Archaeopteryx* is only significant if researchers can demonstrate that the bone evolved from the same pattern of bone loss and bone fusing in each group. The pattern is indeed the same, say two proponents of the ground-up theory, paleontologists Kevin Padian of the University of California at Berkeley and Luis M. Chiappe of the American Museum of Natural History in New York City. Thus, they insist that *Deinonychus* and *Archaeopteryx* are closely related. However, Alan Feduccia and a colleague at UNC, ornithologist Anne Burke, disagree with that conclusion. They argue that the semilunate bones of the two species were not derived from the same bones of the early reptiles. Feduccia and Burke theorize that the common ancestor of *Deinonychus* and *Archaeopteryx* lived much further back in time—probably before the first dinosaurs.

A similar disagreement centers on the *digits,* or fingers, of *Archaeopteryx* and theropod dinosaurs. Once again, the reptilian ancestors had more bones—five digits on each hand. *Archaeopteryx* and most theropods had three similar clawed digits on each hand. Ground-up theorists point to this shared feature as strong evidence of a direct bird-dinosaur relationship. Feduccia and Burke, however, maintain that the *Archaeopteryx* lineage lost the first and fifth digits—the "thumb" and "little finger"— during millions of years of evolution, while theropods lost the fourth and fifth digits—the "ring finger" and "little finger." If their hypothesis is true, then the three-digit similarity would not be relevant to the discussion of bird ancestry. The disagreement over the similarity of hands is

## Different explanations for similar features

Many proponents of the tree-down theory have noted that the similar features of birds and dinosaurs may have evolved independently from an earlier common ancestor.

The hands of *Archaeopteryx* and *Deinonychus,* a theropod dinosaur, both have three clawed fingers and a similar *semilunate carpal,* a wrist bone shaped like a half moon. These resemblances seem to support the ground-up theory, but supporters of the tree-down theory have a different explanation of how these structures evolved.

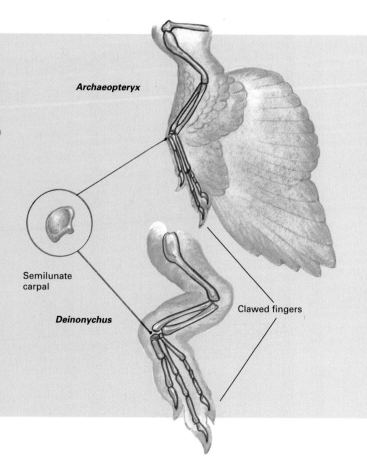

**Archaeopteryx**

Semilunate carpal

**Deinonychus**

Clawed fingers

only a small part of the debate. In fact, almost every characteristic that one group of researchers has cited as a link between birds and theropods has been contested by the other group.

Even when scientists agree on the identity of certain bones, they may draw different conclusions from them. One hotly contested fossil in the proposed bird-dinosaur lineage is a specimen called *Rahona ostromi,* which was found in 1995 in Madagascar fossil beds from the Late Cretaceous Period (about 100 million to 65 million years ago). Paleontologist Catherine Forster of the State University of New York at Stony Brook and her colleagues identified *Rahona* as a bird with dinosaurian features in 1998. Although the fossil included no feathers, the remains had birdlike characters, including little bumps on an arm bone where wing feathers might have been attached. Some researchers criticized the findings, noting that the wing bone might belong to another specimen, of a species called *Vorona berivotrensis,* found at the same site. That fossil, which Forster's team identified as a bird in 1996, had no wings on it. The scientists who believe that the wing associated with *Rahona* really belongs to *Vorona,* contend that *Rahona* was actually a dinosaur.

The bird-dinosaur debate took a slightly different direction in the late 1990's after the discovery of dinosaur specimens with fairly well-preserved internal organs. These finds were significant in themselves be-

Some tree-down theorists have compared the *Archaeopteryx* hand to that of a *Petrolacosaurus*, a common ancestor of birds and dinosaurs. This species had five fingers and two rows of wrist bones. After analyzing the various bones, the researchers concluded that, in intermediate species leading to *Archaeopteryx*, the first and fifth fingers and several wrist bones, *purple,* were lost. According to this theory, the *Archaeopteryx* semilunate carpal, *red,* formed from the bone over the third finger.

When the scientists compared the hands of *Petrolacosaurus* and *Deinonychus,* they concluded that the *Deinonychus* hand evolved from the loss of the fourth and fifth fingers and a different set of wrist bones. The *Deinonychus* semilunate carpal, the scientists argued, formed from the fusion of bones over the first and second fingers.

**Petrolacosaurus**

**Archaeopteryx**

**Petrolacosaurus**

**Deinonychus**

cause soft tissues are rarely preserved in the fossil record. One of the fossils, of a small dinosaur called *Sinosauropteryx prima,* a creature similar to the small theropod *Compsognathus,* was discovered in China's Liaoning province in 1996. Its well-preserved features included the fossilized remains of what appear to have been the lungs, liver, and intestines. The remains of another small theropod species, *Scipionyx samniticus,* found in Italy in 1983 (but not studied and classified until 1998), revealed the liver and colon.

Paleontologist and physiologist John Ruben of Oregon State University in Corvallis studied both of these fossils. He concluded that the size and arrangement of the organs in the specimens suggested that both dinosaurs had a pistonlike breathing mechanism, similar to that of a modern crocodile, in which a diaphragm draws air into the lungs and then forces it out again. Birds, on the other hand, have a more complicated system in which air is continuously drawn through the lungs and through numerous air sacs in the body cavity. Ruben asserted that the breathing mechanism of birds could not have evolved from the pistonlike breathing apparatus of the theropods. Ruben's theory, however, did not convince everyone, because, as many researchers noted, it is difficult to draw conclusions from the fossilized remains of soft tissues.

### Feathered dinosaurs or flightless birds?

In any case, many paleontologists have become less concerned about theories based on the study of internal organs and more interested in another feature—feathers. When *Sinosauropteryx* entered the story, most researchers were struck not so much by the dinosaur's internal organs as by small bristlelike impressions down its back and along the base of its tail. Paleontologists Ji Qiang and Ji Shu-An at the National Geological Museum of China originally described them as primitive feathers.

The following year, the Philadelphia Academy of Sciences sent a team of paleontologists, including John Ostrom and myself, to examine the specimen. Our team did not agree with the Chinese scientists' interpretation, since the structures did not have the characteristic central shaft and branches of feathers. Some team members, however, were willing to identify them as *protofeathers*, structures that in time could have evolved into actual feathers. In any case, many scientists cited the structures—whether primitive feathers or protofeathers—as evidence that birds evolved from dinosaurs. In November 1997, however, Ruben argued that the impressions may actually be the remains of connective tissues under the dinosaur's skin rather than structures on the surface of the skin.

Another candidate for consideration as a feathered dinosaur is *Protarchaeopteryx robusta*, which was also discovered at the Liaoning site. The fossilized remains of this chicken-sized creature included the impressions of feathers that were believed to be attached to its tail. In 1996, Ji Qiang and Ji Shu-An first identified *Protarchaeopteryx* (the name means *Before Archaeopteryx*) as a flightless transitional species between theropods

### One fossil, two interpretations

The fossil of a theropod dinosaur called *Sinosauropteryx prima, left,* which was discovered in China in 1996, has bristlelike impressions along its back and along the base of its tail. After examining the unusual impressions, *below,* scientists drew different conclusions about what the structures were and what the dinosaur looked like.

and *Archaeopteryx*. The Chinese researchers later teamed up with paleontologists Philip J. Currie of the Royal Tyrrell Museum of Paleontology in Drumheller, Canada, and Mark A. Norell of the American Museum of Natural History in New York City to make a closer study of the specimen. In June 1998, that combined team published a cladistic analysis of *Protarchaeopteryx*, describing it as a theropod dinosaur with feathers and presenting the evidence as further support for the ground-up theory.

The scientists reached the same conclusions about *Caudipteryx*. They studied two well-preserved fossils of this species that showed impressions of wing and body feathers. Based on their analysis of both *Protarchaeopteryx* and *Caudipteryx*, the Chinese and North American team concluded that these were flightless theropods. The controversy, they insisted, was now settled: Some theropod dinosaurs evolved feathers to keep warm and later developed the ability to fly, thereafter evolving into birds.

Not unexpectedly, the ornithologists and paleontologists who favor the tree-down theory of bird evolution disagreed with that conclusion. In addition to voicing general concerns about the cladistic analysis of the Chinese fossils, they pointed out particular features of *Caudipteryx* suggesting that it was not a dinosaur at all but rather a flightless bird that evolved from earlier birds that could fly.

The researchers noted, for example, that *Caudipteryx* had only eight small teeth at the front of its bill. These relatively few teeth make *Caudipteryx* appear more advanced than *Archaeopteryx*, which had a mouth full of teeth. The near absence of teeth in *Caudipteryx* actually makes it seem more closely related to the toothless *Confuciusornis*. That species, which

Researchers who support the ground-up theory identified the bristlelike structures as primitive feathers or *protofeathers* (features that might have evolved into feathers). The scientists concluded that *Sinosauropteryx* was covered with these featherlike structures and was related to an earlier—but still undiscovered—theropod that evolved into *Archaeopteryx*.

Other investigators, in the tree-down camp, argued that the bristlelike structures were connective tissues under the skin. They concluded that *Sinosauropteryx* was covered with reptilian skin and that the species did not help explain the ancestry of birds.

# Filling in the evolutionary blanks

Based on the various interpretations of species in the fossil record, researchers have constructed two distinct versions of a bird family tree. Both scenarios begin with creatures called archosaurs, which all experts agree were the ancestors of birds and dinosaurs, as well as modern reptiles.

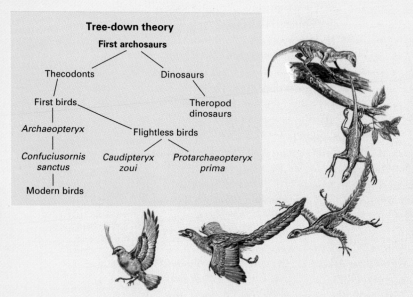

## Tree-down theory

**First archosaurs**

Thecodonts

Dinosaurs

First birds

Theropod dinosaurs

*Archaeopteryx*

Flightless birds

*Confuciusornis sanctus*

*Caudipteryx zoui*

*Protarchaeopteryx prima*

Modern birds

In the tree-down theory, dinosaurs and birds have separate lineages. Because scientists have not yet found fossils linking thecodonts and *Archaeopteryx*, they can only speculate about the physical appearance of the various species in this version of the evolutionary process leading to birds.

The ground-up theorists have drawn a family tree that shows birds as the descendants of theropod dinosaurs. Because the known theropods did not live early enough to be the ancestors of birds, these scientists must also use considerable guesswork in constructing an evolutionary sequence of species leading to birds.

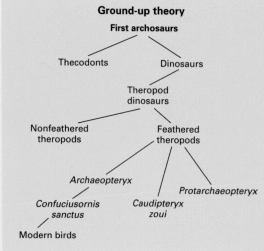

## Ground-up theory

**First archosaurs**

Thecodonts

Dinosaurs

Theropod dinosaurs

Nonfeathered theropods

Feathered theropods

*Archaeopteryx*

*Protarchaeopteryx*

*Confuciusornis sanctus*

*Caudipteryx zoui*

Modern birds

lived about 140 million to 120 million years ago, was more like a modern bird than *Archaeopteryx*. Some scientists also noted that the *Caudipteryx* tail was proportionally about as short as the tail of other birds that came after *Archaeopteryx*. In addition, the claws on the hands and feet of *Caudipteryx* were small and blunt, proportionally smaller than those of *Archaeopteryx* and of theropods that lived as recently as 65 million years ago.

Particular criticism focused on the conclusions about the feathers themselves. *Caudipteryx* had complex feathers, much like the flight feathers of modern birds, attached to its hand and fingers. The critics suggested, as Feduccia had noted before, that these structures were too complex to have evolved only for warmth. They also noted that for a meat-eating, predatory theropod, feathers on the hands would make handling food very difficult. To further their argument that *Caudipteryx* was a bird, they pointed out that the fossil appeared to have a gizzard full of stones. That digestive feature would be expected in an animal that eats plant foods, as many birds do, but not in meat-eating theropods.

Thus, the much-heralded reports of feathered dinosaurs, far from settling the debate about the origins of birds, actually served to rekindle the controversy. Indeed, the research on *Caudipteryx*, *Protarchaeopteryx*, and other species had really just begun in 1999. But as scientists on both sides of the debate studied the fossils, they expected, at the very least, to enhance their understanding of both birds and dinosaurs.

Paleontologists also hoped to unearth new fossils that might finally settle the bird-dinosaur question. That hope seemed justified, especially for the fossil beds in China, which for several years had been yielding an endless stream of surprises. The real solution to the origin of birds may lie further back in time than the fossil evidence available at the end of the 1900's takes us. If theropod dinosaurs and birds are closely related, researchers may find a specimen that predates *Archaeopteryx* and the most birdlike theropods. If the predecessor of birds is an even earlier common ancestor of both birds and dinosaurs, we would need to explore further back in time, where the fossil research is still in its infancy.

## For additional information:

### Books and periodicals

Ackerman, Jennifer. "Dinosaurs Take Wing." *National Geographic,* July 1998, pp. 74-99.

Feduccia, Alan. *The Origin and Evolution of Birds.* Yale University Press, 1996.

Martin, Larry. "The Big Flap." *The Sciences,* March/April 1998, pp. 39-44.

Padian, Kevin and Luis M. Chiappe. "The Origin of Birds and Their Flight." *Scientific American,* February 1998, pp. 38-47.

Shipman, Pat. *Taking Wing: Archaeopteryx and the Evolution of Bird Flight.* Simon and Schuster, 1998.

### Web sites

New Scientist: The Rex Files
    http://www.newscientist.com/nsplus/insight/rexfiles/rexfiles.html
University of California at Berkeley: The Flight Exhibit
    http://www.ucmp.berkeley.edu/vertebrates/flight/enter.html

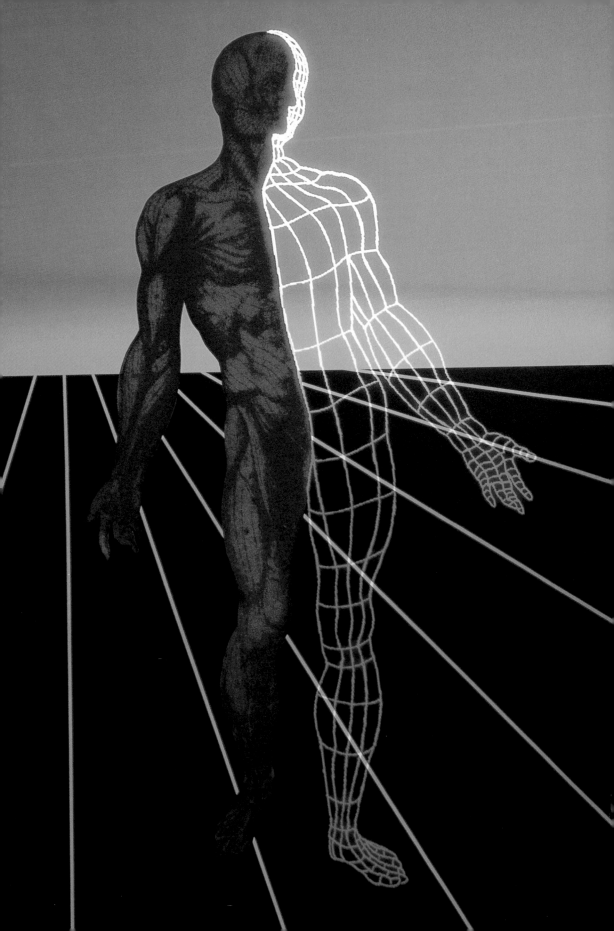

# Tissue Engineering— From Science Fiction to Medical Fact

by Rahul Koka and Joseph P. Vacanti

Researchers are close to achieving the centuries-old dream of replacing damaged or diseased body parts, an advance that promises to change medicine forever.

In July 1994, a 12-year-old Massachusetts boy named Sean McCormack became a part of medical history when surgeons implanted a laboratory-grown section of chest in his body. Sean had been born with a rare disorder that prevented the development of bone and cartilage tissue on the left side of his chest. As a result, the left side of his torso, covered only by skin, was unprotected by a rib cage. As Sean grew, his parents and doctors became increasingly concerned that he might suffer a serious injury. A strong blow to the chest—from a baseball, for example—could have been fatal to him.

Inquiries about Sean's case led the family to Harvard Medical School in Boston, where a team of researchers was doing pioneering work in tissue engineering. This is a new field of medical science that seeks to repair or replace body parts that have been damaged by disease or injury using living human tissues grown in a laboratory. After examining Sean, the Harvard physicians concluded that they could protect the boy's chest with a "shield" constructed from cartilage cells taken from his *sternum* (breastbone).

In the first phase of the procedure, surgeons extracted cartilage cells from Sean's sternum. The cells were then "seeded" onto a porous *matrix* (mold) made of a specially designed *polymer* (a substance, such as nylon, whose molecules are formed by long chains of simpler molecules). The polymer matrix, fashioned into the shape of Sean's chest, acted as a scaffold on which the cartilage cells could multiply. The seeded mold was then bathed in a mixture of nutrients to encourage cell growth. When the new cartilage had grown to a thickness of several millimeters, Sean underwent another operation to implant the shield into his chest.

**The authors:**

Rahul Koka is a research technician at the Laboratory for Tissue Engineering and Organ Fabrication at Massachusetts General Hospital in Boston. Joseph P. Vacanti is the John Homans Professor of Surgery at Harvard Medical School and the director of the Laboratory for Tissue Engineering and Organ Fabrication at Massachusetts General Hospital.

Within a few months, the polymer scaffolding had disappeared—it had been designed to gradually dissolve as it was overgrown by living tissue. All that remained of the implant were Sean's own cartilage cells, which had begun to fuse with his partial ribcage.

The procedure carried out by the Harvard researchers was hailed as a medical milestone. Since ancient times, medical practitioners have dreamed of replacing missing or defective body parts with healthy new ones. Over the centuries, doctors had given people a variety of *prostheses* (artificial parts), ranging from wooden "peg legs" to steel-and-plastic joints, but no prosthesis could adequately substitute for live, functioning tissue. By the late 1990's, medical technology had advanced to the point where several restorative procedures—organ transplants, tissue grafts, and techniques to reattach severed limbs—were relatively commonplace. But these procedures, too, often had their drawbacks.

Tissue engineering (sometimes known as bioengineering) promises to usher in a new era of medicine in which replacing missing or faulty body parts with perfectly functioning new ones is routine. These replacement parts will either be assembled in medical laboratories or regenerated in the body using the body's own natural ability to restore damaged tissues. This technology has the potential to lengthen the human life span and improve people's quality of life.

## The limitations of traditional replacement techniques

All parts of the human body age and eventually degenerate. Almost every part of the body is also vulnerable to damage from accident or disease. In addition, many people are born with *congenital* conditions, medical problems present at birth, ranging from deformities to abnormal— or even missing—tissues and organs. Compensating for such defects has been one of the major challenges facing modern medicine.

Drug treatments can sometimes make up for a malfunctioning internal organ. For instance, diabetes mellitus, a disease caused by the body's inability to remove *glucose* (sugar) from the blood, can be managed with insulin injections. Insulin is a hormone that controls blood sugar. Although daily insulin injections may be inconvenient, most diabetics lead otherwise normal lives.

In cases where drug treatment is not applicable, treatment often consists of surgery—for example, a coronary bypass operation for a person with advanced heart disease or an organ transplant for someone whose liver or kidneys are failing. Although organ transplants are usually quite successful, many patients who need one have difficulty obtaining a suitable organ due to a critical shortage of donor organs.

Because of the scarcity of donated organs, patients often spend months or even years on waiting lists before receiving a transplant. According to the United States Department of Health and Human Services, in January 1999 there were 64,580 registered patients on waiting lists to receive an organ transplant in the United States but only 9,286 registered organ donors. About 55 patients in the United States receive an organ transplant every day, but 10 others on a waiting list die.

What's worse, the shortage of organ donors shows no signs of improving. In fact, many experts expect the demand for donated organs to continue rising, even as the supply remains steady or declines.

Prostheses are designed to last a long time without breaking or triggering an immune response, but they have many limitations. A prosthesis can never fully replace the original, living part in either function or appearance. Also, since prosthetic devices do not grow, they pose a particular problem in treating children because they can distort or fracture as a child's body grows, requiring replacement. In addition, even though a prosthesis is made of biologically nonreactive materials, a patient's immune system may still recognize it as foreign and slowly break it down.

Prosthetic replacements for internal organs have also been tested. The best known of these was the mechanical artificial heart developed by American physician Robert K. Jarvik, which was tested in the 1980's as a way to preserve the lives of critically ill heart-transplant candidates who could wait no longer for a donor. However, the device had a tendency to form blood clots, which caused strokes in several test subjects. In addition, the implant had to be connected to a large external power unit.

The field of tissue engineering emerged as a possible, and significant, improvement on existing therapies. Because of its potential to give patients a natural, permanent replacement for a lost limb or diseased organ, tissue engineering was seen as a vastly better way to treat many conditions. For example, a tissue-engineered pancreas that could produce insulin would be a cure for some types of diabetes, and a tissue-engineered arm or leg would be far superior to any prosthesis. Engineered tissue can grow, reshape itself, or otherwise alter its function as the body requires. This capability is of particular importance in young patients, who need organs or limbs that will develop along with the rest of their body. But the most important benefit of tissue engineering may be the possibility of creating stores of replacement parts in the laboratory, thereby eliminating dependence on organ donation. By early 1999, more than 30 companies and a dozen academic institutions had become involved in efforts to engineer virtually every type of organ and tissue, including skin, blood vessels, heart valves, livers, kidneys, bladders, connective tissues, breasts, and spinal-cord nerves.

## The origins of tissue engineering

The origins of tissue engineering can be traced to the late 1960's and early 1970's, when scientists first began experimenting with the idea of growing cells on a three-dimensional matrix in order to create replacement parts using living material. One pioneer in this research was Carl F. W. Wolf, an American physician and biomedical engineer. In the late 1960's, Wolf experimented with the use of hollow textile fibers as "artificial capillaries" in an attempt to imitate the waste-filtering process that takes place in the capillaries of a kidney. Then, as now, the only treatment for kidney failure was a transplant or *hemodialysis,* a procedure in which the blood is filtered through a machine to remove wastes. Hemodialysis, which takes several hours to complete, is usually done three times a week. Wolf believed that he could build a filter made of fiber strands that could be implanted in a patient's body to reduce or eliminate the need for mechanical dialysis. Continuing this research while working as a pathologist at the New York Blood Center in New York City, Wolf began experimenting with growing living cells on and within the fiber strands. Cells on the inner walls of the fibers formed living tissue that enabled blood to pass through without clotting, while cells on the exterior simulated the living tissue that would surround the device if it were implanted.

In 1972, Wolf and Richard Knazek, a researcher at the National Institutes of Health, independently developed artificial-capillary devices that imitated the function of the liver and of various *endocrine* (hormone) glands. Wolf's device consisted of *hepatocytes* (liver cells) grown within and around a bundle of hollow fibers. The device was connected to the circulatory system of a laboratory mouse whose natural liver function had been disrupted and successfully purified the animal's blood. At about the same time, William L. Chick, a physician at the University of Massachusetts Medical School, began growing pancreas cells on a matrix made from meshes of hollow fibers woven together and implanting them to treat diabetes in dogs. Similar efforts continued throughout the 1970's and 1980's.

The term "tissue engineering" was first used in 1988 at a National Science Foundation workshop. In 1994, the Tissue Engineering Society was organized by Charles A. Vacanti, chairman of the department of anesthesiology at the University of Massachusetts Medical Center; his brother, Joseph P. Vacanti, a transplant surgeon and professor at Harvard Medical School; and Robert S. Langer, a chemist and biomedical engineer at the Massachusetts Institute of Technology (MIT). The stated purpose of the society was "the advancement of the science and technology of tissue engineering in both its basic and applied aspects."

## Simple idea, difficult execution

Although the fundamental idea in tissue engineering—growing cells on an artificial scaffolding—is fairly simple, in practice it is quite complex. That is because every tissue and organ contains many different types of cells. For example, a human leg bone is made up of two parts—a hard shell that surrounds a spongy, soft marrow. The outer layer, called compact bone, consists of various kinds of bone cells, while the softer marrow contains blood vessels, connective tissue, and blood-forming cells. The first step in a tissue engineering project is identifying each of the cell types that make up the original body part and studying how they all fit and work together.

Some tissues are less structurally complex than others, and so their fabrication is less difficult. Because skin has a relatively simple structure—the epidermis at the surface, the dermis just below that, and the subcutaneous layer at the base—it was one of the first organs to be reproduced in the laboratory. In the 1970's and early 1980's, surgeon John F. Burke of Massachusetts General Hospital and mechanical engineer

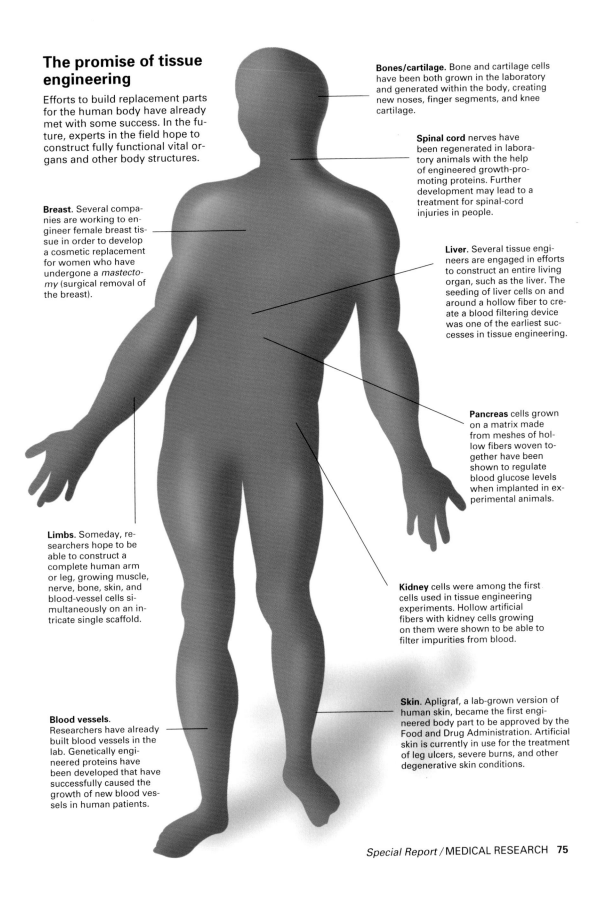

# The promise of tissue engineering

Efforts to build replacement parts for the human body have already met with some success. In the future, experts in the field hope to construct fully functional vital organs and other body structures.

**Bones/cartilage.** Bone and cartilage cells have been both grown in the laboratory and generated within the body, creating new noses, finger segments, and knee cartilage.

**Spinal cord** nerves have been regenerated in laboratory animals with the help of engineered growth-promoting proteins. Further development may lead to a treatment for spinal-cord injuries in people.

**Breast.** Several companies are working to engineer female breast tissue in order to develop a cosmetic replacement for women who have undergone a *mastectomy* (surgical removal of the breast).

**Liver.** Several tissue engineers are engaged in efforts to construct an entire living organ, such as the liver. The seeding of liver cells on and around a hollow fiber to create a blood filtering device was one of the earliest successes in tissue engineering.

**Pancreas** cells grown on a matrix made from meshes of hollow fibers woven together have been shown to regulate blood glucose levels when implanted in experimental animals.

**Limbs.** Someday, researchers hope to be able to construct a complete human arm or leg, growing muscle, nerve, bone, skin, and blood-vessel cells simultaneously on an intricate single scaffold.

**Kidney** cells were among the first cells used in tissue engineering experiments. Hollow artificial fibers with kidney cells growing on them were shown to be able to filter impurities from blood.

**Blood vessels.** Researchers have already built blood vessels in the lab. Genetically engineered proteins have been developed that have successfully caused the growth of new blood vessels in human patients.

**Skin.** Apligraf, a lab-grown version of human skin, became the first engineered body part to be approved by the Food and Drug Administration. Artificial skin is currently in use for the treatment of leg ulcers, severe burns, and other degenerative skin conditions.

Ioannis V. Yannas of MIT pioneered the development of artificial skin for the treatment of severe burns.

In severe burns, the dermis is destroyed. Although epidermal tissue regenerates well, the regeneration relies on the presence of underlying dermal tissue, and destroyed dermis does not regrow. The synthetic skin Burke and Yannas developed was designed to encourage dermal regrowth by providing a framework in which such growth would occur. The matrix was a highly porous polymer sheet made from cowhide and shark cartilage. As the dermis returned, the epidermal layer could begin to grow as well. Additional patches of epidermis from elsewhere on the patient's body would then be transplanted over the rebuilt dermal layer, leading to complete epidermal regrowth.

Solid organs like the liver, kidneys, and heart, however, are much more complex than skin, with many cell layers, various cell types, and a high degree of *vascularization* (penetration by blood vessels). This complexity must be recreated in an engineered organ, with cell types carefully chosen and arranged, for the organ to perform properly.

Tissue engineers, whether making an organ or limb, must also select the most appropriate material for the structural scaffold for the organ or tissue. Although the matrix can sometimes be permanent, such as with engineered heart valves, it is usually designed to dissolve over time, thereby reducing the chance of an immune response. In choosing the scaffolding material, scientists consider such factors as its physical strength and how fast it will disappear after implantation. It is also critical to ensure that the material will not be *toxic* (poisonous) or cause any sort of harmful reaction when placed in the body. A variety of engineered tissues—including heart valves—have been successfully implanted in laboratory animals. Human trials are on the horizon.

## New polymers lead to engineering advances

Some tissue scaffolds are made from biological materials. For example, collagen, a major protein component of connective tissue that is found throughout the human body, is sometimes used as a matrix. But synthetic polymers are the materials most commonly used today. Polymers have several superior properties, including the ease with which they dissolve in the body and their ability to be shaped into a variety of structures. The most useful of these materials for tissue engineering are called branching polymers. They are so named because their chains of molecules link together in a way that mimics the branching growth pattern found in the cells of many living things.

During the 1980's when branching polymers first became available, Joseph Vacanti and Langer teamed up to build a liver. Influenced by the efforts of Burke and Yannas to grow skin, Vacanti and Langer set out to grow liver tissue on a three-dimensional matrix.

By this time, tissue engineers had learned that if cells seeded onto a matrix are placed close enough together, they tend to form structures. This phenomenon, known as neomorphogenesis, shows that cells are "intelligent" enough to organize themselves into specific arrangements

## Seeking a better way to restore function

During the 1900's, compensating for a lost or malfunctioning limb or organ involved transplants or mechanical solutions. But advances in biotechnology in the 1980's and 1990's showed that it would be possible to create replacement body parts from human tissues.

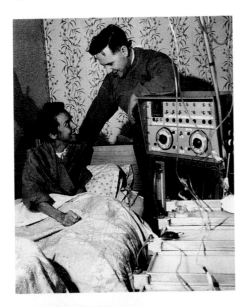

For decades, people with kidney failure have needed to have their blood filtered by a dialysis machine, *above, left,* to remove wastes. People with missing arms or legs have traditionally received artificial limbs of metal and plastic, *above.* In a significant step toward practical tissue engineering, researchers in Boston announced in 1995 that they had built a model of a human ear using living cartilage cells. The cells were grown on an ear-shaped polymer mold that was attached to the back of a laboratory mouse, *left.*

entirely on their own. However, the placement of these cells must be done very precisely. If the cells are crowded too close together, many will die from lack of oxygen and nutrients. If, on the other hand, they are placed too far apart, organized tissue growth will not occur.

Realizing that nature frequently solves this same problem by using branching structures to pack cells efficiently within a given space, Vacanti and Langer hit on the idea of imitating nature by building a matrix with branching polymers. Although they were unable to make a functioning liver, the scientists became convinced that branching polymers were the ideal material for creating complex scaffolds. By the end of the 1980's, branching polymers were being used in experimental efforts to grow a wide variety of other human organs and tissues.

Tissue engineering made rapid progress in the 1990's. In a striking demonstration of the potential benefits of this emerging technology, a team headed by Langer, Joseph Vacanti, and Charles Vacanti announced in October 1995 that they had built a structure resembling a human ear out of living cartilage. A photograph of the ear—growing on the back of a laboratory mouse—was published worldwide. The ear was created from cartilage cells seeded onto a polymer scaffold and then surgically attached beneath the skin of the mouse. The animal's body served as a living incubator, providing the seeded matrix with the proper conditions for development.

More publicity followed in August 1998, when Charles Vacanti and University of Massachusetts surgeon John V. Shufflebarger used samples of a patient's own bone cells to regenerate a segment of his thumb bone that had been severed in an industrial accident. Shufflebarger surgically attached the regrown segment to the stump of the thumb. The researchers planned to grow additional bone, as well as cartilage and skin, to fully recreate the man's thumb. In January 1999, doctors in Florida used Apligraf, the first artificial skin product approved for routine use in the United States, to treat a baby born with a degenerative skin-cell disease.

## Manufacturing human skin

Apligraf, a skin-graft product, in May 1998 became the first engineered human tissue approved for routine medical use in the United States. After implantation, Apligraf, which contains no pigments, becomes virtually indistinguishable from the patient's own skin.

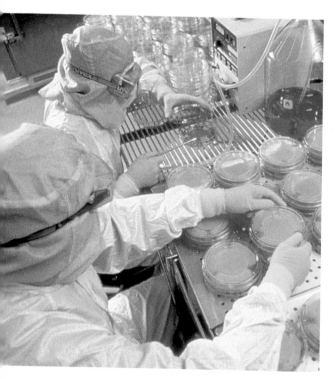

Apligraf is made from the donated foreskins of circumcised babies. The cells in a single piece of foreskin the size of a postage stamp can be used to produce 200,000 similar-sized units of Apligraf.

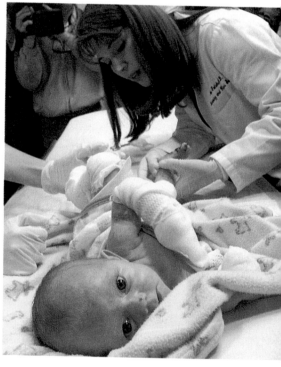

In January 1999, an 8-week-old Florida girl with epidermolysis bullosa, a rare skin disorder, received extensive grafts of Apligraf. The treatment worked so well that several other children with the same disease soon received the same treatment.

**An early success story**

Sean McCormack, *left,* received an implant of engineered cartilage on the left side of his chest in 1994. An X-ray image, *below,* shows a cross-section of Sean's chest cavity after the implantation procedure. The implant (at lower right) is thinner than the natural chest wall but provides protection for the heart, lungs, and other internal organs.

## An important breakthrough—growing human stem cells

In November 1998, scientists at the University of Wisconsin at Madison and Johns Hopkins University in Baltimore announced a biological milestone with potentially huge implications for the future of tissue engineering. The researchers independently announced that they had, for the first time, isolated human *primordial stem cells* and grown them in the laboratory. Primordial stem cells are immature cells that have not yet become any of the specialized types of cells that make up the body. The researchers harvested these cells from tiny human *embryos* (organisms at an early stage of development before birth) obtained from fertility clinics. By exposing laboratory-cultured stem cells to certain hormones and chemicals, a number of investigators speculated, it should be possible to "instruct" the cells to become whatever kind of cells are desired. Thus, if stem cells can be produced in large numbers, they could be stored in their immature state and used to create any type of body tissue on demand. Because they are not completely developed, some experts speculate that primordial stem cells would produce tissues that are less likely to provoke *rejection* when transplanted than tissues grown from mature cells. Rejection is a condition in which the patient's immune system attacks the transplanted organ as it would an invading organism.

The harvesting of stem cells from human embryos raised ethical concerns for many people, even though the embryos were just clusters of cells and would not have developed further. Fertility clinics routinely create a number of embryos for couples seeking to have a baby but im-

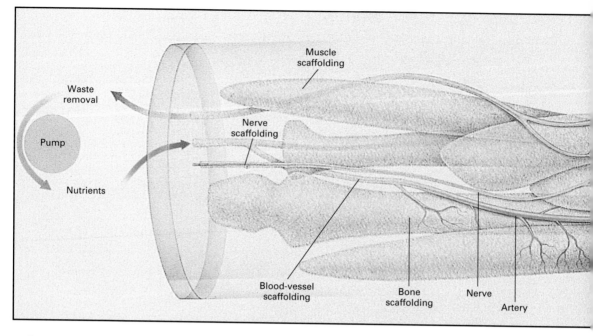

Waste removal

Pump

Nutrients

Muscle scaffolding

Nerve scaffolding

Blood-vessel scaffolding

Bone scaffolding

Nerve

Artery

## Constructing human limbs

In 1999, tissue engineers were developing the technology to build a complete replacement arm or leg in the laboratory out of living tissue. The biggest remaining challenge was finding a way to grow each of the different tissue types—nerves, bones, and blood vessels—simultaneously and having all the components work together as they do in a natural limb.

plant just a few of them in the woman's womb. The others are frozen, destroyed, or donated for research. Nonetheless, the fact that every embryo represents a potential human life makes their use in research ethically questionable to many in the public.

An apparent way out of that moral dilemma came in April 1999, when scientists at Osiris Therapeutics, a small biotechnology company in Baltimore, announced that they had grown bone, cartilage, fat, tendon, and muscle tissue from a type of stem cell found in all people. These cells, called human mesenchymal stem cells, were found in bone marrow. Mesenchymal cells offer a much easier way to grow tissues from a patient's own cells, and to do so without controversy.

## New treatment options—and current challenges

Even though tissue engineering is still a new field, decades of research have already led to treatments using engineered human tissues to treat certain conditions. For example, physicians in 1999 were using engineered cartilage to correct urinary tract problems in both children and adults. Surgeons use a patient's own cartilage cells, multiplied in the laboratory, to strengthen weakened areas of the urinary tract.

While the fabrication of engineered limbs and organs will not be possible until well into the 2000's, many experts in the field expect a variety

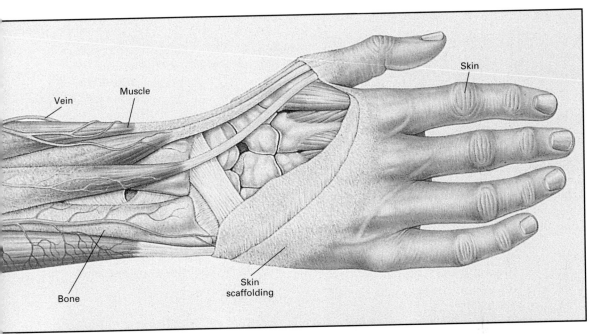

Vein

Muscle

Skin

Bone

Skin
scaffolding

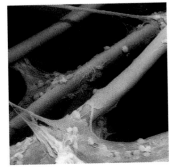

of significant treatments to develop out of their research by about 2010. By that time, heart-attack patients may be treated with engineered growth factors that promote the development of new heart muscle, which does not regrow naturally. Mesenchymal stem cells may be useful in helping some cancer patients recover from treatment. For example, many leukemia patients must undergo a bone-marrow transplant after receiving radiation therapy to destroy their diseased bone marrow. Physicians could use engineered bone marrow grown from mesenchymal stem cells to boost the effectiveness of bone-marrow transplants after the radiation treatment is completed.

Because it is still in its infancy, tissue engineering faces a number of challenges. For one thing, creating all the different types of body tissues has proved quite difficult. Researchers have been particularly frustrated in their efforts to build a functioning liver because liver cells are very difficult to keep alive and healthy in the lab. Therefore, a large part of the research effort in this area has focused on improving methods to preserve liver cells in culture.

In an artist's conception of an arm being constructed on a polymer scaffold, *top* (with time advancing from left to right), several kinds of tissues grow into a fully functioning limb. The arrows represent nutrients being supplied to the developing arm and waste products being removed. Highly magnified views of a polymer scaffold, *above,* reveal that as cells multiply, they slowly absorb the polymer material until it eventually disappears.

Some creatures, such as the five-lined skink, *above,* have the natural ability to regrow a severed tail and others, such as the starfish, can replace a lost limb. A branch of tissue engineering called tissue regeneration is seeking ways to stimulate the human body to renew itself in the same way.

Another problem has been keeping engineered tissues alive and functioning after implantation. Implanted cells tend to die if they do not have a steady supply of oxygen and nutrients. This is especially true in the case of tissues that are more biochemically active, such as those of the heart and liver. These tissues must *metabolize* nutrients (convert them into energy) at a much faster rate than most other cells do. Therefore, efforts to engineer metabolically active tissues concentrate on ways to supply adequate nutrients to all the cells of the transplanted tissue. One approach has been to seed the matrix with a large number of blood-vessel cells to encourage a high degree of blood-vessel growth. Another method has been to reduce the total number of cells in the engineered tissue to enable nutrients to penetrate it more easily.

Such long-range goals as creating replacement limbs will require many more years of development before becoming achievable. Growing complex structures—including skin, muscles, blood vessels, and nerves—separately but at the same time within a single matrix remained in 1999 a distant frontier.

## Another approach: tissue regeneration

While some researchers worked to perfect tissue-replacement techniques, others focused on tissue regeneration, which involves stimulating the body to repair or regrow its own tissues. The natural regenerative ability of some animals inspired many scientists to study how such regeneration takes place. A starfish, for example, can regrow a severed arm and a salamander, a severed tail. Many researchers thought that it may be possible to induce the regeneration of human tissues that do not normally renew themselves, such as nerves, heart muscle, and bone. A number of tissues in the human body—skin, hair, nails, and bone marrow—regenerate naturally, either by a continuous cycle of cell death and replacement or in response to injury or disease. However, most human tissues respond with tissue repair, not regeneration. For example, while a small cut on a finger will eventually heal over with new skin, a broken bone will mend with the formation of scar tissue at the injury site.

Because researchers were still a long way in 1999 from knowing how to induce the complete regeneration of tissues, they worked on less ambitious repair methods. One promising technique that was being investigated involves taking genetically defective cells from a patient and inserting new genes into them. The repaired cells are then returned to the patient's body in the hope that they will multiply and carry out the function that the defective cells are failing to provide. This approach was being studied as a possible treatment for diseases in which genes play a role, such as Parkinson disease, a degenerative condition caused by the destruction of brain cells.

In a related regeneration method, damaged tissues are seeded with specialized "messenger cells" that secrete signaling molecules to help induce the growth of new tissues. A variation of this method involves introducing the signaling molecules themselves to the regeneration site. To restore function to a severed spinal cord, for example, scientists could place a matrix in the gap and seed it with cells that secrete nerve-growth factors or inject the growth factors directly into the gap. Experiments with laboratory rats using this approach have met with some success.

Despite recent advances in tissue regeneration, however, there were still several problems to be overcome. For one thing, normal regeneration is often a slow process, so if a patient were experiencing sudden organ failure, the gradual repair of the organ would not save the person's life. In addition, researchers still had much to learn about the factors and mechanisms involved in the regrowth of tissues.

While basic research continued, tissue regeneration methods were already being used experimentally to treat several conditions. For example, physicians were combining tissue engineering and regeneration methods to treat foot ulcers—persistent open sores on the feet—in patients with diabetes. In clinical trials, doctors reported several cases in which Dermagraft, a bioengineered skin graft that contains lab-grown dermal cells, generated new dermal tissue that promoted regrowth of the skin on a patient's foot. These results boosted researchers' confidence that tissue regeneration would lead to widely used treatments.

Although tissue engineering still had far to go in 1999, researchers were increasingly confident that renewing the human body with new or regenerated tissues, organs, and limbs was an achievable goal. Perhaps one day, a doctor in any hospital will be able to give a dying patient an "off the shelf" liver or regrow the severed spinal cord of a car-crash victim. Physicians would then have the kind of power for enhancing people's lives that for most of human history could only be dreamed of.

## For additional information:

### Books and periodicals

Arnst, Catherine. "Biotech Bodies." *Business Week,* July 27, 1998, pp. 56-63.
Hirshberg, Charles. "The Body Shop." *Life: Special Issue: Medical Miracles for the Next Millennium,* Fall 1998, pp. 51-57.
Langer, Robert S. and Vacanti, Joseph P. "Artificial Organs." *Scientific American,* September 1995, pp. 130-133.
Langer, Robert S. and Vacanti, Joseph P. "Tissue Engineering: The Challenges Ahead." *Scientific American,* April 1999, pp. 86-89.

### Web sites

Tissue Engineering Society
    http://www.t-e-s.org/
Pittsburgh Tissue Engineering Initiative Inc. (PTEI)
    http://www.pittsburgh-tissue.net/

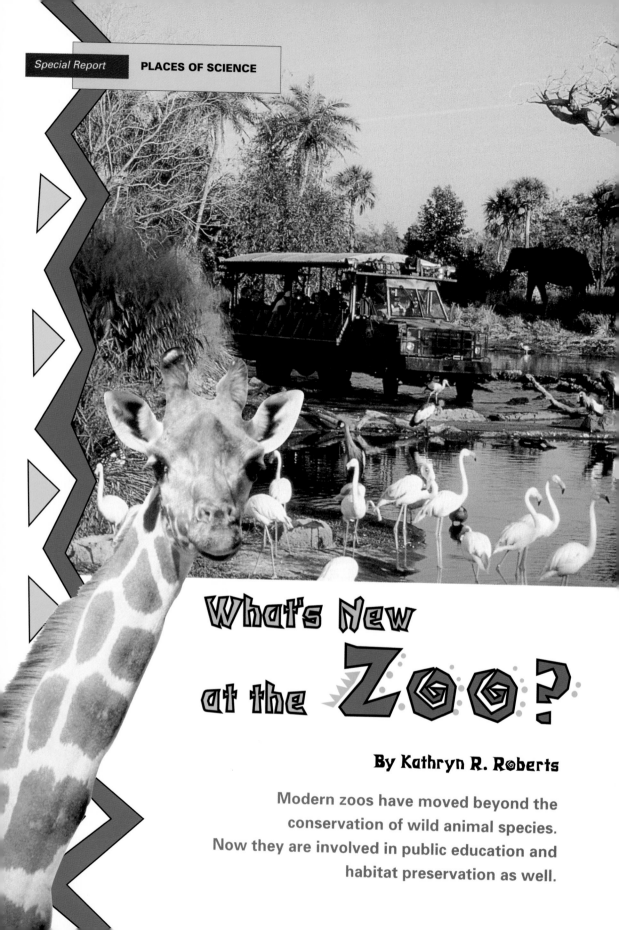

# What's New at the Zoo?

## By Kathryn R. Roberts

Modern zoos have moved beyond the
conservation of wild animal species.
Now they are involved in public education and
habitat preservation as well.

A canopied truck takes vacationers on safari to view the plains of Africa and watch animals roaming freely. Elephants forage, giraffes scratch themselves on trees, and birds call shrilly to one another. Suddenly, the vacationers spot a group of poachers at work, and their driver begins a high-speed chase to frighten the thieves away.

Have these people journeyed to the wilds of Africa? No—they have entered the Walt Disney Company's Animal Kingdom in Lake Buena Vista, Florida, the newest zoo in North America. Animal Kingdom, which opened in 1998, covers about 200 hectares (500 acres), making it one of the largest of the nearly 200 zoos in the United States and also one of the largest in the world.

Animal Kingdom represents a new generation of zoos. Rather than confining single animals to cages, as was once common, the best modern zoos place family groups of animals in spacious settings resembling their native habitats. Also, instead of displaying as many different animals as possible, as earlier zoos did, the new zoos feature a smaller number of species. Limiting their animal populations enables these zoos to give the animals better care, so that they are more likely to reproduce. The emphasis has changed from quantity to quality.

Leading zoos have adopted many other changes as well. They now participate in public-education, conservation, and research-oriented programs. And instead of competing with one another for animals, zoos around the world have begun to work together, combining their resources to save endangered species and preserve animals' natural habitats. How did such sweeping changes come about? Many of them have taken place only since the early 1970's, though people have collected living animals for more than a thousand years.

Visitors in a canopied truck, *opposite page,* observe animals of the African savanna at Walt Disney Company's Animal Kingdom in Lake Buena Vista, Florida. Disney's Animal Kingdom, which opened in 1998, is one of the largest and newest zoos in North America.

## The first zoos

The earliest animal collections, often called menageries, were only for the rich and royal. Chinese and Aztec emperors and Egyptian pharaohs all boasted about their exotic and fantastic collections of animals. As early as 700 B.C., menageries had been established in most Greek city-states. In England in the A.D. 1100's, King Henry I kept a large menagerie that historians believe was established by his father, William the Conqueror. In 1519, the Spanish explorer Hernando Cortes, conqueror of the Aztecs of Mexico, admired a menagerie belonging to the Aztec Emperor Montezuma. The animals were kept in ornate bronze cages.

The modern era of zoos began in the late 1700's, when European cities began to put animals on display for the public. Vienna founded a zoo in 1752, and Paris followed in 1793. The London Zoological Society opened Regent's Park Zoo in 1828. The society coined the term *zoological garden* but soon shortened it to *zoo*.

These first zoos served as models for other European cities. In the mid-1800's, the German cities of Berlin, Frankfurt, Hanover, and Cologne all constructed zoos. Animals were housed in large, ornate buildings constructed in a variety of architectural styles, such as baroque, Moorish, and Arabian. The buildings were situated in spacious, landscaped grounds, allowing visitors to stroll through gardens while viewing the animals.

Well-traveled Americans who visited Europe's zoos liked what they saw, and they began to recreate these institutions at home. The first zoo in the United States was established in Philadelphia in 1859, but it did not open until 1874 because the Civil War (1861-1865) delayed its construction. The Philadelphia Zoo was modeled after London's Regent's Park Zoo. In quick succession, zoos opened in cities across America. New York City's Central Park Zoo began its collection with donations of animals that it displayed in a small area of the great park during the 1860's. The Buffalo (New York) Zoo and Chicago's Lincoln Park Zoo opened in the 1870's. In 1889, the U.S. Congress established a zoo in Washington, D.C. The zoo's stated purpose was the "advancement of science and the instruction and recreation of the people."

The administrators of these early zoos rarely gave much thought to the needs of the animals. The value of a zoo's collection in attracting visitors was determined by the number of animals it contained, so zoos exhibited as many odd and interesting animals as they could. Single animals were housed in as small a space as possible so that more species could be shown. Animals were typically kept in concrete cages with bars at the front of the enclosure. These features made exhibits easy to keep clean and allowed people to view animals at very close range.

When zoos needed to replace animals that were ill or had died, or to add new animals to their collections, they simply captured them from the wild. Breeding animals was not a priority, so very few animals were born in zoos.

## Better conditions for animals

In the early 1900's, cities of every size built zoos for their communities. As zoos multiplied in number, the public responded with tremendous interest and attendance soared. At the same time, zoo designers began to think more seriously about the animals' needs.

One of the first to do so was Carl Hagenbeck, a German animal trainer and dealer who collected animals from the wild for his own menagerie and for zoos and circuses throughout Europe. In 1907, Hagenbeck opened a new kind of zoo in Sellingen, a suburb of Hamburg,

## Terms and concepts

**Biodiversity:** The genetic variety that exists among different types of animals and plants.

**Immersion exhibit:** A type of zoo exhibit that allows visitors to enter an animal's natural habitat.

**Inbreeding:** The breeding of animals that are closely related, a practice that can produce weak offspring.

**Menagerie:** An early term for a collection of animals.

**Taxon:** A group of related species of animals.

**The author:**

Kathryn R. Roberts is the Vice President of Programs at the Minneapolis Foundation in Minnesota and a former director of the Minnesota Zoo in Apple Valley.

## The first zoos

Zoos have existed in various forms for more than 2,000 years. However, the modern era of zoos began only in the 1700's, and the most significant advances in zoos took place beginning in the mid-1900's.

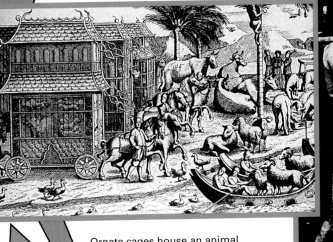

Ornate cages house an animal collection as it is transported into Persia in 1670. Such collections, often called menageries, were owned by royalty and not accessible to the general public.

Visitors to New York City's Central Park Zoo in the early 1900's observe a lion through the bars of its small concrete cage. Though such enclosures allowed zoos to house many different types of animals in a minimum amount of space, the animals were not able to live as they did in the wild, so they tended to be less healthy and rarely produced offspring.

Visitors at the Tiergarten in Sellingen, near Hamburg, Germany, in the early 1900's enjoy an unobstructed view of polar bears and antelope in an enclosure that resembles their natural habitats. The Tiergarten, opened by animal trainer and dealer Carl Hagenbeck in 1907, was the first zoo to dispense with traditional cages. By the 1930's, many European and American zoos had begun to imitate its exhibit design.

A gorilla named Willie B sits alone in his concrete cage at Atlanta's Grant Park Zoo in the 1960's, *top*. As zoo professionals learned more about the needs of the animals in their care, they improved the conditions under which the animals were kept. *Above,* Willie B observes one of his offspring, born after the renamed Zoo Atlanta constructed a new, natural habitat for the gorilla in the late 1980's.

Germany. Hagenbeck's Tiergarten was the first zoo without bars. Large, open displays replaced cages. Moats, greenery, and rocks were used to create enclosures that looked like the animals' natural habitats. Animals that lived together in the wild were displayed together in the same enclosure. The new exhibit design benefited the zoo's visitors as well as its animals. For the first time, people could view wild animals in a zoo without looking through bars and see animals in relationship to their natural environment.

By the 1930's, some U.S. zoos had begun to imitate Hagenbeck's exhibit designs. In addition, over the next 20 years, veterinary medicine became more sophisticated. New drugs allowed veterinarians to tranquilize animals to treat injuries and diseases that had been impossible to treat a generation earlier. By the early 1960's, changes in zoo management had also led to the successful breeding of many animals. The public was enthralled with the newborn animals, and as a result, people visited and supported zoos in growing numbers.

## Zoos confront new challenges

At the same time, zoos began to find themselves in a bind. Animal-rights activists, a growing force in the United States and Europe, demanded even better conditions for animals in zoos and a halt to collecting animals in the wild. Even if zoos wanted to collect wild animals, however, many popular species were becoming scarce in their native habitats. A number of countries began to consider their native animals as national treasures and closed their borders to collectors. Zoos were forced to find other ways to acquire new animals and to invest more effort in maintaining the health of the animals they already had.

To meet these challenges and encourage the general improvement of zoos, a group of zoo professionals came together in 1972 to form the American Association of Zoological Parks and Aquariums, headquartered in Wheeling, West Virginia. The group later moved to Silver Springs, Maryland, and changed its name to the American Zoo and Aquarium Association (AZA). The AZA developed an accreditation program for zoos (and aquariums). To be accredited, a zoo had to meet certain standards, including providing excellent animal care and undertaking sophisticated public-education programs and conservation efforts. For many years, accreditation was not mandatory for membership in the AZA. However, many zoos voluntarily undertook improvements to gain accreditation. Zoo professionals could develop necessary skills at the School for Professional Management Development for Zoo and Aquarium Personnel, established in 1975 by the AZA in Wheeling. In addition, the member institutions of the AZA adopted a strict code of ethics in 1976 to guide their daily operations and decision making.

In an effort to help zoos learn how to breed animals successfully, researchers at the National Zoo in Washington, D.C., in the late 1970's reviewed the records of zoo animals throughout the world. They compared the records of animals born in the wild with those of animals born in zoos. The researchers found that significantly fewer of the animals born in zoos survived to adulthood. In addition, zoo-born animals were more prone to disease and birth defects. It became obvious to the scientists that *inbreeding* (breeding animals that are closely related) among zoo animals was causing them to produce weak offspring.

## Species survival plans

Armed with this insight, the AZA began in the early 1980's to implement a cooperative breeding program based on a collection of guidelines called Species Survival Plans (SSP). An important goal of the program was to increase the number of healthy offspring born to zoo animals. However, even more importantly, it was aimed at preserving species that were disappearing in the wild. Many zoo professionals had come to believe that zoos should act as modern-day equivalents of Noah's Ark. That is, zoos should save as many threatened species as possible by maintaining populations of those species in captivity, "two by two," until the time comes when the animals can be reintroduced into the wild. Zoo professionals hoped that animal habitats—dwindling from the pressures of human population growth—would someday rebound and once again be able to support large numbers of species.

Under the AZA's program, each species has its own survival plan, managed by an expert coordinator, such as a zoo curator. A record of the parentage and inherited traits of each animal of that species living in captivity in zoos participating in the program is stored in a computer. When a zoo wants to expand its collection of a particular species, the SSP coordinator helps determine which animals of that species should mate to produce the sturdiest offspring. The coordinator looks for traits that will not only result in the birth of a healthy animal but will also improve the genetic diversity of the species as a whole. The selected animals are then transferred to the zoo needing them, either permanently or temporarily, until a healthy baby has been born. In 1982, the AZA simplified the process by requiring all zoos to use the same system for maintaining their animal records. By 1998, 86 SSP's were in place, for a variety of mammals, birds, fish, and reptiles. Animal breeding in zoos had become so successful that more than 90 percent of the new mammals being added to zoo collections had been born there.

In the 1970's, many member institutions of the AZA had voluntarily made changes to meet the organization's standards of animal care, public education, and conservation. However, in 1980 AZA accreditation became mandatory for all new members of the organization, and in 1985 it became mandatory for existing members. During the 1980's, many zoos also made changes in response to criticisms from animal-rights groups.

## Zoos as educators

With nearly 120 million annual visitors in the mid-1990's, zoos have had a unique opportunity to teach people about the plight of the natural world. As late as the 1970's, zoo educators had considered their primary mission to be teaching visiting schoolchildren about the animals on display. In later years, the scope of zoos' educational programs expanded greatly to include programs for adults as well as for children. In addition to providing information about the animals, zoos began to teach visitors about such conservation issues as the preservation of habitats, the importance of *biodiversity* (the genetic variety that exists among different types of animals and plants), and the interdependence of all the organisms in a particular environment.

Zoos also moved beyond the use of such traditional educational tools as graphics, labels, and exhibit design. For example, Brookfield Zoo near Chicago introduced an interactive, life-sized board game called Quest to help visitors learn about the environment. People could walk on stepping stones through the "Bog of Habits" and the "Path of Temptation," making decisions about real-life environmental choices, such as whether to use plastic or permanent tableware.

In addition, zoos encouraged teachers to use the zoo grounds as classrooms. Using curriculum guides designed by zoo staff members and tailored to particular grade levels, teachers could send children to the zoo to fulfill class assignments, such as reports on threatened species and environments.

To fulfill their goal of becoming accredited by the AZA, zoos that had continued to exhibit animals in small, bare enclosures began to install larger, more naturalistic ones. Many of these exhibits also helped to educate the public about the role of species in their natural environment. In one new type of exhibit, zoo designers demonstrated biodiversity by placing a variety of species that normally live together in the wild in the same enclosure—separated by unintrusive barriers, such as moats—so that people could better understand how much space animals need to live in harmony and what land features are necessary for the physical well-being of animals.

Other exhibits, called *immersion exhibits*, allowed visitors to enter animals' natural habitats, or at least simulations of them. A leading model for these exhibits was the Arizona-Sonora Desert Museum, which opened in 1952 near Tucson. The museum created paths, lined with artificial rockwork and native

Giraffes, ostriches, and zebras share an enclosure without barriers at the North Carolina Zoo near Asheboro. Beginning in the 1970's, many zoos constructed such exhibits, in which animals that co-exist in the wild are grouped together in enclosures that simulate their natural habitat, to give people a better idea of how the animals really live.

plants, through a small section of the Sonora Desert. Along the paths, naturalistic enclosures housed native desert animals.

Some zoos created immersion exhibits in which visitors could experience life in a tropical rain forest. The Topeka (Kansas) Zoo developed the first such exhibit in 1974. Although this exhibit was rather small, it inspired other zoos to construct more complex and dramatic recreations of rain forests. These included the National Zoo's Amazonia, Lied Jungle at Omaha (Nebraska's) Henry Doorly Zoo, and the Denver (Colorado) Zoo's Tropical Discovery. Many species live together in these exhibits, which are filled with lush tropical vegetation and usually feature a simulated thunderstorm or gentle rain shower. Guides or visual displays explain the rate at which the fragile rain forests are being destroyed, the numbers of rain forest species that are threatened with extinction as their habitat disappears, and what people can do to help preserve the remaining forests. Zoo directors hoped that such exhibits would give visitors a better understanding of the animals and their needs and would prompt people to support efforts to conserve natural habitats.

Although many people think of zoos as places to see land animals, in the 1990's, as ocean habitats became ever more polluted, zoos became increasingly concerned about the fate of animals that live in the water. As a result, many zoos built major aquatic exhibits. Those popular new attractions included the Minnesota Zoo's Discovery Bay and the Columbus (Ohio) Zoo's Discovery Reef. Zoo professionals hoped they could use their extensive knowledge about the diet, breeding, and behavior of the marine mammals and birds in their collections to help species in the wild. Exhibits and programs about Earth's aquatic ecosystems also became important parts of most zoos' public-education efforts.

### Seeking further improvements—and more space

Exhibit design continued to evolve throughout the 1990's, with an even greater emphasis on meeting the animals' needs. Many zoos began including so-called behavioral enrichment devices in their exhibits. At the Los Angeles Zoo, exhibit designers added an artificial termite mound to the primate area. Chimpanzees could use sticks to "fish" in the mound for treats, such as honey and applesauce, just as they "fish" for termites in the wild. Animal behavior specialists believe that such devices increase the activity of animals in the exhibit, making the animals happier, healthier, and more interesting for people to watch.

Similarly, many new exhibits in the 1990's included areas for training animals in behaviors that can help improve the quality of their care. A new dolphin exhibit at the Minnesota Zoo included a "slide-out" area, a shallow pool in which the dolphins can be examined by zoo veterinarians. The animals were trained to present a fluke for blood sampling and to provide urine for a pregnancy test. The slide-out area made it unnecessary for the dolphins to be netted and restrained for physical examinations.

Although the improvements in animal enclosures and breeding and education programs proved to be a great success, they also contributed to an emerging problem in the late 1990's: Space for zoo animals was running out. The main reason for the space crunch was that zoo construction had virtually come to a halt by the 1960's, as the cost of opening new facilities soared. Prior to the opening of Disney's Animal Kingdom in 1998, the last major zoos to be established in the United States were the Minnesota Zoo in Apple Valley, near Minneapolis, in 1978 and the North Carolina Zoo, near Asheboro, in 1979. Both zoos were established by state legis-

Visitors watch a polar bear dive at the North Carolina Zoo. Throughout the 1970's and 1980's, exhibit designers created enclosures in which animals could engage in natural behaviors and people could observe them at close quarters with minimal barriers.

## Getting creative

Beginning in the 1970's, zoos developed "immersion exhibits" and other novel attractions to bring people even closer to the animals and teach them about habitats and conservation issues.

Visitors at the Bronx Children's Zoo in New York City get a close-up look at life in a prairie dog colony.

Brookfield Zoo's "Quest" exhibit, an outdoor, life-sized game board, allows visitors to consider conservation issues in their daily lives as they choose whether to follow paths to "The Easy Way" or "The Harder Way."

A boy crawls in a simulated opossum den in the "Explore" exhibit at Brookfield Zoo near Chicago. Zoo professionals designed the exhibit to allow children to experience how opossums and other animals live.

latures, which provided the money to build them.

In addition, the new types of exhibits actually reduced the space available in zoos for the animals. In order for zoos to house different animals in the same enclosure, they had to build structures such as moats or hidden fences to keep the animals apart. To allow the public to see how animals live in the wild, zoos had to plant trees, build pools, and include rocks for the animals to climb, all of which took up space. Features that encouraged certain animal behaviors, such as the termite mounds for chimpanzees and slide-out areas for dolphins, also took up space.

*A keeper at the Minnesota Zoo in Apple Valley, near Minneapolis, trains a dolphin to present its fin on command. Teaching such behaviors allows zoo professionals to perform necessary medical and maintenance procedures, such as the taking of blood samples, with less stress for the animals.*

Competition for the limited space was inevitable. The coordinators of each species survival plan attempted to obtain as much space as they could for the long-term survival of their species. However, zoo directors also realized the impossibility of preserving all of the hundreds of endangered animal species in their Noah's Ark. They needed a mechanism to help them determine which species to save and how much space to allocate for a particular species or for the *taxon* (group of related animals) to which that species belongs. Zoos needed help, for example, in deciding how much space to allot to the taxon of bears, which includes such species as sun bears, grizzly bears, and polar bears.

To address this problem, the AZA created the Taxon Advisory Groups in 1990. These groups establish target population sizes for the taxon covered by their committee to ensure that available space is used for the maximum benefit of the taxon as a whole. The groups also address such difficult questions as which species should be maintained and which should not. Popularity, availability, and personal preference on the part of zoo directors, curators, and keepers all play an important role in the selection of species. In some cases, the groups must convince curators of the need to remove some of the species they have in their collections. Despite the hard job facing these committees, zoos remained optimistic about chances for their success.

### Protecting wildlife habitats

The space problem of the 1990's also prompted zoo professionals to expand their efforts beyond preservation of animals into conservation of habitats. They realized that unless zoos became involved in efforts to protect habitats, there might eventually be no suitable wild areas into which animals could be released. In 1993, the AZA established a Field

Conservation Committee to help its member zoos protect animal habitats around the world, and many zoos were soon actively involved in habitat preservation. A few zoos "adopted" a national park in another country. The Minnesota Zoo, for example, began supporting a national park in Indonesia, a nation that is home to many of the animals in the zoo's collection. The Minnesota Zoo provides supplies and expertise to the Indonesian park to help it protect its own wildlife.

Other zoos, such as the Wildlife Conservation Society's Bronx Zoo in New York City, the San Diego Zoo, and Brookfield Zoo, have staff members working full time in the field on species and habitat preservation. From 1990 to 1998 alone, the Wildlife Conservation Society helped establish nearly 53 million hectares (130 million acres) of new reserves and parks in Belize, Bolivia, Brazil, Cameroon, Chile, Congo (Brazzaville), Congo (Kinshasa), the Dominican Republic, Madagascar, Papua New Guinea, and Tibet.

Some zoos raised money to purchase land for private nature reserves. Often these habitat protection programs were linked to captive breeding programs. Certain animals in the zoos, including pandas, snow leopards, and black rhinos, became "ambassadors" for the protection of their own habitats.

## Reintroducing animals to the wild

By the 1990's, zoo professionals had also concluded that it was best not to wait for some ideal future date to reintroduce captive animal species to the wild, and the number of reintroduction programs increased significantly. Until then, there had been only a handful of such programs. In one well-known project in the 1960's, the Arabian oryx—a white antelope—was bred at the Phoenix and San Diego zoos and reintroduced to the deserts of Oman. In another much-publicized effort, monkeys called golden lion tamarins were taught skills that would help them survive in the wild, such as banana peeling, at Washington, D.C.'s, National Zoo. They were then released into protected reserves in eastern Brazil. In the 1990's, more than a dozen reintroduction programs were underway, including one in which *herpetologists*

A technician at the San Diego Zoo prepares to freeze eggs, sperm, and embryos from endangered animals. Geneticists hope that as techniques of artificial insemination and surrogate motherhood improve, these genetic samples will ensure that endangered species can survive for many generations to come.

A staff member from the North Carolina Zoo (NCZ) speaks with co-workers on a World Wide Fund for Nature site in Cameroon. As part of the zoo's habitat preservation program, NCZ staff help track African elephants in Cameroon in an effort to find ways for the herds to coexist with local farmers.

(reptile experts) at the Riverbanks Zoo in Columbia, South Carolina, were preparing a threatened rattlesnake species for reintroduction to the Caribbean island of Aruba.

The reintroduction programs at some zoos shifted from breeding exotic animals for return to their native countries to breeding animals native to the United States. The Bronx Zoo, for instance, helped breed bison and successfully reintroduced them to the American West. Other native American species being bred at U.S. zoos for reintroduction to the wild included a wide variety of frogs, snakes, lizards, birds, and mammals.

As they approached the year 2000, zoos were justifiably proud of what they had achieved. They had transformed themselves from centers for idle diversion into educational and conservation institutions that benefit people, animals, and the environment alike.

## Zoos of the future

What will zoos of the future be like? Zoo architect Jon Coe of Coe Design, Incorporated, in Philadelphia foresees the development of exhibits that incorporate elements commonly associated with amusement parks. Ride simulators may become teaching tools, taking people on imaginary journeys through environments that zoos cannot re-create. Visitors may follow in the footsteps of a mother gorilla, for example, as she cares for her young, forages for food, and prepares her nest. Or they may ride along in an underwater vehicle as it examines the condition of coral reefs and counts the populations of tropical fishes.

More importantly, zoos will face difficult decisions about the fate of the animals in their care. As the human population continues to grow at a rate of about 80 million people a year, natural habitats will continue to shrink and to disappear. As the only conservation organizations that care for generation after generation of exotic animals, zoos will need to decide which species they will try to save and which are beyond saving. And with zoo animals living longer—and experiencing more illnesses and conditions associated with age—zoos will have to determine what to do with elderly, sickly, or surplus animals.

Finally, if animals cannot be returned to the wild and are destined to live out their lives in captivity, zoos must ask themselves, "What is the purpose of saving these creatures in the first place?" For many zoo professionals, the answer to that question lies in the simplest interactions that take place in zoos every day: the feeling of wonder that people experience in face-to-face encounters with exotic animals. The opportunity for such encounters, many experts agree, must never disappear, for without them people will quickly lose interest in safeguarding the few wild places that are left in our world.

## For additional information:

### Books and periodicals

*New Worlds, New Animals: From Menagerie to Zoological Park in the Nineteenth Century.* Ed. by R. J. Hoage and William A. Deiss. Johns Hopkins University Press, 1996.

Nichols, Michael, and others. *Keepers of the Kingdom: The New American Zoo.* Lickle Publishing, 1996.

### Web sites

American Zoo and Aquarium Association Web pages—http://www.aza.org

Bagheera Endangered Species Web pages—http://www.bagheera.com

Wildlife Conservation Society Web pages—http://www.wcs.org/wild

■ ··· ■ ··· ■

## Questions for thought and discussion

Suppose you are the curator of gibbons (the smallest members of the ape family) in a city zoo. Your exhibit space is limited, so you must make wise use of it. However, your department has received a budget increase this year. You've decided to spend the money on redesigning the enclosure so that the black-striped gibbons—the most popular species with visitors at your zoo—would have a larger territory. Two of the black-striped females are about to give birth, and if you do not provide more room for the group, you will need to give the babies away to a zoo that has more space. Before you can begin the work, the coordinator of the Kloss's gibbon species survival plan calls to ask whether you can house the last two captive members of this endangered species at your zoo. If you cannot take the animals, the species is likely to become extinct, because their habitat is on the verge of being destroyed.

**Questions:** Would you refuse the Kloss's gibbons and hope some other zoo will take them? Or would you accept the Kloss's gibbons, use your additional funds to build them an enclosure, and give away the black-striped babies? Can you think of other alternatives that would help both species of gibbons and also benefit your zoo?

# The Wireless World

By James R. Asker

Placing a telephone call—
even from remote regions of
the Earth—became easier in
1999 thanks to advances in
satellite communications.

## Terms and concepts

**Analog signal:** A signal that varies continuously and smoothly.

**Digital signal:** A signal that is translated into a string of 0's and 1's.

**Footprint:** The area in which a satellite signal can be received.

**Gateway:** A surface base station through which satellite transmissions are sent and received.

**Geosynchronous orbit:** An orbit above the equator at an altitude of about 22,300 miles (35,900 kilometers). A satellite in this orbit, moving in the same direction that the Earth rotates about its axis, stays above the same point on the Earth's surface.

**The author:**
James R. Asker is the Washington bureau chief for *Aviation Week & Space Technology* magazine.

magine a scene in which a television producer needs to make a phone call from a remote location. He is in the desert in Morocco taping a team competition that combines sports, nature, and adventure. The producer wants to call network executives in California to tell them that the event has begun with a camel race. Some of the participants were immediately tossed from their saddles, and the athletes are chasing the humped beasts across the barren landscape. It should make a great television special.

But how can the producer phone home with the news? There isn't a telephone pole for miles, and even cellular phones won't work in such a faraway place. The nearest city, Marrakech, is some 120 kilometers (75 miles) away. Without missing a beat, the producer flips out a portable telephone, keys in the TV studio number, and moments later he's chatting away. The producer is one of the first users of a new type of satellite system—one in which handheld telephones transmit directly to, and receive signals from, satellites in low Earth orbits.

In early 1999, one such system, a $5.5-billion network called Iridium, had 66 satellites in orbit. Competing satellite phone systems were also in development in 1999, including Globalstar, which planned a $3.26-billion system with 48 satellites. A third system under construction, Teledesic, was designed to connect users to the Internet. Teledesic, requiring nearly 300 satellites to operate, was expected to make almost any type of information available anywhere at any time of day.

By 1999, there were already some 400 active communications satellites in orbit. A 1998 study by the United States Federal Aviation Administration projected that 1,200 new satellites would be launched into low Earth orbits through 2010. And as work progressed on the various satellite systems, new types of ground-based wireless systems were also in development. Thus, as the 2000's approached, it was apparent that a new wireless world of communications was being born.

## The beginning of wireless communication

Some of the technology being used in these systems is not new. Commercial communications satellites have been in use by telephone companies, television networks, and major corporations since the 1960's. By the 1990's, satellites were sending high-quality television signals directly to homes via dish antennas about the size of a large pizza. But a growing desire on the part of consumers for telephone and data services that offer mobility and global coverage seemed destined to make wireless technology explode even further.

Even before the new satellite systems had been conceived, two-way wireless communications had become an ordinary marvel for many people. In the United States and other technologically advanced countries, it had become common by the 1990's to see a person talking on a phone while walking along the street or even driving a car. But such phones use ground-based equipment rather than satellite relays.

The first wireless phones, which were introduced in the 1920's for transatlantic calls, were basically radio transmitters and receivers and worked like a citizens band (CB) radio. They converted a user's voice into an *analog* radio signal (a signal that varies smoothly and continuously). Such devices transmitted directly to each other, without a relay station in between, and used a single radio frequency for a two-way conversation.

Beginning in the 1950's, the regional affiliates of the American Telephone and Telegraph Company (AT&T) began offering local mobile telephone service in many metropolitan areas. These early mobile phone systems used a single, centrally located transmitting and receiving tower or base station. This was similar to the radio systems that are still used for dispatching taxicabs and tow trucks. The base station connected users' wireless handsets to the hard-wired phone system. Since there was just one base station in a calling area, the early radio-telephone units had to be powerful enough to send signals up to 40 kilometers (25 miles). In addition, the systems used only a few dozen channels, so not many subscribers could use an area's system at the same time. As a result, the number of mobile phones that could be operated was strictly limited, and monthly fees were very high.

## A better idea for local service—cellular phones

Commercial cellular telephone systems, which began operating in 1983 in the United States, greatly expanded communication possibilities. Rather than using a single base station for a city, cellular systems blanket a region with hundreds of base stations, each serving a much smaller area. These service areas, called cells, range in size from about 3 to 16 kilometers (2 to 10 miles) in diameter. Since a user is never far from a base station, phones do not have to be nearly as powerful as before, and so they can be made smaller. Also, because an area is divided into many cells, the same channel can be used at the same time by two subscribers in different locations.

Cellular networks must be able to pass calls from one cell to another as a user moves through an area. To do so requires the ability to analyze the signal strength of each user's phone to determine when a caller is nearing the edge of a cell. When the caller moves into a different cell, the system usually must switch the call to a different channel, since the caller might be nearing another subscriber who is using the same frequency. A pair of conversations on the same channel in a single cell can interfere with each other. In the first cellular systems, if such a switch was not done rapidly, the conversations were interrupted. Cellular communication engineers later created ways for several conversations to take place on the same channel without interference by allowing multiple access to a single communications channel.

The biggest limitation to such technology is that cellular phone companies cannot provide global service. For example, a base station cannot be constructed in the middle of the ocean. By 1999, less than 20 percent of the world's land areas had cellular service. A desire for global wireless communications provided the impetus for the new satellite systems.

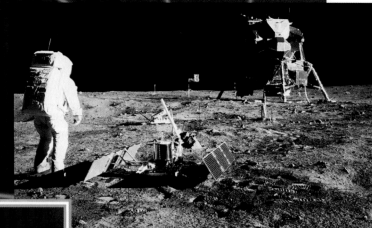

## Early satellite communications

The International Tele-communications Satellite Organization (Intelsat) was established in the 1960's to stimulate the development of satellite communications. Intelsat satellites made it possible to beam telephone messages and television signals worldwide.

In 1969, an Intelsat satellite provided live global television coverage of the Apollo XI moon landing.

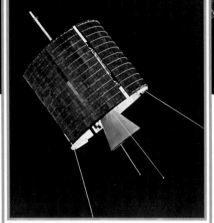

The organization's first satellite, Intelsat I, launched in 1965, could carry only 240 simultaneous telephone calls or one television channel. But there were many improvements to the technology, and by the late 1990's Intelsat satellites were able to simultaneously transmit 120,000 telephone calls and 32 television channels.

## Geosynchronous satellites

Despite the rapid evolution of satellite telecommunication systems in recent years, the general idea is not new. In 1945, the British science fiction author Arthur C. Clarke proposed placing a spacecraft in a special orbit to serve as a communications station. Clarke knew that the period of a satellite—the time it takes to complete one orbit around the Earth—increases with its distance from the planet. It occurred to him that at a certain distance, which he calculated to be 35,880 kilometers (22,300 miles), the orbital period would be 24 hours. Because the Earth rotates on its axis once every 24 hours, a satellite orbiting above the equator at that height, and in the same direction as the Earth rotates, would appear to be stationary over a single point on the planet's surface. A satellite in such an orbit—called a geosynchronous orbit—would make a stable target for relaying communications from the ground to another point partway around the world. Clarke said it would take only three geosynchronous satellites spaced 120 degrees apart in space to provide wireless communications for the entire globe.

It would be years before Clarke's idea could be realized. The first American satellites, launched in response to the Soviet Union's Sputnik in 1957, were in much lower orbits. In 1960, the United States Department of Defense launched Courier 1B, the first "active" communications satellite—one that amplifies a signal before redirecting it. (In contrast, a "passive" satellite can only reflect a signal.) Courier was able to receive and store messages as it passed over a base station on Earth and then transmit them on command. Courier was followed in 1962 by the first private telecommunications satellite, Telstar I, launched by

AT&T. Both of those satellites were in variable orbits taking them to a maximum height of about 6,000 kilometers (3,600 miles). Telstar transmitted voice, television, and data, but its signals could be relayed between two ground stations for periods of only about 10 minutes, because the satellite would soon pass out of view.

The first high-altitude, geosynchronous-orbit satellite, part of the Syncom series, was launched in 1963. The Syncom spacecraft, which at last allowed continuous satellite communications, were developed by the Hughes Aircraft Corporation for the National Aeronautics and Space Administration (NASA). Many experts consider the use of Syncom III to transmit televised events from the 1964 Summer Olympics in Tokyo to the United States as the true beginning of the age of satellite communications.

## Intelsat and Inmarsat

To aid in developing the potential of communications satellites in the 1960's, the United States helped create the International Telecommunications Satellite Organization (Intelsat), headquartered in Washington, D.C. Intelsat is the largest operator of geosynchronous communications satellites. In 1969, an Intelsat satellite provided live global television coverage of NASA's Apollo XI moon landing to some 500 million people worldwide. By 1999, Intelsat had more than 140 member nations and a network of 19 active satellites.

In 1979, a similar organization, the International Maritime Satellite Organization (Inmarsat), was formed to offer satellite communications to ships at sea and to people needing emergency assistance in isolated areas. By 1999, Inmarsat had become a private company with corporate or government shareholders in 86 nations.

Inmarsat also offers worldwide mobile telephone service, but its system has several disadvantages. For example, the smallest Inmarsat phone unit is the size of a briefcase. A user must open the case and unfold panels on the lid to form an antenna. Because the unit must send a signal to a satellite some 36,000 kilometers away, it must be fairly powerful. As a result, an Inmarsat user must be careful not to walk in front of the antenna when the unit is transmitting, since it produces a signal strong enough to damage human tissue. Also, to assure that the signal is as strong as possible when it reaches the satellite, the user must point the antenna at a specific satellite.

The Inmarsat satellites—and all geosynchronous satellites—have another limitation that may annoy some telephone users. Radio waves, like all forms of electromagnetic radiation, travel at the speed of light, 299,792 kilometers (186,282 miles) per second. With a geostationary satellite, a radio signal from a user must travel so far up to the satellite and then so far back down to the other person in a conversation that there is a slight time lag—approximately half a second. Although such a delay is considered insignificant in relaying television signals or data, in phone conversations it can lead to echoes or to people "stepping" on each other's words.

## Two approaches to satellite telecommunications

Most satellite phone networks use a "bent-pipe" system, so called because a signal is redirected like water flowing through a pipe. But a new kind of system passes a signal from one satellite to another.

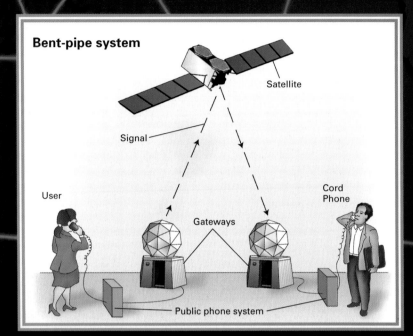

**Bent-pipe system**

Satellite

Signal

User

Cord Phone

Gateways

Public phone system

In a typical bent-pipe system, a call is sent through the public phone system to a satellite base station, called a gateway. At the gateway, an antenna sends the signal to a satellite, which transfers it to another gateway. From there, the call is routed into the public phone lines. The Globalstar wireless network, which was due to be completed in late 1999, uses a bent-pipe system, but its phones send signals directly to a satellite.

## Turning communications systems upside down

The Iridium and Globalstar phone systems were designed to overcome the limitations of geosynchronous satellites. The idea was to use low-altitude satellites but to have so many of them that several would always be available from any point on Earth.

Iridium got its start in the mid-1980's, when Bary Bertiger, an engineer with the Motorola Corporation in Arizona, was preparing to go on vacation to an isolated part of the Bahamas with his wife. His wife, a real estate agent, was trying to complete a real estate deal and was frustrated that she would be unable to use her cellular phone. She challenged her husband to devise a system that would work in such places.

Bertiger and his colleagues hit on the solution: They would turn a cellular phone system "upside down" and put the cells in space, with a satellite serving as the base station in each cell. With satellites in low Earth orbit, their antennas would be able to pick up low-intensity signals from small handsets. The phone's transmitters would be *omnidirectional* (sending signals in all directions) so they would not have to be pointed at the satellites.

The Motorola engineers devised a scheme in which 77 satellites would orbit the Earth at the same altitude, 780 kilometers (485 miles). They named the system after the 77th element in the periodic table of

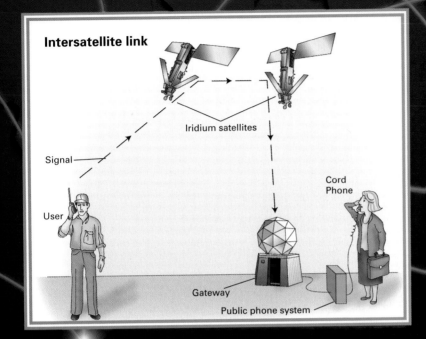

**Intersatellite link**

Iridium satellites

Signal

User

Cord Phone

Gateway

Public phone system

Another system is being used by the Iridium network, which began operation in late 1998. Iridium satellites use a process called an intersatellite link, which enables them to communicate with one another. A signal from an Iridium wireless phone is transmitted to the nearest satellite and is then transferred from one satellite to another until it reaches one that is near a gateway. That satellite then transmits the signal to the gateway, where the call is directed into the public phone lines.

the elements—iridium. The engineers later developed a more efficient design for the satellites that boosted their power and reduced the required number of satellites to 66, but they did not change the name. The system was completed and began operation in November 1998.

Iridium differs from traditional communications satellite systems in the way it sends and receives messages. Traditional communications satellites are referred to as bent-pipe systems. Like a pipe redirecting a flow of water, a satellite of this type changes the direction of a transmission beamed to it and sends it to another point on Earth. A transatlantic telephone call in such a system begins in a surface network and is routed to a base station, called a *gateway*. There a large antenna "uplinks" the call to the satellite. An antenna on the satellite "downlinks" it over a broad area, known as the antenna's *footprint*. On the other side of the ocean, somewhere in that footprint, another gateway receives the signal and routes the call into the land lines of the public phone system.

Such a routine would not work with the Iridium system because its satellites were designed to produce a much smaller footprint, each of which would require its own gateway at all times. Providing that many ground stations would be too costly. Furthermore, some ground stations would have had to be built in remote areas and at sea.

The engineers reasoned, however, that if the Iridium satellites could communicate not just with the ground but also with one another, a call could be relayed from one satellite to another until it reached one whose footprint had a gateway within view. That idea was adopted, making Iridium satellites the first commercial satellites with the ability to communicate with one another, a process called an intersatellite link. Even though the satellites' footprints are constantly moving as the satellites themselves move, these links enable continuous coverage.

Like all satellite phone systems, Iridium uses digital technology. Digital technology transforms a radio signal into a string of 0's and 1's. Each 0 or 1 is a *bit* (binary digit) that acts as a single unit of information. Digital technology enables the satellite to handle larger amounts of data than is possible with a fluctuating analog signal. It also increases the efficiency with which information is transmitted. Digital technology was not a new idea—it had long been used in computers and had been adopted by some cellular phone systems in the late 1990's. Iridium, however, was the first system to use onboard "packet" switching, a technique that divides a telephone conversation into many short strings of digital data. Each such string begins with a short code that identifies the collection of data as a packet and tells the system where to route it. This is followed by a piece of the telephone conversation and finally by a signal to the system that it has reached the end of the packet. Packet switching makes it possible to greatly expand a system's data transmission capacity.

Iridium also relies on "phased-array" technology, allowing an antenna on a satellite to focus its power in multiple beams simultaneously.

An Iridium satellite enables users of an Iridium phone, *right,* to place a call from anywhere in the world.

This allows a satellite to communicate with many telephones at the same time. The 66 operational Iridium satellites, along with six spare spacecraft, are arranged in six orbital planes almost perpendicular to the equator. Each plane contains 11 satellites, which follow each other along a nearly circular path around the planet, crossing near both poles, like a huge Ferris wheel in space. The satellites were launched in 15 flights from three continents in just over one year.

An Iridium phone is slightly larger and heavier than a typical cellular phone, mainly because it has a larger antenna. Communication experts, however, consider the Iridium technology a big step in the miniaturization of satellite mobile phones. When an Iridium phone is switched on, it sends a signal identifying the subscriber that is picked up by the nearest Iridium satellite. That satellite sends the signal to a gateway within the Iridium network, at which the caller is registered for subscription verification, call-routing, and billing. Unless the person being called is another Iridium user, the call is routed through one of the system's 12 gateways constructed worldwide and into the public phone network. Iridium designed its service mainly for international business travelers who want a small, portable phone to use anywhere in the world. Service for the system in 1999 cost about $3 per minute.

## Other competitors in the wireless marketplace

Competitors in the wireless world emerged at a rapid pace in the 1990's. Globalstar, a satellite telephone project based in San Jose, California, and led by Loral Space & Communications of New York City and Qualcomm Incorporated of San Diego, was in the final stages of preparation in mid-1999. Globalstar consists of a 48-satellite constellation, plus four nonoperational spare satellites. Six satellites are spaced evenly in each of eight orbital planes, circling the globe in a low Earth orbit 1,414 kilometers (878 miles) high.

Globalstar developed a patented system that virtually eliminates the problem of a signal being blocked by a building or geographical features. The system gives a caller access to two or three satellites simultaneously, any one of which can handle the transmission. Globalstar planned to attract a clientele consisting of both mobile phone users and customers living in areas that are not served, or are poorly served, by land-based phone systems.

Unlike Iridium, Globalstar decided not to put its satellites into polar orbits. Instead its satellites are in orbits slanted 52 degrees to the equator. As a result, the system will be unable to provide global coverage, though it will still offer service to the majority of the world's population. Also unlike Iridium, Globalstar uses a bent-pipe system, so even though signals go straight from a user's phone to a satellite, call processing and switching occurs on the ground. Thus, the Globalstar system needs more ground stations than its competitor. Plans called for 60 gateways, five times the number used by Iridium. Globalstar began launching satellites in 1998 and planned to complete the system in late 1999. Calls were expected to cost between $1 and $1.50 per minute.

Not all new ventures into satellite-based communications in the late 1990's involved new business ventures. Inmarsat, the pioneer in mobile satellite communication, created ICO Global Communications to build a new, nongeosynchronous system to work with smaller, lighter handsets. ICO's 10 satellites, transmitting through 12 ground stations, were to circle the planet in a medium Earth orbit 10,390 kilometers (6,460 miles) high. From that altitude, each satellite would be able to cover about 30 percent of the planet's surface. With two to four satellites always in view of a user on the ground, ICO was also expected to have alternative paths available for each call. Charges for calls were predicted to cost about $2.50 per minute. The company expected to have its system operational in late 2000.

## Big Leos, Little Leos, and broadband

The various low- and medium-orbit satellite systems being developed in the late 1990's are referred to collectively as Leos, for low Earth orbit. In the telecommunications industry, however, satellite systems are defined not by the orbits they use but by the part of the radio spectrum used by the satellites' communications equipment. Although all radio waves travel at the speed of light, a radio signal can have different frequencies and wavelengths. The higher the frequency, the shorter the wavelength. Low-frequency/long-wavelength signals such as those of AM and FM radio stations are not very susceptible to interference, enabling them to pass through buildings and to remain relatively unaffected by weather. The highest-frequency/shortest-wavelength radio waves can be blocked by buildings and disturbed by rain or fog.

The portion of the radio spectrum carrying signals for such low- and medium-orbit systems as Iridium, Globalstar, and ICO is relatively high frequency, so those systems are often referred to as the Big Leos. Big Leo systems transmit at 1 billion and 3 billion bits per second. A band with a somewhat lower frequency will be used by low-orbit satellite systems called Little Leos for paging devices and short-text messages. Little Leo systems transmit data at less than 1 billion bits per second. There were several satellite systems being planned in 1999 that would make use of frequencies even higher than the band being used by the Big Leos. They would offer so-called broadband wireless services, especially access to the Internet. Broadband systems are capable of transmitting 20 to 30 billion bits per second.

One such broadband system under development in 1999 was Teledesic. Plans for the Teledesic system called for using 288 low-orbit satellites—24 satellites in 12 planes orbiting 1,375 kilometers (855 miles) high. Teledesic officials predicted the system would cost approximately $9 billion, though some industry analysts said that the cost could easily reach $15 billion.

Teledesic's system was not expected to be available to customers until 2003. Rather than market directly to consumers, the company planned to first become a sort of "Internet in the sky," providing broadband communications capabilities to other companies. Teledesic

## Major new satellite communications systems

| System | When operational | Number of satellites | Altitude of orbit | Uses |
|---|---|---|---|---|
| Iridium | 1998 | 66 | 780 kilometers (485 miles) | Phone service |
| Globalstar | Late 1999 | 48 | 1,414 kilometers (878 miles) | Phone service |
| ICO Global Communications | Late 2000 | 10 | 10,390 kilometers (6,460 miles) | Phone service |
| Teledesic | 2003 | 288 | 1,375 kilometers (855 miles) | Internet service |

expected future users to range from information industry employees who don't want to be confined to the often-congested telephone lines to people in less-developed nations to multinational corporations seeking to link their offices. The service is not intended to be mobile. Instead, customers would link to the network by installing a Teledesic terminal on the roof of their office, home, or school.

Medium- and low-orbit satellite systems in operation or development in 1999 will give consumers several options for wireless telephone service and also make wireless access to the Internet available to people around the world.

### A booming industry—but an uncertain one

Experts predicted that the market for all kinds of satellite-based communication would continue to grow rapidly in the early 2000's. Some telecommunications industry analysts predicted in the late 1990's that by 2007 there would be 30 million users of satellite phones. Globalstar executives estimated that at least that many customers could be found in developing countries alone. Some analysts said the worldwide market for broadband satellite systems would reach 150 to 200 million users by about 2010.

Satellite systems were not the only wireless technologies undergoing rapid development in the late 1990's; ground-based systems were also advancing. Technologies becoming available in 1999 included wireless cable, which offered the broad bandwidth of a cable television system but without the wiring. One of the most promising of these systems was called Local Multipoint Distribution Service (LMDS).

LMDS uses high-frequency microwaves. Unlike AM and FM radio

signals, however, microwave radio signals cannot pass through walls. A signal from a transmitter must therefore pass unobstructed to the receiver. Also, the signals do not travel very far, which means such systems must be cellular. Each cell can be no more than 5 kilometers (3 miles) wide, though some tests have indicated that the most effective cell size may be closer to 2 kilometers (1.25 miles).

Within that limitation, LMDS can transmit huge amounts of digital data. The bandwidth of a typical system is enough to provide rapid access to the Internet. Since microwaves are absorbed by water, LMDS systems must boost their power when it rains. If the power is not increased, then the cell size will shrink. In 1999, tests showed that only about 60 percent of homes within an LMDS cell received clear reception. With overlapping cells, the proportion of U.S. homes obtaining a strong signal could be boosted economically to 85 percent.

Some critics of ground-based technologies like LMDS argue that satellite communications systems have the power to provide any service needed. Television service provides an instructive example. Cable television offered viewers dozens more channels and better reception than most homes could pick up with a roof-top antenna. By the 1990's, some 60 million homes in the United States were cable subscribers. By the mid-1990's, however, high-powered satellites were capable of beaming digital TV signals to home satellite dishes. The digital signals made television reception better than that provided by cable, and satellite systems offered hundreds of channels. By 1999, there were about 9 million U.S. households with satellite TV dishes.

Companies in 1999 were competing to expand satellite communications systems to radio. Potential customers ranged from American motorists seeking uninterrupted radio signals as they travel cross-country, to people in less-developed nations who simply have no conventional radio stations from which to choose.

The creation of so many satellite communications systems offered exciting possibilities, but it was also an economic and technological gamble. For example, the Iridium company estimated that the system would need 600,000 subscribers in order to begin paying off its debt. By May 1999, fewer than 11,000 had signed up. Company officials nonetheless predicted that the system would have 5 million subscribers by 2010. Many financial analysts, however, were wary of such predictions. Some cautioned that the marketplace in the 2000's would be too small, at least initially, to support several satellite communications systems. Others warned that if major systems such as Iridium or Globalstar stumbled—either technically or financially—it could hinder future investments in such communications ventures.

How these developments would play out in the end was anybody's guess. All that could be said for certain as the 1990's drew to a close was that communications technology was changing more rapidly than ever. Experts predicted that in the 2000's, wireless networks, whether ground-based or in space, would be able to handle all sorts of applications, including voice, video, and data, and almost all networks would become interconnected. But who would be the winners and losers in the new wireless world? Some investors may lose fortunes, but it seems safe to say that consumers will find themselves in the winner's circle.

## For additional information:

### Books and periodicals

Elbert, Bruce R. *Introduction to Satellite Communication.* Artech House, 1998.

Jamalipour, Abbas. *Low Earth Orbital Satellites for Personal Communication Networks.* Artech House, 1998.

Rees, David W. E. *Satellite Communications: The First Quarter Century of Service.* John Wiley & Sons, 1990.

Vizard, Frank. "Space Calls." *Popular Science,* October 1998, pp. 94-97.

### Web sites

Globalstar Web page—www.globalstar.com
ICO Web page—www.ico.com
Inmarsat Web page—www.inmarsat.org
Intelsat Web page—www.intelsat.com
Iridium Web page—www.iridium.com
Teledesic Web page—www.teledesic.com

## Introduction

A new millennium dawns just once every thousand years, so the year 2000, marking the beginning of the third millennium (or the end of the second for mathematical sticklers), is a noteworthy date. Of course, the years of a millennium are completely arbitrary, because different calendars have different starting points. The Gregorian calendar, which has tallied the years in the Western world for centuries, takes the birth of Christ as its starting point. But it could just as easily date from the founding of Rome, in which case we would be embarking on the year 2753.

Be that as it may, people the world over in 1999 were looking forward to the new millennium and recalling the achievements of the one coming to an end. The celebratory atmosphere surrounding the year 2000 contrasted sharply with the widespread feeling of foreboding that historical sources tell us was abroad in the land as the year 1000 approached. Many people in that superstition-ridden time feared that the approach of a new millennium heralded not a beginning but an end—of the world.

Of course, there were still superstitions, and apprehensions about the future, in 1999. To most forward-looking people, though, the year 2000 appeared to be a threshold beyond which lay a wondrous "World of the Future" that had been long predicted but slow in arriving—a world of genetic medicine, routine space travel, advanced computers and robots relieving us of life's tedious chores, and scores of other marvels.

In observance of the new millennium, and to look back on the one passing into history, the editors have prepared this Special Section. In four features, we present the story of science's development over the past 1,000 years, a sampling of predicted technological wonders that failed to materialize (but may be on the way), a futurist's look at the decades to come, and a few of the great questions that science still seeks to answer.

# A Thousand Years of Discovery

**Once almost extinguished by ignorance and superstition, scientific inquiry has become a triumph of the human mind.**

**By James Trefil**

The demon-haunted world that the people of the Middle Ages thought they inhabited is portrayed in an engraving from the 1400's. At the time the engraving was made, however, the superstitions that had ruled human society for centuries were starting to be challenged by a new spirit of scientific inquiry.

Try to imagine what your world would be like if not for the scientific discoveries of the past millennium. If you are reading this indoors, you wouldn't have artificial lights. A thousand years ago, the best you could have done was some candles or torches. Is the building you are sitting in heated or air-conditioned? You wouldn't have those luxuries, either. A thousand years ago the only heat people had to stave off the cold of winter came from an inefficient fireplace, and in the summer they could find relief from the heat only by fanning themselves.

Look at your clothes. Anything that isn't cotton, wool, silk or leather is the result of the chemical industry that came into being in the 1900's. And the food you eat is the result of enormous advances in our understanding of agriculture. Finally, if things today were still the way they were 1,000 years ago, you would be lucky to live past age 25. Medical science and public health are both developments of the last millennium.

It's hard for most people to imagine a world without constant progress in science and technology. Society has become so used to constant progress that it is now taken for granted, and we forget the long, hard path our predecessors followed to give us the modern world. As we approach a new millennium, it may be instructive to pause and consider how science developed and grew. Doing so might give us a stronger appreciation for the contributions that science has made to everyday life.

A Chinese star map from the 1100's, painted on the ceiling of a tomb found near Beijing, depicts the major constellations. China, where astronomers had been observing the heavens for centuries, was one of several parts of the world where scientific investigation continued, while in the Western world it stagnated.

## Science in the ancient world

Science is not entirely a recent development. In the ancient world, Greek and Roman civilization had reached a high level of scientific sophistication and spread its influence throughout most of southern Europe and the Middle East. For example, around A.D. 150, the Greek astronomer Ptolemy, who lived and worked in Alexandria, Egypt, took centuries of astronomic observation and combined them into a single view of the universe. Ptolemy's scheme envisioned the planets moving on crystal spheres around a stationary Earth. Although this idea was completely wrong, it helped ancient astronomers predict eclipses and the positions of the planets. In fact, Ptolemy's theory explained the universe as well as it could be com-

**The author:**
James Trefil is Clarence J. Robinson professor of physics at George Mason University in Fairfax, Virginia.

prehended at that time, and it was accepted as authoritative throughout Europe until the mid-1500's.

Other Greeks made advances in the field of mathematics and
medicine. Most of the geometry that is taught today in high school
was systematized in the works of the Greek philosopher Euclid in
about 300 B.C. In medicine, the great Greek physician Hippocrates
showed in the 400's B.C. that disease had natural, rather than divine,
causes. And another famous Greek physician, Galen, who practiced
medicine in Rome in the A.D. 100's, experimented on animals to develop theories on anatomy, disease, and physiology that influenced
doctors for nearly 1,500 years.

Despite the influence of Ptolemy and other Greek thinkers, the
Romans, who dominated the Mediterranean world from the 200's
B.C. to the A.D. 400's, were less interested in pure science than in
architecture and engineering. Concerned mostly with *pragmatic*
(practical) problem solving, the Romans became known as great
builders of roads and aqueducts.

Internal power struggles began to weaken the Roman Empire in
the late A.D. 300's, causing it to be split into the West Roman Empire
and the East Roman Empire. By the mid-400's, Germanic peoples
had invaded much of the West Roman Empire, which they divided
into several kingdoms. The decline of the West brought an end to
scientific inquiry, and intellectual pursuits were soon confined to the
study of theology.

## Keeping the flame alive

As scientific curiosity stagnated in Europe during these times—
often called the Dark Ages—the Arab world kept the flame of learning alive, especially in the areas of astronomy and mathematics. Arabs
in the Middle East preserved much of the ancient Greco-Roman science and accurately translated many Greek and Roman texts into
Arabic. In the 800's, the Arab mathematician al-Khowarizmi organized and expanded algebra to include a number system developed in
India. This system used place values and also introduced the concept
of zero.

Arabs also made advances in the field of medicine. One prominent
Arab physician, Avicenna, who lived from 980 to 1037, accurately described meningitis, tetanus, and other diseases and published his
findings in a medical encyclopedia titled the *Canon of Medicine*.

In the Far East, China pursued its own course in science and technology. Observations made by Chinese astronomers over hundreds of
years proved invaluable to Western astronomers more than 1,000
years later in tracing long-term occurrences of such celestial phenomena as comets, eclipses, meteor showers, and sunspots. During
the Song dynasty, which ruled from 960 to 1279, the Chinese invented gunpowder, the magnetic compass, and movable type for printing.

Meanwhile, in Europe, the Dark Ages were not a period of complete ignorance. Despite the stagnation of science, several technolog-

ical advances did occur. The moldboard plow, which allowed the cultivation of the heavy soils of northern Europe, the mechanical clock, and the widespread use of water wheels and windmills were all developments of the Middle Ages.

## The rebirth of scientific inquiry

In the 1100's, Europeans began to rediscover the works of the ancient Greeks and Romans that had been preserved by the Arabs and by Western monasteries. Many Arabic texts—which included original works as well as Arabic versions of ancient writings—were translated into Latin, the language of learning in the West. As a result, the writings of such scholars as Ptolemy once again became known in the West after an absence of hundreds of years.

Many historians credit the Spanish King Alfonso X, known as Alfonso the Wise, with promoting the rebirth of scientific thought in Europe in the 1200's. Among the scientific advances made by Alfonso's court scholars was a table of planetary motions based on Arabic sources but updated by Spanish astronomers.

The intellectual climate of Europe had changed radically by the 1400's with a revival of learning called the *Renaissance,* a cultural movement that began in Italy during the early 1300's and spread to other European nations by the late 1400's. During this period, a new breed of researchers—engineers in the mold of the ancient Romans—who were less interested in abstract knowledge than in solving specific problems, began to appear. If one had to point to a particular place and person that characterized this rebirth of pragmatism, a good case could be made for Milan, Italy, in the early 1500's and the work of an engineer and mathematician named Niccolo Tartaglia.

In the early 1500's, artillery was just being introduced in Europe, and the Duke of Milan asked a simple question: "At what angle should I set the barrel of a cannon to get maximum range for a given amount of powder?"

To find an answer, Tartaglia did not pore over ancient texts in search of information that might help him solve the mathematical aspects of the problem. Rather, he opted for a more creative approach. Tartaglia took a cannon into a field and fired it at a number of different angles. After analyzing the results, Tartaglia determined that the maximum range was attained when the cannon barrel was at an angle of 45 degrees. The experiment was one of the first to require physical proof before a theory could be accepted as true. At a time when the rediscovered teachings of the Greek and Roman philosophers were unquestioned by most Europeans, Tartaglia's experiment represented an important first step toward scientific investigation.

The rebirth of science in Europe continued to grow in the mid-1500's with the work of Nicolaus Copernicus, one of the great names in the history of astronomy. Copernicus was an administrative official

The Polish astronomer Nicolaus Copernicus theorized in the mid-1500's that the sun, and not the Earth, was at the center of the universe.

Danish astronomer Tycho Brahe, center, and his assistants calculate celestial movements. Brahe's observations were among the most accurate of the 1500's.

at the Cathedral of Frauenburg in Poland, and he also pursued astronomy as a hobby. He had a small observatory in which he was able to study the night sky, but his most important work was using Ptolemy's records of the varying positions of the planets and stars to develop a new theory of the universe. In the Copernican theory, the sun, and not the Earth, was at the center of the universe. His book, *On the Revolutions of the Heavenly Spheres* (1543), was itself to cause a revolution. But revolutionary or not, there was no denying the fact that the Earth did indeed revolve about the sun. By the early 1600's, astronomers such as Tycho Brahe and Johannes Kepler had built on Copernicus's work to explain the motions of the planets.

## Galileo and experimental science

One of the greatest scientists of that period, the Italian astronomer and physicist Galileo Galilei, studied the heavens with the newly invented telescope and also became convinced that Copernicus was right. Galileo, enlarging on the new approach pioneered by Tartaglia, also conducted many painstaking experiments in physics, for which he is referred to as the father of experimental science.

The systematic approach he pioneered for acquiring knowledge about the natural world developed into the scientific method, the cornerstone of modern science. Following the scientific method, if we want to learn something about how the world works, we must either make detailed observations of a natural phenomenon or conduct laboratory experiments. Once we have accumulated a sufficient body of information about the question being investigated, it should be possible to summarize the information in a compact set of statements, or laws, from which an overall theory—usually based on mathematics—can be formulated. The final step in the cycle is to use this theory to make predictions about things that have never been measured, and then to conduct further experiments and see if those predictions are correct. Thus, the scientific method is a way of approaching the truth and disproving incorrect notions about the world.

Galileo employed this questioning attitude in his investigation of the motion of objects being accelerated by gravity. Before the 1600's, people believed that if two objects of different weights were dropped from the same height at the same moment, the heavier object would fall faster and hit the ground first. Galileo conducted various experiments and concluded that all objects fall at the same speed (disregarding air resistance), and as they fall their speed increases at a regular rate. His results became known as the law of falling bodies.

## The contributions of Newton

Although Galileo laid the foundation for the scientific method, it was the English physicist and mathematician Isaac Newton who developed this investigative method to its complete form. For this reason, historians of science consider Newton one of the greatest scientists who ever lived. Thanks to Newton, when scientists in the 1800's began investigating the properties of electricity, magnetism, and heat, they didn't have to start from scratch—they knew how to proceed.

Newton's work included extensive experiments with light, in which he demonstrated that white light is a mixture of a spectrum of colors. But his most important discovery was that the forces and laws that operate on Earth also govern the motions of heavenly bodies.

Newton recounted that he came to this understanding quite suddenly while he was walking in an apple orchard and saw an apple fall from a tree. In the daylight sky beyond, he saw the moon. He realized that for the moon to stay in orbit about the Earth, a force had to be acting upon it, and that for the apple to fall, a force had to be acting on it as well. His conclusion, that in both cases the same force—gravity—was involved, was the foundation of what became known as "Newton's Law of Universal Gravitation." Newton's discoveries, which also included laws explaining other forms of motion, were published in 1687 in a book called *Principia Mathematica*, which is considered to be one of the greatest single contributions in the development of science.

Newton's insight that a single force could explain the motions of both a falling apple and the orbiting moon was the beginning of our understanding of the solar system. Scientists for the first time could explain what keeps the planets in their orbits—the gravitational pull of the sun. If the

In the 1600's, the Italian astronomer and physicist Galileo Galilei experimented with the newly invented telescope and used it to study the heavens.

Isaac Newton, best known for his theory of gravitation, also experimented with light. By passing a beam of sunlight through a prism, he showed that white light is a mixture of many colors.

The French chemist Antoine Lavoisier, shown with his wife, explained combustion and founded the modern science of chemistry.

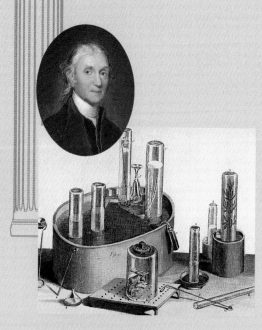

Using an improvised apparatus, *above,* the English chemist Joseph Priestley, *inset,* discovered the element oxygen.

sun's gravity were somehow turned off, the planets would fly off into space.

Newton's picture of the solar system was confirmed by the work of Edmond Halley, an English astronomer of the same period who became curious about why comets sometimes appear in the sky. Astronomers had always regarded comets as unpredictable phenomena. Halley, however, realized that Newton's laws must apply to comets as well as to planets. In the early 1700's, he collected data on a large number of comets that had been sighted in the past and used Newton's laws to determine their orbits. From his calculations, he discovered that a comet he had observed in 1682 followed the same orbit as a comet astronomers had seen in 1531 and 1607. Halley concluded that what had been observed was not three separate comets but a single comet returning about every 76 years. He predicted it would return in 1758 and, when it did, its appearance confirmed Newton's view of the universe. The comet was later named Halley's Comet.

Thus, Newton's legacy to science was a picture of a very ordered and predictable universe. This view spread through the Western culture of the 1700's, an era known as the Enlightenment or the Age of Reason, and it influenced not only the scientific world but the world at large. For example, when the framers of the United States Constitution debated how best to found a just and lasting nation, they worked within the Newtonian framework. They believed that there were natural, God-given laws governing the actions of human beings, just as there were laws controlling the motions of the planets, and they fashioned the Constitution to take those laws into account.

## The quest for knowledge continues

The 1700's were marked by a growing thirst for knowledge and scientific discovery. Using the scientific method, researchers in a number of fields, including such emerging disciplines as biology, geology, and chemistry, continued to enlarge science's understanding of the natural world.

The Swedish botanist and naturalist Carolus Linnaeus, second from right, devised a system for classifying plants and animals. He is pictured classifying the plants in the garden of his estate.

In biology, researchers began cataloguing the incredible diversity of living things. The Swedish naturalist and botanist Carolus Linnaeus developed the first modern system for naming and classifying plants and animals. The scientific study of the Earth got its start with James Hutton, a Scottish chemist who is often called the father of modern geology. In 1795, Hutton published a theory that the Earth was immensely old and that the planet's features were constantly, though very gradually, changing. His theory challenged a belief that the Earth was just a few thousand years old and that only huge natural disasters, particularly the Great Flood recorded in the Bible, altered its appearance.

In chemistry, the English chemist Joseph Priestley and Swedish chemist Carl Scheele independently discovered the element oxygen. The French chemist Antoine Lavoisier conducted experiments on combustion, proving that it involved the combination of a substance with oxygen. Lavoisier's *Elementary Treatise on Chemistry,* published in 1789, was the first modern chemistry textbook.

Chemists made great strides in the 1800's. In 1803, the British chemist John Dalton proposed a theory that matter is made up of tiny units called atoms. Unlike the ancient Greek philosophers who had earlier proposed the idea but made no attempt to demonstrate it experimentally, Dalton based his version of the atomic theory on observation. He and other chemists knew that most materials could be broken down by chemical reactions, such as burning, but that once a material was reduced to a certain point, it could be broken down no further. So although burning would produce charcoal, or pure carbon, no chemical reaction could further break down the carbon. Carbon was thus recognized as an elemental substance.

Dalton theorized that each chemical element—of which about 16 had then been identified—was made of a specific kind of atom. In his

In the 1830's, the British naturalist Charles Darwin served aboard the H.M.S. *Beagle, right,* —shown in the Strait of Magellan in South America—on a scientific voyage around the world. Darwin's observations on that trip led him to develop the theory of evolution, the idea that all life developed from simpler forms.

In the late 1800's, the Scottish mathematician and physicist James Clerk Maxwell combined existing theories of electricity and magnetism into a unified mathematical theory.

theory, atoms were indivisible but could be put together in different ways to make all of the materials in the world.

Chemists in the late 1800's, using Dalton's atomic theory, tried to discover the laws that govern the interactions of atoms. In 1869, the Russian chemist Dmitri Mendeleev and the German chemist Julius Lothar Meyer independently announced the discovery of a *periodic law.* They observed that when elements were listed according to their atomic weights, elements with similar properties appeared at regular intervals, or periods. From this finding, the chemists organized the elements into a periodic table of the elements that soon became a standard fixture in the world of chemistry. The table imposed order on the growing number of known elements, but why the elements could be arranged in such an orderly way was something the chemists of the 1800's could not explain. This problem would not be solved until the 1900's.

Meanwhile, great strides were made in the biological sciences. Beginning the 1870's, Louis Pasteur, a French chemist, and Robert Koch, a German physician, established a new theory of disease. Through their studies, Pasteur and Koch proved that infections and many diseases are caused by microscopic organisms. This "germ theory" of disease revolutionized medicine and public health. Pasteur benefited humanity by applying this basic discovery to important practical problems. Among his major accomplishments was the use of heat to kill microbes in wine, beer, milk, and food, in a process that became known as pasteurization. Pasteur also demonstrated the value of vaccination against microbes and developed several vaccines, including a rabies vaccine.

## Darwin and the theory of evolution

An event important to humanity's understanding of nature—and of itself—occurred in 1859. This was publication of the book *The Origin of Species* by the British naturalist Charles Darwin. Darwin theorized that the multitude of species on Earth had evolved from earlier species by means of a mechanism he called natural selection. He began his argument by noting that people could produce diversity within a given population of animals by selective breeding. Darwin, who was also a pigeon breeder, discussed at great length the different kinds of pigeons that breeders had produced by selective breeding of their stock over a period of just a few hundred years. He called this process artificial selection, and he argued that if human beings could produce such major changes over relatively short periods of time, then nature, acting over enormously longer spans of time, could produce even greater changes.

This process of natural selection, Darwin said, depends on two things: first, the fact that in any population there will always be natural variations among individuals and, second, that there is always competition among members of a species for food, mates, and other things necessary to survive and reproduce. Darwin argued that in any given environment, some members of a species will have certain genetically inherited traits that give them a slight advantage in surviving and reproducing. Because these individuals will tend to have more surviving offspring than the average for their species, over many generations the advantageous traits will come to be shared by more and more of the population. Through natural selection—often called the survival of the fittest—the entire population will eventually be made up of individuals that have the advantageous traits. If enough changes accumulate, then a new species will emerge.

Darwin's theory of evolution provided a coherent unifying theme for all the life sciences. Before Darwin, a researcher studying the distribution of butterfly species in Argentina and a scientist studying the distribution of fish in the Arctic Ocean had no common framework in which to discuss their work. The theory of evolution showed that all organisms are to some degree relat-

The French chemist Louis Pasteur, shown at work in his Paris laboratory, proved in the late 1800's that infections and most diseases are caused by microorganisms. This "germ theory" of disease led to the use of vaccines to prevent illness in people and animals. Pasteur himself developed several vaccines, including one to protect people from rabies.

The German physicist Albert Einstein revolutionized science with his theory of relativity, which created new concepts of time and space.

Marie Curie, a Polish-born French physicist, isolated the element radium in 1898, helping advance scientists' understanding of radioactivity.

ed because they all arose from the process of natural selection, a mechanism that continues to dominate life on Earth.

The idea of natural selection had a major impact on Western culture. It implied that human beings were produced by the same evolutionary process as other animals. In fact, some biologists said, humans are almost certainly descended from apes. Such a theory shocked many people in the 1800's. The majority of Europeans at that time believed that each species had been created by a separate divine act in the Garden of Eden and that human beings had been made in God's image. Although many theologians accepted the theory of evolution, some denounced it. Nonetheless, its acceptance spread quickly in the scientific community and among most educated lay people.

## The pace of discovery accelerates

By the late 1800's, scientific inquiry was exploding. The Scottish physicist James Clerk Maxwell extended the known laws of electricity and magnetism and combined them into a single set of four equations. The equations, published in 1865, described how electric and magnetic fields arise and interact. In 1897, the British physicist Joseph

American biologist James Watson, at left in photo, and British biologist Francis Crick, in 1953 discovered the structure of DNA (deoxyribonucleic acid, the molecule genes are made of), an achievement that greatly advanced scientists' understanding of human genetics.

Thomson discovered that atoms contain negatively charged particles, which he called electrons. The following year, the Polish-born French physicist Marie Curie and her husband, French physicist Pierre Curie, isolated the element radium, thereby advancing the study of radioactivity. In the late 1890's, astronomers were busily cataloguing stars through ever-larger telescopes, and biologists were developing an increasingly coherent view of life on Earth. Nevertheless, no one living at the close of the 1800's could have predicted the advances in science that would change the world in the century to come.

One of the first major scientific achievements of the 1900's came in 1905, when the German physicist Albert Einstein published his special theory of relativity. The theory, a major conceptual and theoretical event in physics, is based on the notion that the laws of physics are the same in all frames of reference—that is, no matter where you are or how you are moving, you will see the same laws of physics operating. Einstein's theory was based on the idea that different observers of the universe would see a given event differently, depending on how they were moving, and it made predictions that ran against human intuition. For example, the theory predicted that a moving clock will appear to run slower to a stationary observer than to a person traveling with the clock. As Einstein's predictions passed scientific tests, his theory of relativity became as essential a part of physics as Newton's laws of motion and changed the way scientists looked at the universe.

At the same time that Einstein was doing his most famous work— which also included a general theory of relativity, explaining gravity as a warping of space—other researchers were learning more about

Because of science, the world since 1900 has seen technological triumphs such as U.S. landings on the moon, *left,* and been blessed with many advances, such as modern medicine, *above right,* that have greatly improved human life. Science has also brought new perils to the world, including nuclear weapons of enormous destructive power, *above.* But most people would undoubtedly say that the good far outweighs the bad and would prefer to live in today's world than in a world without science.

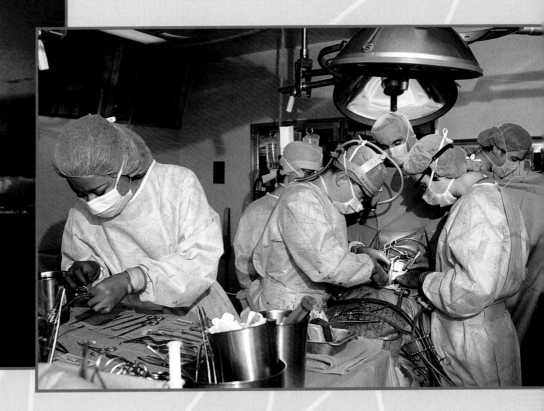

the atom. One important development took place in 1911 when the British physicist Ernest Rutherford performed an experiment that revolutionized scientists' understanding of atomic structure. Thomson had pictured atoms as tiny balls with electrons interspersed all through them. Rutherford showed, however, that an atom has a well-defined structure, with most of its mass locked inside a very tiny, positively charged nucleus around which the negatively charged electrons whirl at tremendous speeds.

Rutherford's experiments did not explain how electrons were arranged around the nucleus. It was not until 1913 that the Danish physicist Niels Bohr suggested that electrons are confined to particular orbits, or "shells" around the nucleus. Bohr also theorized that when an electron jumps from an outer orbit to an inner one, it emits a quantum, or unit, of energy. His theory became a part of quantum mechanics, a field of physics that explains the behavior of atoms and the subatomic particles of which they are made. Quantum mechanics explained why the elements could be arranged in an orderly way in the periodic table. According to quantum theory, similarities in the behavior of a group of elements result from similarities in the structure of the atoms of those elements.

From then on, much of the research in physics in the 1900's involved attempts to understand how an atom's nucleus is structured and how it works. Scientists had discovered in the late 1800's that positively charged particles called protons are one of the particles that make up the nucleus. Another particle, the uncharged neutron, was discovered in 1932. Beginning in the 1930's, researchers learned to conduct experiments in which they smashed fast-moving particles into nuclei and examined the debris to learn more about what's at the heart of an atom. They discovered that there are many other kinds of "fundamental particles" that make up protons and neutrons, and that all of these supposedly elementary particles are made of even smaller things. In 1964, the American physicists Murray Gell-Mann and George Zweig independently proposed that all nuclear particles are composed of smaller building blocks called quarks. By 1999, investigators had discovered evidence for six kinds of quarks, which combine in different ways to form all of the nuclear particles known to science.

Equally dramatic changes occurred in the 1900's to alter our perceptions of the universe. Astronomers had known since the 1860's that the stars are very distant, but they still had many questions about the heavens. In the 1920's, American astronomer Edwin Hubble, working at the Mount Wilson Observatory in California, was able to solve an old problem in astronomy. The puzzle concerned whether our own galaxy, the Milky Way, constitutes the entire universe or whether other "island universes" exist. By showing that many hazy "nebulae" are actually huge collections of stars far beyond the boundaries of the Milky Way, Hubble determined that our own galaxy is just one of billions in the universe.

Hubble also ascertained that these many other galaxies are moving away from us and from one another. This finding revealed that the universe is expanding, which indicated that it began in an explosive event—a "big bang"—at a specific time in the past. Astronomers later calculated the age of the universe to be between 10 billion and 20 billion years.

### Understanding the molecular basis of life

Other scientific advances led to changes in our understanding of living systems. Even after Darwin offered a general framework within which to understand life, biologists continued well into the 1900's to be concerned with examining and cataloging organisms. But beginning in the late 1950's, that focus changed abruptly as life scientists realized that they could at last understand the molecular basis of life. Biologists knew, based on discoveries made by Gregor Mendel, an Austrian Monk, in the mid-1800's, that physical characteristics are produced by basic hereditary units that transmit traits from generation to generation. But the event that led to a new turn in biology was the 1953 discovery of the structure of DNA (deoxyribonucleic acid—the molecule that genes are made of) by biologists James Watson of the United States and Francis Crick of Great Britain. Science's understanding of human genes exploded in the following years.

Advances in biology and medicine have increased the average life span by 50 percent since the beginning of the century, and by 1999, a new day in medicine—the era of genetic medicine—was in the offing. Genetic medicine, which focuses on the molecular basis of disease, may lead to gene therapy, or the replacement of faulty genes with new genes to cure many diseases. Discoveries in all the major fields of science have transformed human society during the 1900's. Physics, for example, has given us aviation, computer technology, nuclear power, space travel, and telecommunications, not to mention such everyday amenities as electricity.

## Concerns and hopes

Of course, not every scientific advance has been viewed as beneficial by all parts of society. For example, though scientists and engineers found ways to harness nuclear energy as a power source, critics of science contend that nuclear plants are an environmental hazard. Furthermore, they say, the development of the atomic bomb in the 1940's and the hydrogen bomb in the 1950's created a threat of global annihilation that existed for nearly 50 years. In the late 1990's, critics of science expressed considerable opposition to the newfound ability to *clone* (make exact genetic duplicates of) living creatures. This astonishing technology, though it promised great benefits in agriculture, could presumably be used to create human clones, a prospect viewed with horror by many people.

Despite such valid concerns, few people would deny that science has brought humanity great progress in the last millennium. And it can undoubtedly carry us to even greater heights in the next millennium. However, as history has shown, continued progress in science and technology is not guaranteed. Some observers in 1999 warned that science was being threatened by a rising tide of irrational beliefs among the public. Proponents of science argue that the quest for knowledge is necessary for the survival of civilization, and that humanity cannot take for granted that the future will be a continuation of the present. Only science, they say, can ensure that there will be no new Dark Age.

## For additional information:

### Books and periodicals

Porter, Roy. *The Greatest Benefit to Mankind: A Medical History of Humanity.* W. W. Norton & Company, 1998.
Silver, Brian L. *The Ascent of Science.* Oxford University Press, 1998.
Trefil, James. *Reading the Mind of God.* Scribner's, 1989.
Watson, James. *The Double Helix.* New American Library, 1991.

An underwater jet-propelled vehicle carries people from a domed city beneath the sea in a vision of things to come from 1962. The creation of such a massive complex on the deep-sea floor—capable of withstanding tons of pressure from the waters above—remained only a dream in 2000.

# The Way Things Were Supposed To Be

**In the mid-1900's, people had great expectations for the world of the year 2000. As the new millennium dawned, most of those hopes remained unfulfilled.**

## By Al Smuskiewicz

One morning in the year 2000, Jim Smith, an average American business executive, hops into his jet car and flies from his solar-powered, completely mechanized home to the atomic-powered rocket factory where he works. When he arrives, one of the factory's intelligent robots greets him with a pleasant "Good morning, Jim." The robots are busy manufacturing the company's latest spaceship, to ferry humans on yet another trip to Mars. Combined with the Martian colony, several lunar colonies, and numerous space stations orbiting Earth, there are now approximately 1 million people living and working in space.

But wait! You say you don't recognize this world of the year 2000? That's because, obviously, it doesn't exist. Still, such visions of things to come were widely shared by many scientists, science-fiction writers, and futurists in the mid-1900's. Optimism in humanity's destiny to conquer the world with technology grew from the 1939-1940 New York World's Fair, which highlighted such new wonders as television and long-distance phone calls, and was reaffirmed by the technological triumphs of the dawning space age in the 1950's and 1960's.

For a while, future possibilities seemed limitless. And, in fact, a number of midcentury predictions have come true. Among these are the interstate highway system, a worldwide network linking computers, robotic assembly lines, cellular telephones, fax machines, microwave ovens, and synthetic materials in everything from homes to clothes.

However, the flashier predictions have fizzled. There are still no jet cars in people's garages. Robots and computers endowed with artificial intelligence are nowhere in sight. Humans have not yet set foot on Mars. And the glowing prophecies for solar and atomic energy no longer shine as brightly as once envisioned. Nevertheless, a number of these predictions were scientifically valid when they were made, and they remain so. Many scientists believe that the reasons they did not come true by the year 2000 have more to do with economics and politics than science. So perhaps the technological wonders on these pages still await us in the future.

The author:
Al Smuskiewicz
is a *Science Year*
senior editor.

Mechanized fields stretching for miles, cattle in skyscraper pens, and grain flowing through long pneumatic tubes would feed America's growing population in the year 2000, according to this 1970 vision. But over the next 30 years, only minimal changes occurred in agricultural methods.

In the classic 1968 motion picture *2001: A Space Odyssey*, two astronauts hide in a soundproof pod to discuss shutting down their ship's erratically functioning computer "HAL." HAL, however, read their lips through the pod window and tried to sabotage their plan. Scientists were not even close to achieving such artificial intelligence at the start of the new millennium.

Thousands of square kilometers of desert land covered with huge solar panels would meet all of the energy needs of the United States by 2000, according to some futurists in the 1970's. But these "solar farms" never materialized, in large part because researchers failed to find a cost-effective way to convert the sun's energy into electricity. Nonetheless, scientists made considerable progress in the development of affordable and efficient solar panels.

In a 1955 illustration, a soldier fires his "ray gun" at the enemy, who have scored hits on his rocket ship. Although electromagnetic radiation was being used for a number of purposes by 2000, including killing cancer cells, scientists had not found a way to concentrate such destructive energy into a hand-held weapon.

*Year* **2000**

A man of the future gets his air car serviced in a scene envisioned in the late 1960's. Flying cars were one of the most common mid-1900's predictions for 2000. One of the reasons they were not developed is that air traffic controllers raised concerns about the chaos that large numbers of such vehicles might cause in the skies. In addition, building a flying car that almost anyone could operate proved to be a formidable challenge.

Personal rapid transit vehicles (PRT's) respond to the press of a button to pick up passengers and take them nonstop to their destinations. Among the most popular predictions of the 1970's, the inefficient PRT system had achieved only limited application by 2000.

# Year 2000

In a 1955 depiction of life in the year 2000, personal grooming robots do a woman's hair and nails while her boyfriend waits impatiently by the videophone. Although smaller-sized videophones were available by the end of the century, people still had to do their own grooming.

In the early 2000's, conservationist-astronaut Freeman Lowell plays poker with the robot "Huey" in the 1971 movie *Silent Running*. Huey, along with his robopals "Dewey" and "Louie," could also perform mechanical, surgical, and gardening jobs. In the real world of 2000, robots were capable of only simple, repetitive tasks, such as industrial assembly-line work.

An enormous cylindrical space colony, *right,* rotating to produce artificial gravity, provides suburban comfort to adventurous space travelers. Some scientists in the mid-1970's speculated that 1 million people might be living and working in space by 2000, a wildly optimistic prediction.

Astronauts "walk" in deep space, *above,* while tethered to their circular space station, a design widely predicted in the 1950's that bears no resemblance to the awkward-looking International Space Station on which construction began in 1999.

Mining operations on the moon send valuable minerals to Earth, *right,* in a peek at the future painted soon after the Apollo missions ended in the 1970's. The main reason such ambitious space projects failed to become a reality by 2000 was the incredible expense that they would have entailed.

# *Eyes on the Future*

A physicist and science fiction writer looks ahead to the world of the 2000's and tells what he sees for science and society.

**An interview with Gregory Benford, conducted by David Dreier**

Gregory Benford—*physicist, science fiction writer, and futurist—has been thinking about the world of tomorrow for most of his life. In both his fictional works and a large collection of scientific essays, Benford has speculated on what the future may hold for humanity. In contrast to science fiction writers who don't worry too much about technical accuracy when telling a tale, Benford is what is known as a "hard" science fiction writer—he bases his stories and novels on solid scientific principles. His approximately 20 novels and 120 short stories have brought him many honors, including two Nebula Awards, science fiction's equivalent of the Pulitzer Prize. Benford also devotes considerable time to teaching and research. He has been a member of the physics faculty of the University of California at Irvine (UCI) since 1971 and a full professor since 1980. In addition, he has been a visiting fellow at Cambridge University in England and has served as an adviser to the White House Council on Space Policy, the National Aeronautics and Space Administration, and the Department of Energy. Despite his busy schedule, Benford always makes time for writing. His latest book,* Deep Time, *a nonfiction work about the possible impact of present-day technologies on the far future, was published in January 1999. He is now looking forward to new projects—and, of course, to the future.*

**Science Year:** As a century—and a millennium—comes to an end, people are eager for predictions about what life may be like in the decades ahead. Accurate predictions about the future, of course, are very difficult to make. People in the late 1800's had no idea that automobiles, airplanes, computers, and the power of the atom would transform society, and, likewise, there may be advances in coming years that will take us by surprise. Nonetheless, there are many current trends in science and technology that offer clues about what the future may hold. In looking ahead, what do you as a scientist and a visionary think will be the major scientific and technological advances that will shape people's lives in the 2000's?

**Benford:** I think we'll see tremendous progress in a number of areas, including computer technology, telecommunications, biology and medicine, and transportation, to name but a few. I think biology, in particular, will dominate the coming century in much the same way that physics—which enabled the development of aviation, space travel, nuclear power, television, and many other technologies—has dominated the 1900's. For that reason, I predict that the 2000's will be known as the Biological Century.

**Science Year:** Then let's start with biology. What are some of the particular advances you foresee?

**Benford:** I think biotechnology will develop into an extremely powerful tool. It will transform agriculture, and it will also be widely used in manufacturing. We will use biology to produce directly many of our goods, rather than just the raw materials for those goods.

**Science Year:** Could you explain how that might be done?

**Benford:** Well, for instance, we could alter the genes of tree cells so that they grow according to a new blueprint. It should be quite possible, in this way, to grow a wooden house. At present, a wooden house is made by raising trees, cutting them into pieces, and then putting the pieces together. We could eliminate those steps and grow walls and other wooden structural members in place. Then we'd simply have to add windows, plumbing equipment, and what not.

**Science Year:** That would certainly give new meaning to the term "tree house." What else will the Biological Century bring us?

**Benford:** It will transform the human prospect. We will gain control of our genetic destiny. We're on the verge of that now, with the several projects aimed at deciphering the human genome [all the genetic material in a human cell], which should be completed within a few years.

But the ability to guide human genetics will pose profound philosophical dilemmas. The control of biology so that it does not strip society of its ethical and social underpinnings will be the big issue of the coming decades. Would it be immoral, for example, for parents to want children who are uniformly tall, slim, and good-looking? Because we all know that such people tend to do better in life than people who are short, overweight, and homely. That's just one of the issues we will have to confront in the future.

**Science Year:** What do you foresee in the medical sciences?

**Benford:** We'll continue to make progress in curing heart disease

**The author:**
David Dreier is the managing editor of *Science Year.*

and cancer. Heart disease is mostly a plumbing problem and, therefore, will be easier to treat. Basically, if you could clear out arteries with nonsurgical methods, you could solve most heart problems. There's no particular reason why we shouldn't be able to knock the fatty deposits off the walls of arteries. In fact, there's a research program right here at the University of California at Irvine aimed at doing just that with sound waves. So far, it works in mice and pigs.

Cancer is a far more fundamental problem and it will take longer to solve. But within, I hope, about 30 years, we should largely have found our way through that thicket, and in perhaps 50 years we'll have conquered all forms of cancer. Cancer is essentially a genetic disease, so once we have an intimate understanding of the human genome, we should also have an understanding of cancer and how to cure it. Genetic medicine is the wave of the future for the treatment of cancer, as well as for many other diseases. We'll have gene therapy so advanced that presently fatal diseases like cancer will be treated on an out-patient basis—the doctor will inject you with some new genes, and you'll be on your way.

*I think biology will dominate the coming century. For that reason, I predict that the 2000's will be known as the Biological Century.*

You have to understand, though, that if we succeeded in eliminating heart disease and cancer, we would increase the average human life span by only about five years. People start dying of all kinds of things as they get older.

**Science Year:** Then what about conquering the diseases of old age, or even aging itself?

**Benford:** Most conditions associated with extreme old age will undoubtedly be cured eventually, and as a result I expect that in the 2000's, medical science will accomplish what it did in the 1900's—that is, increase people's life span by 50 percent. In 1900, the average man or woman in the United States lived to the age of 50. In 2000, the average life span in this country will be roughly 75, actually a little better than that. If we achieve that again—with new medicines, genetic therapies, and other advances—we will add something like 35 or 40 years to the human life span by the year 2100, and people will routinely live to be more than 100 years old.

By that time, it may simply be aging itself that finally does people in. People today who stay healthy past the age of 100 finally reach a point where their bodies fail—an age of about 125 seems to be the limit. Will that always be true? Maybe not. Sometime in the 2000's, medical science may well find a way of halting or reversing aging. There's a lot of research now underway on the aging process, and scientists may find that aging is a disease like any other, and is therefore curable.

**Science Year:** That's what some people think will happen. The well-known science fiction writer Arthur C. Clarke, with whom you have collaborated, has predicted that people born after the year 2000 won't die, except in accidents.

**Benford:** He could be right. Of course, that possibility doesn't do much for people born before the year 2000, but perhaps there's a chance for them, too.

**Science Year:** What do you mean?

**Benford:** I think cryonics—the freezing of human bodies after death in the hope of bringing them back to life at a later time—offers people the possibility of extended life. People who are frozen now could plausibly be revived in 50 or 75 years, and they would awaken in a world that is able to manipulate matter at the molecular level with great assurance and restore a body to a youthful life. Cryonics is the only way I know to possibly defeat death. So for an investment of $50,000 or more with a cryonics company, you can try to grab that brass ring hanging up there half a century or more away. And I think it's a very reasonable hope. Some biologists have derided cryonics as a ridiculous idea, a pipe dream, but I think they're wrong. Nature shows that it can be done. There are several species of frogs in Canada that freeze at the bottom of lakes throughout the entire winter and then revive in the summer.

**Science Year:** How would a frozen person be brought back to life?

**Benford:** Well, we're talking about very futuristic technologies, so there's really no way of saying. But I think nanotechnology—the coming ability to manipulate matter at its smallest levels—may hold part of the answer.

**Science Year:** In what way?

**Benford:** This technology will make it possible to build microscopic robotic devices to circulate through the body by the billions doing repair work to keep us healthy. And if they can mend the cells of a living person, they should be able to do the same for someone who has been frozen. A frozen body will need a lot of repair work at the molecular level, because the freezing process is likely to rip cell membranes and cause various other kinds of damage.

*I think cryonics—the freezing of human bodies after death in the hope of bringing them back to life at a later time—offers people the possibility of extended life.*

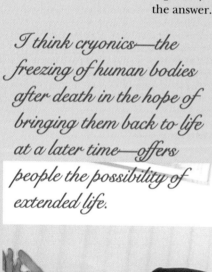

**Science Year:** Nanotechnology sounds as though it will be an incredible technology. What does it involve?

**Benford:** It's the ultimate goal in chemistry—the ability to place individual atoms and molecules in precise arrangements to make whatever we want from them, including many kinds of goods. But putting large numbers of

*By the end of the 2000's, manipulating matter at the molecular and atomic level to manufacture everything from clothing to automobiles will be routine.*

atoms and molecules exactly where you want them is a tough task, so I think it'll be several decades before nanotechnology really begins to be a doable thing, at least for the large-scale production of commercial products. After that, though, we'll get better and better at it. By the end of the 2000's, manipulating matter at the molecular and atomic level to manufacture everything from clothing to automobiles will be routine.

**Science Year:** So that's the Holy Grail for chemistry. What will it be for physics?

**Benford:** It will be a theory unifying all the fundamental forces of nature. We will then have an intimate understanding of the structure of matter at its tiniest dimensions, the way particles interact with one another, and the nature of space at the smallest and largest scales. Achieving such an understanding is quite possibly the closure of the true agenda of physics. It doesn't mean we'll be able to predict everything that's going to happen in the universe, but it does mean, I think, that we'll achieve a deep understanding of why the universe is the way it is. Then we should have an answer to the question posed by [the German-born physicist Albert] Einstein: whether God had any choice when he made the universe.

**Science Year:** What did Einstein mean by that?

**Benford:** He was pondering whether a universe like ours could have been constructed in any other way, with different physical laws, different masses for particles, and so forth. And perhaps it couldn't have been. There may be only one way to make a universe so that it gives rise to intelligent life. That's a very profound question, but there's a good chance we'll be able to answer it. When you consider that the goal of answering such a question wasn't even perceptible to scientists 100 years ago, you can appreciate what amazing progress physics has made in this century.

**Science Year:** Let's touch on a few other issues related somewhat to physics, starting with computer technology. How do you see computers developing over the next 25 to 50 years, and what capabilities do you think they will have?

# Inhabiting Two Worlds

Books have played a central role in Gregory Benford's life. Not only has he written many books, but, like most authors, he was first an avid reader. And it was two books that he chanced upon in his boyhood that, more than any other influence, launched him on his twin careers as a science fiction writer and physicist.

Benford encountered the first of those life-changing books in about 1950, while living with his family—mother, father, and twin brother, Jim—in Japan. His father, James, a career Army officer, had been transferred to Tokyo to serve on the staff of General Douglas MacArthur, who was heading the U.S. military occupation of Japan after World War II (1939-1945). At the school library, 9-year-old Gregory found a novel titled *Rocket Ship Galileo* by the American science fiction writer Robert Heinlein. This tale of space travel fired the boy's imagination and soon he was an ardent fan of science fiction. Although many youngsters go through a science fiction phase,

Benford, a long-time resident of Laguna Beach, California, enjoys sailing when his busy schedule allows it.

for Greg it was the start of a lifelong passion.

His brother, too, acquired the science fiction bug, and in 1954 the two of them started a "fanzine"—a science fiction fan magazine—called *Void.* By then, after a short stay in Atlanta, Georgia, the family had been moved to West Germany.

In 1957, the Benfords returned to the United States for good, settling in Dallas. While attending high school, Greg had another fateful encounter with a book. He chanced to read *Atoms in the Family,* by Laura Fermi, the wife of the late Nobel Prize-winning Italian-American physicist Enrico Fermi. Inspired by the book's account of Fermi's life as a nuclear physicist, Greg decided that physics was the career for him.

In 1959, he and Jim—who had made the same decision—won scholarships to the University of Oklahoma. They graduated in 1963 and then both went to the University of California at San Diego for graduate work.

While pursuing his graduate studies in physics, with an emphasis on the physics of solids, Greg met and began dating Joan Abbe of Boston, an art teacher at a girls' school in La Jolla, California. They married in 1967. That same year, Greg received his Ph.D.

By that time, he had begun trying his hand at science fiction, writing short stories whenever he could take time away from his studies. Much to his surprise, the stories began to sell. In 1969, while working as a research physicist at the Lawrence Radiation Laboratory in Livermore, California, Benford got a contract to write a science fiction novel, *Deeper Than the Darkness,* which was published the following year. With that success under his belt, he knew that from then on he would be pursuing a double career as both a physicist and a writer.

At the Lawrence laboratory, Benford worked in a program aimed at developing a power reactor using controlled nuclear fusion. Fusion is the process by which stars and hydrogen bombs produce energy, and many scientists were hopeful that fusion could be harnessed for generating electricity. But it soon became apparent that the technical problems involved were formidable, with no probability of success for decades. His interest in fusion dwindling, Benford left Livermore in 1971 to take a position as an assistant professor of physics at the recently established University of California at Irvine (UCI). He and Joan went to live in the nearby resort town of Laguna Beach, where they raised two children—a daughter, Alyson, and a son, Mark.

Benford found his new life agreeable. He taught physics classes and conducted research in theoretical astrophysics, including studies of the enormous jets of matter that stream from the center of many galaxies. At the same time, he continued his writing.

In 1974, Benford won his first Nebula Award, an annual literary prize conferred by the Science Fiction Writers of America, for the novelette *If the Stars Are Gods*. That book was coauthored with another science fiction writer, Gordon Eklund, with whom he would collaborate on several other fictional works in the 1970's.

Professional recognition as a writer was followed in 1980 by financial success with the publication of *Timescape*, which was to be his most popular novel. *Timescape* is the story of scientists who communicate with researchers in the past with the aid of tachyons, particles that travel faster than light and backward in time. The book became a best seller and won Benford his second Nebula Award. He was also named the recipient of the 1981 John W. Campbell Award, an honor bestowed each year on the best new science fiction writer by the World Science Fiction Convention.

Benford's most ambitious literary project was a group of six novels that became known as the *Galactic Series*. The series, which took him 25 years to complete, involves intelligent machines that human astronauts discover to be the dominant life form in the Milky Way Galaxy. The last of the *Galactic* novels, *Sailing Bright Eternity,* was published in 1995.

Between writing each of the *Galactic Series* books, Benford worked on other novels and short stories. With the novel *Artifact,* published in 1985, he switched gears and spun a tale of suspense about an archaeological excavation in Greece. Because his publisher refused to treat the book as anything but another Gregory Benford science fiction novel, Benford decided to take a pen name for his next foray into the suspense genre, *Chiller.* That 1993 novel, written under the name Sterling Blake, was based on the technology of cryonics—freezing bodies after death in the hope that doctors of the future will be able to restore them to life. But Blake's true identity was soon leaked, so Benford discarded the pseudonym.

Benford's other projects have included a stint as consultant and scriptwriter for "A Galactic Odyssey," an eight-part series on modern science and the Milky Way Galaxy produced in the late 1980's by Japanese National Broadcasting. At about the same time, he collaborated with the

In 1991, Benford visited science fiction writer Arthur C. Clarke in London after collaborating with him on a novel.

famed British-born science fiction author Arthur C. Clarke on *Beyond the Fall of Night* (1990), a sequel to Clarke's 1953 novel *Against the Fall of Night*. And a few years later, in an homage to Isaac Asimov, another of the great names in science fiction, Benford wrote *Foundation's Fear* (1997), an addition to Asimov's *Foundation* series of novels. Even before embarking on that last project, and before completing his *Galactic Series,* Benford had received yet another prestigious writing award, the 1990 United Nations Medal in Literature.

Throughout his writing career, Benford has continued as a faculty member at UCI, though in 1994, because of federal funding cutbacks and a heavy workload, he ceased doing laboratory work. Since then, he has confined his academic duties to teaching and theoretical research, principally on plasmas, extremely hot gases in which atoms have been stripped of their electrons.

In 1997, in an autobiographical essay, Benford summed up his twofold career as author and scientist: "The comforts of writing are many, as are those of physics. The two worlds rub against each other uneasily in my life. I shall probably always cycle between them, seeking the ideas and experiences that lend a vibrancy to life." [D.D.]

**Benford:** They'll of course get increasingly fast, but they'll also get increasingly small. Eventually, many computers will actually be microscopic in size. That's something we can expect from nanotechnology.

**Science Year:** How will computer technology affect people's lives in coming years?

**Benford:** Computers will be everywhere and, because they will be so small, they will be mostly embedded—that is, they will be integrated into our surroundings, and they'll inhabit everything we own or use. Computers will make everything "smart." We'll have self-adjusting shoes, for example, and desks that carry out the functions of a secretary. Within the next 20 years, the entire environment people inhabit will be smart: smart houses, smart cars—smart cities, in fact.

When we get to that point, people will be immersed in an artificial world that anticipates their needs and does everything possible to satisfy them. Say, for example, that you drive to a shopping mall. As you get out of your car—which drove itself to the mall on an intelligent roadway and then identified itself to the parking lot—the entire side of a building lights up in a large smile and says, "Hello, there, Eleanor. We're happy to see you back. You haven't been here for three weeks, but in that time, we've had a number of sales and, in fact, there are three of them going on now at your favorite stores, which I see from your buying record are Bloomingdale's, Barnes & Noble, and The Limited." At home, your bedroom will talk to you and adjust everything to your liking—temperature, mood music, flat-screen photographs on the wall of the Alps or a tropical island, and so on.

**Science Year:** How will computers affect the kinds of work people do?

**Benford:** They'll take over many difficult and routine and tedious jobs, a trend that is already well underway. Within a century in the advanced nations, the only jobs of any substance for most people will be in either personal services or the creative arts. The present growth of services in the American economy is an indicator of what is to come—an explosion of more personalized, custom services of every imaginable variety.

**Science Year:** Speaking of personal services, what do you anticipate in the development of household robots that will do people's routine chores for them?

**Benford:** The 1950's vision of helpful robots doing our heavy lifting and dirty work will happen, though not until we've made further advances in the development of artificial intelligence. But most household robots, when they do arrive on the scene, probably won't look like humans; they'll look like gadgets. And I don't think they'll walk on legs like humans do. We're the only species on the planet, other than birds, that walks by starting to fall and then catching itself. That, in essence, is how we get around, and it's a tricky form of mobility. Building a robot that walks on two legs would probably be quite difficult. So a household robot is likely to roll around on wheels, and it'll be some kind of cute little thing, like Sojourner on Mars. People will consider it a sort of pet. I think most people would feel uncomfortable

*Computers will make every-thing "smart." Within the next 20 years, the entire environ-ment people inhabit will be smart: smart houses, smart cars—smart cities, in fact.*

if these machines looked like weirdly constructed humans—it would be too much like owning slaves.

Still, for those who like the idea of having personal servants, and are willing to pay for the technology, extremely subservient humanlike robots will undoubtedly be available. In fact, if you want it and can afford it, you will probably be able to get a whole house—robots and other electronic technology—that treats you like a Roman emperor. We'll largely be able to tailor our personal environments to our individual needs and wants.

**Science Year:** Let's turn to another physics-related topic: communications technology. What do you foresee there?

**Benford:** Well, for one thing, people will be free of their offices and able to interact with high-quality wireless "visiphones"—phones that transmit images as well as sound. Most people, or at least most people working in the information industries, will be able to get the bulk of their work done at home or at the beach.

**Science Year:** How do you see the Internet evolving?

**Benford:** I think it will evolve into something that I've called in a couple of my stories the Mesh. A mesh is more finely divided than a net, so I see the Mesh as being far more pervasive than the Internet. At the moment, you can log onto the Internet at work or at home or at a few other places. Pretty soon, though, you'll be able to log on anywhere, and you'll be able to do so with a voice mike. We'll have voice-activated computers that are integrated into our homes, our clothing, our cars. You'll be able to access your e-mail, and have it read to you, over your car radio—beamed to you from a satellite link—while you drive home from work. And you'll be able to respond to it effortlessly. So let's say you're cruising down the road, and you get a recorded message from your pal Fred. He says, "Hey, what do you say we get together next Wednesday for dinner?" And you reply, "Love to. Let's make it 6:30 at the Club Mediterranean." Your smart system will even be able to call the restaurant and make the reservation and to remind you of it on Wednesday morning. That sort of thing will be routine in another 10 years. It'll just be so easy to get things done and to stay in touch with people wherever you go.

**Science Year:** What do you foresee for the use of virtual reality in the future? Right now this is a fairly crude technology in which people don large goggles and other clumsy equipment for the experience of immersing themselves in a simulated environment. Do you think this technology will someday make virtual environments and experiences nearly indistinguishable from the real thing?

**Benford:** I think so. You'll be able, for example, to explore a virtual cave, and you'll smell the wet rock, feel the clamminess of the air, hear the echo of objects falling far back in the cave. Exactly how the technology will develop is not easy to predict at this stage. You may wear some sort of very advanced body suit that stimulates all your senses, along with much-improved goggles or glasses that present lifelike images to your eyes.

An even better technique, however, might be to bypass the eyes and other sensory organs and tap directly into the brain and central nervous system. We could plug in and receive sensory data through a special implant—a jack in the back, so to speak. With this approach, virtual reality would be a sort of waking dream. But it would seem absolutely real.

**Science Year:** On a related subject, do you foresee any melding of the human brain with computers to increase human intelligence? Some scientists have suggested that this could be done by implanting computer chips in people's brains.

**Benford:** The problem with that idea is that information processing by computers is utterly different in style and speed from that of the brain. The brain operates at very low frequencies, computers at very high frequencies. So even if we develop tiny implantable computers, it's going to be very hard to couple them to the brain. It's an interface problem. It's a problem that many computer scientists and brain researchers think will be solved, though at this point they don't know how. It could take 50 years or more for this technology to become a reality.

When people do get computers in their heads, I believe, it will mostly enable them to remember things better and faster. You might say to yourself, for instance, What was the name of that person I met in Washington last week?, and you'll immediately access that person's name with the aid of an implanted computer that keeps track of all your activities. Or you might need to know the capital of Kansas and, boom, it's there in your mind. But this ability wouldn't change the quality of your thinking—it wouldn't make you an Einstein. Truly original thinking involves a lot more than just moving information around quickly.

**Science Year:** Let's turn to a subject that has always been of great interest to people and that has a major impact on their daily lives: transportation technology. During the 1900's, society has seen the development of the automobile and the airplane. What can we look forward to in the 2000's?

**Benford:** For one thing, we'll be seeing sleeker, bigger, and much faster airplanes. There will be suborbital aircraft—planes that take you into space for a large part of your trip—flying at maybe 10 times the speed of sound. So, for example, you could get from the United States to Australia in two hours or so. That's something we'll see by perhaps the year 2020. And by about the same time, we'll have orbital hotels where you can go and live in a zero-gravity environment for a week. Delivering people on private spacecraft to the Orbital Hilton will be an everyday thing by 2020 or 2030.

**Science Year:** What advances in ground transportation do you think we can expect?

**Benford:** There will be much faster trains, including maglevs—trains using magnetic levitation instead of wheels. This is a proven technology, but it's being held back by extremely high construction costs. Something I hope to see is the development of so-called gravity maglevs, which would travel deep underground and at essentially zero fuel cost. You would go from, say, Los Angeles to San Diego by falling along a deep curve connecting the two cities that goes about a mile underground. The train would simply fall to the bottom of the curve, accelerating to a speed of maybe 300 miles an hour, and then use its momentum to rise back to the surface at the far end. If you evacuated all the air from the tunnel so there's no wind resistance, you basically would get free transport between two cities, as long as they're at the same altitude.

**Science Year:** What about personal transportation—what do you predict there?

**Benford:** I think automobiles running on four wheels will be with us for a long time to come. But they'll be smarter and made with lighter and more durable materials. Also, they'll probably be cheaper than today's cars, thanks to highly advanced manufacturing processes.

**Science Year:** Is there anything else you foresee for the future of transportation?

**Benford:** Well, I'll tell you something I'd like to see happen for the benefit of ocean traffic, and that's the construction of a sea-level canal through Nicaragua. It would replace the Panama Canal. The building of that waterway was a tremendous achievement, but the Panama Canal is now hopelessly obsolete. It simply can't handle enough traffic. The waiting time for a ship to get through the canal is between two and four weeks, and that delay is a major impediment to trade between the Atlantic and Pacific economic communities.

A sea-level canal across Nicaragua would pass though Lake Nicaragua, and there would be no locks [chambers for raising and lowering ships to different water levels]. It would simply be a wide, unimpeded waterway between the Atlantic and Pacific. There would be some environmental concerns about mixing life forms between the two oceans, but that has been going on for a long time, anyway, particularly with the dumping of ballast water from ships. From a strictly engineering standpoint, however, it could be done, and I think it would be a perfect opportunity for the peaceful use of nuclear energy.

*We'll have orbital hotels where you can go and live in a zero-gravity environment for a week. Delivering people on private spacecraft to the Orbital Hilton will be an everyday thing by 2020 or 2030.*

**Science Year:** How so?

**Benford:** The best way to cut a path across Nicaragua—particularly through a fairly good-sized mountain that stands in the way of the shortest route—would be with the use of small nuclear charges. It would show the world that "nuke" doesn't necessarily mean radioactivity. There are nuclear devices from which all traces of radioactivity are gone in a week. I don't suppose the idea of using nuclear explosions to dig a ditch could be sold to the public or the international community, but it would in fact be the quickest and easiest way to build a sea-level canal.

**Science Year:** Speaking of nuclear energy, do you foresee much of a future for the use of nuclear power in generating electricity?

**Benford:** No. The antinuclear community will probably hold electrical generation with nuclear-fission power plants to roughly present levels, which in the United States is only about 20 percent of the total energy produced annually. So we'll continue to generate most of our electricity with oil and coal, just like today. There aren't any other alternatives. Solar power and wind power will never be able to generate the huge amounts of energy needed by the United States and other advanced nations.

**Science Year:** What about nuclear fusion? The controlled fusion of hydrogen nuclei to release energy has long been predicted as the answer to all of our energy problems.

**Benford:** Lots of luck. I worked in the U.S. fusion program, so I know the difficulties first hand. There's a saying in the physics community: Fusion is 20 years away from realization, and it always will be. Eventually, fusion will become a commercially viable way of generating energy, but we're talking decades in the future. Until then, it's going to be fossil fuels.

**Science Year:** Aren't we going to run out of those resources sometime in the 2000's?

**Benford:** Deposits of liquid oil, yes, but then we could turn to oil-bearing shale, which we have in abundance. With the processing of shale to extract its oil content, we could extend oil production for several more centuries—though at a much higher price. And then there's coal, of which the United States has at least a 500-year supply. So fuels may become more expensive, but we're not going to run out of them.

**Science Year:** Let's look away from the Earth and into outer space. What do you foresee happening in the exploration of the solar system in coming years?

**Benford:** I think a manned exploration of Mars would be a great adventure that would fire everyone's imagination. It would also be an excellent enterprise from a scientific standpoint, because the really big scientific issue, the question of whether there is life on Mars, simply cannot be settled by a bunch of robot rovers. It's got to be done by field biologists working on the Martian surface. I think this is a project that the United States will undertake, probably by the year 2030.

Something else we could do in space, and which I also expect will happen, is the mining of asteroids and moons for minerals. In fact, the exploitation of the resources of the whole inner solar system will be the big new bonanza for generating great wealth. It probably won't happen in the early part of the 2000's, but within 100 years it's a certainty.

Mars and the asteroids are the frontier for the coming century. But farther out in the solar system, things get tough. The outer planets have ice moons that are swept by a "sleet" of protons [positively charged particles from atomic nuclei], trapped by strong magnetic fields, that would kill a human being in a day. So those moons will have to be explored and mined by machines.

**Science Year:** What about the possibility of mining resources on our own moon?

**Benford:** There's no point in going to the moon for raw materials, because there just aren't any that would justify the expense of getting them and bringing them back to Earth. With one exception: helium 3, a form of helium with one neutron in the nucleus, rather than the usual two. Helium 3 is rare on Earth but fairly common in the top three feet or so of the lunar soil. The solar wind [a stream of charged particles from the sun] slams into the moon and leaves helium-3 nuclei in the surface. Helium 3 would be a good fuel for controlled nuclear fusion, if and when that technology is perfected. A ton of it, distilled from moon soil, would be worth a billion dollars.

Aside from obtaining that one resource, however, I see no reason for going to the moon, except to satisfy our curiosity about it and perhaps to build an astronomical observatory on the far side. But it might also be a vacation destination someday. I could see the appeal of visiting the moon and staying in a hotel built under a huge transparent dome. The moon's mild gravity—about one-sixth that of the Earth's—would allow people to engage in some very entertaining activites. For instance, within an enclosed dome with air you could strap on wings and fly like a bird!

**Science Year:** One last question. Since we're starting a new millennium, let's look forward 1,000 years. What do you see when you look toward the end of the third millennium?

**Benford:** The solar system will have been thoroughly explored and settled. The nearby stars and their planets will surely also have been explored—by robots and probably also by people. The human species, from the influence of genetic manipulation and the effects of living on other planets, will have diversified into several subspecies. And the human life span will be measured in centuries. But the people of the year 3000 will still be faced with philosophical problems centering on the meaning of human life and of humanity's place in the universe. What specific questions they will be asking themselves we can't begin to guess at, just as the issues we are now grappling with—such as how we should use our new-found ability to alter our genetic makeup—would have been inconceivable to our ancestors of 1,000 years ago. We can only speculate on what the human experience will be in the far future. The reality will undoubtedly be something that we could not possibly even envision.

# Great Unanswered Questions of Science

Although the 1900's were a period of unparalleled scientific accomplishment, scientists at the end of the century were still trying to answer many profound questions about the universe, living organisms, and the history of life.

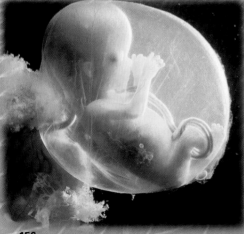

## Introduction

People living during the 1900's witnessed an explosion of scientific knowledge unsurpassed in history. A person born around 1900 could have read the first newspaper reports of early century physicists explaining the nature of gravity and describing the behavior of subatomic particles. In midcentury, that person might have heard on the radio that scientists had split the atom and, later, discovered the structure of *DNA* (the molecule that makes up genes). If the person lived to a ripe old age, he or she could have watched the news on television that researchers had identified numerous genes that cause disease in humans and created a *clone* (genetic duplicate) of a sheep.

Despite this dizzying pace of discovery, much of the world remained a mystery to scientists as the 1900's came to a close. Scientists in many different fields were grappling with a number of big questions that defied solution. For example, what kinds of exotic entities make up space at its tiniest dimensions? What will ultimately happen to the universe in the most distant future? How do all the complex characteristics of a human develop from a single fertilized cell? What exactly is the mind, and how is it generated? Why do we age, and is it possible to conquer death? Why have large numbers of plants and animals vanished from the Earth at various times in its history?

Although these six mysteries are explored in the following pages, they are by no means the only great unanswered questions of science. Scientists of the 2000's will also be engaged in such daunting tasks as trying to learn how the universe began, what is the source of high-energy cosmic rays, how life on Earth originated, if there is life on other planets, and how genes and environmental factors interact to influence human behavior. With so much left to learn and with the tools of science becoming ever more sophisticated, scientists in 1999 were confident that the pace of scientific inquiry and accomplishment would continue to accelerate throughout the new century.

# What Is the Fundamental Nature of Space?

## By Gary T. Horowitz

What does the space between this book and your eyes consist of? Although it looks like nothing at all, just empty space, we know that there are air molecules floating around. However, what if we were to get rid of all the air molecules and create a complete vacuum between the book and your eyes? Then surely that volume of space would be truly nothing—an empty void with no properties of its own. Right?

It may surprise you to learn that this "common sense" view of space is, in fact, wrong. Modern physics has shown that space—even an absolute vacuum—has a number of surprising properties. In fact, a vacuum is full of activity, with various kinds of particles winking in and out of existence and tiny bundles of energy interacting with each other. In the late 1990's, researchers were trying to gain a more complete understanding of the bizarre conditions that characterize the fundamental nature of space, one of the great remaining mysteries of science.

Our usual notion of space goes back to Aristotle and the other ancient Greek philosophers, who believed that space is a uniform void with three dimensions—length, width, and height.

**The author:**

Gary T. Horowitz is a professor of physics at the University of California at Santa Barbara.

The long scientific process that ultimately changed this view began in the 1600's, with the work of the Italian physicist and astronomer Galileo Galilei and the English mathematician and scientist Sir Isaac Newton. Galileo described the motion of free-falling objects, and Newton expanded Galileo's findings into a universal theory of gravitation.

Newton discovered that the force of gravity between two bodies, such as the sun and the Earth, depends on the mass of the bodies and the distance between them. But he did not know how gravity works, beyond being some mysterious force that acts between objects. One reason why Newton was unable to discover how gravity works was that he accepted the classical notion of space as a uniform void that cannot change. It was not until the early 1900's that scientists abandoned this concept of space.

### Einstein's view of space

In 1915, the German-born scientist Albert Einstein developed the general theory of relativity, which explained gravity in terms of curved spacetime. Spacetime was a concept derived from Einstein's 1905 special theory of relativity, which demonstrated that time is a relative concept. For example, the theory implied that time would appear to move slower for a spaceship full of people traveling near the speed of light (299,792 kilometers [186,282 miles] per second) than for a stationary observer that the spaceship passes. The theory also implied a four-dimensional universe,

Physicists believe that space and time are unified in a single entity called spacetime. Researchers in the late 1990's sought to learn how space is constructed at its smallest dimensions.

in which the three dimensions of space are unified with the dimension of time.

Einstein's general theory stated that a massive object, such as the sun, curves space-time around itself, much as a bowling ball would form a depression in a rubber sheet. Smaller objects, such as Earth and the other planets, move around the large object along paths in the curved spacetime, just as marbles would roll around the depression in the rubber sheet.

Therefore, gravity is not a mysterious force, as Newton thought it was. Rather, general relativity shows that the gravitational attraction between objects is actually a geometric effect. These objects fall toward one another along the curves that their masses produce in spacetime.

**Matter and forces from vibrating strings**
According to string, or superstring, theory, all particles of matter and all force-carrying particles are made up of tiny vibrating, linelike objects called strings.

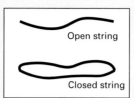

Open string

Closed string

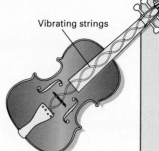

Vibrating strings

Physicists believe that strings exist in two basic forms—open and closed. Strings can interact so that open strings become closed, and closed strings become open. In addition, one string can break into two, and two strings can join to become one.

Just as violin strings can vibrate at different frequencies and produce varying musical notes, strings in spacetime can vibrate in many ways to create all the different types of elementary particles that make up atoms.

## Quantum mechanics

Although general relativity explains the effects of gravity around planets, stars, galaxies, and other large masses, it cannot account for phenomena that occur at very small distances, in the realm of atoms and subatomic particles. These effects are described by quantum mechanics, a branch of physics developed in the early 1900's by a number of physicists, including Niels Bohr of Denmark, Erwin Schrodinger of Austria, and Werner Heisenberg of Germany. Quantum mechanics explains how atoms absorb and give off units of energy and momentum called quanta. Quanta act as both particles and waves. Quantum mechanics also explains the behavior of the various particles that make up atoms.

Physicists have developed a quantum theory, known as the Standard Model of Particle Physics, to describe all of the fundamental particles that make up atoms. According to this theory, protons and neutrons, which make up an atom's *nucleus* (core), are themselves made up of tiny particles called quarks. There are six known kinds of quarks, varying in mass and electric charge. In addition, atoms contain negatively charged particles, called electrons, that orbit the nucleus.

The Standard Model also describes the forces at work in the subatomic realm, including the electromagnetic force (which holds atoms and molecules together), the strong force (which holds protons and neutrons together), and the weak force (which causes certain forms of radioactive decay). The forces are transmitted by particles called bosons. For example, bosons known as photons carry the electromagnetic force. The Standard Model does not include a description of gravity, which in the subatomic realm is much weaker than the other forces.

Another particle in the Standard Model is the Higgs boson, which had not yet been observed as of 1999. Particle physicists believe these bosons are generated by a field—the Higgs field—that permeates the vacuum of space, giving mass to all particles of matter.

Scientists thus have one theory (general relativity) to describe the behavior of large masses such as planets and another theory (quantum mechanics) to explain the behavior of atoms and subatomic particles. What is needed, physicists say, is a single theory that would somehow combine these two separate concepts. This "unified theory" would describe the fundamental principles that physicists believe underlie the workings of both general relativity and quantum mechanics.

## Superstrings

Physicists began searching for such a theory soon after the development of quantum mechanics. However, progress was limited until the early 1980's, when John Schwarz of the California Institute of Technology in Pasadena and Michael Green of Queen Mary College in England began

According to the theory of quantum geometry, the tiniest areas of space consist of a complex fabric of interwoven threadlike entities, like the weaves of a cloth. The interconnections and interactions of the threads of the weave give rise to the spacetime *continuum* (continuous whole).

to develop an idea called string, or superstring, theory. This theory not only combined general relativity and quantum mechanics, it also sought to describe exotic entities that make up all the particles in the universe.

String theory states that quarks, electrons, and other particles of matter, rather than being point-like, are actually tiny linelike objects called strings. These strings are incredibly small. Each is approximately the size of the *Planck length*, the smallest possible distance in spacetime—less than one billionth of one billionth the radius of an electron. Just as the strings of a violin can vibrate at different frequencies and produce varying musical notes, the tiny strings in spacetime can vibrate in many ways to create different types of elementary particles.

Besides making up all particles of matter in spacetime, according to the theory, strings also make up all the force-carrying particles that act on matter. Furthermore, the theory proposes that strings move within a curved spacetime. Therefore, string theory naturally incorporates the explanation of gravity provided by general relativity.

Physicists theorize that strings exist in two basic forms—open and closed. An open string has two free ends, while a closed string forms a loop with no free ends. Strings interact by splitting and joining, so that one string can break into two, and two strings can combine to form one. In addition, open strings can close up, and closed strings can break open. These interactions and motions determine the kinds of particles and forces the strings give rise to.

One prediction that string theory makes is that spacetime has more than four dimensions. In fact, the mathematical description of string theory doesn't work unless physicists assume that strings vibrate in 10 dimensions. But where are these other dimensions, and why can't we see them? Physicists working on string theory propose that we cannot see the extra dimensions because they are

wrapped up in tiny balls. Each ball may be the size of the Planck length, and there is a ball at each point in the three-dimensional space that we observe. Physicists believe that the properties of these rolled-up dimensions determine some of the properties of the elementary particles that occur in nature.

## Quantum geometry

Another approach to combining quantum mechanics with general relativity is known as quantum geometry, a theory begun by physicist Abhay Ashtekar at Syracuse University in Syracuse, New York, in the mid-1980's. Unlike string theory, which attempts to describe the underlying quantum nature of all the particles and forces that exist in spacetime, quantum geometry describes the fabric of spacetime itself as having quantum characteristics. According to this theory, space at its smallest distances resembles a weave of cloth with interlocking threads. The interconnections and interactions of the threads of the weave give rise to the spacetime *continuum* (continuous whole) at the heart of general relativity. Physicists are not sure how the behavior of these energetic threads of spacetime might relate to particles of matter or force-carrying particles.

Because the threads themselves make up spacetime, the tiny areas between the threads are regions where space and time do not exist at all. This is a concept that boggles the minds of even physicists!

Quantum geometry does not require the existence of extra dimensions to describe spacetime, as string theory does. Instead, it sees spacetime as a four-dimensional background on which elementary particles exist. However, the theory does not explain what makes up the particles, as string theory does. Some physicists speculate that the ideas of string theory and quantum geometry might eventually be incorporated into a single theory, though, as of 1999, physicists did not know how this might be accomplished.

As the year 2000 approached, physicists presented the world with a view of space almost unimaginably different from what most people take for granted. Theorists expected this conception to evolve further, leading to an ever greater understanding of the fundamental nature of space. It will be a challenge to the physicists of the new millennium to arrive at such an understanding, which could help us to better comprehend the universe as a whole.

# What Is the Ultimate Fate of the Universe?

## By Robert P. Kirshner

About 15 billion years ago, scientists believe, an unimaginably dense, hot speck of matter and energy suddenly erupted in a "big bang" to give birth to the universe. Astronomical observations indicate that ever since that time, the universe has been expanding and cooling. As the universe has expanded, innumerable stars—many like the sun—have burned brightly and then died out, only to be replaced by countless others. It is a cycle that is still going on and will continue for billions of years to come.

What remains unclear is what will happen to the universe in the even more remote future. Will it expand forever, with all cosmic objects getting farther and farther apart throughout endless time? Or will the expansion eventually reverse itself, resulting in everything in the cosmos contracting together into a "big crunch"? During the late 1990's, research suggested an answer to these questions, but at the same time, it raised new questions. As long as those uncertainties remain, the ultimate fate of the universe will continue to be one of the great unsolved mysteries of science.

## Discovery of expansion

The quest to discover the fate of the universe has gone through several steps. It began with the work of the American astronomer Edwin Hubble. In the 1920's, Hubble (for whom the Hubble Space Telescope is named) discovered the universe's expansion. He did so by studying the distances to and movements of galaxies, large concentrations of stars grouped together by gravity.

Hubble determined the distances to various galaxies by examining certain stars, called Cepheid variables, in the galaxies. Cepheid variables are giant stars that undergo very regular periods, or cycles, of brightening and dimming. Another American astronomer, Henrietta Swan Leavitt, had discovered that the periods are directly related to the *luminosity* (energy output) of the stars. She learned how to calculate a Cepheid's luminosity by measuring how long it takes the star to brighten and dim.

Once the luminosity of a Cepheid is known, astronomers can determine how far away it is by comparing its brightness to other Cepheids with the same period. In other words, Cepheids are somewhat like candles. Because the average luminosity of candles is well known, the distance to a candle in the night can be estimated from its apparent brightness. The dimmer the candle, the farther away it is. For this reason, astronomers refer to Cepheids as "standard candles."

There are billions of galaxies in the universe, and they are almost all moving away from us. Astronomers interpret this observation to mean that the universe is expanding.

After Hubble had measured the distances to the galaxies, he sought to discover whether they were moving in relation to our own Milky Way Galaxy. In this process, a device called a spectrograph spread the light from each galaxy into its component colors. All light is composed of the colors of the rainbow in a spectrum, ranging from red to blue. The faster a light source is moving away from an observer, the more its light is *red shifted*—that is, the more that features in its spectrum are displaced toward the red end of the spectrum. Conversely, if a light source is moving toward an observer, features in its spectrum are *blue shifted*, or displaced toward the blue end.

Hubble found that practically all the galaxies he looked at were red shifted, meaning that they were retreating from the Milky Way. Moreover, he

**The author:**
Robert P. Kirshner is a professor of astronomy at the Harvard-Smithsonian Center for Astrophysics in Cambridge, Massachusetts.

found that the more distant a galaxy was, the greater was its red shift—that is, the faster it was receding from us. This finding was the first evidence that we live in an expanding universe.

The phenomenon discovered by Hubble—that the velocity of a galaxy away from the Milky Way is directly proportional to its distance—might be better understood by imagining a room full of astronomers playing on a magical jungle gym that continually and uniformly grows in every direction. Each astronomer would see his or her neighbors moving farther away as the jungle gym increases in size, stretching the spaces between everyone. Consider astronomer X watching as astronomers Y and Z move away. When the jungle gym has doubled in size, astronomer Y, who had been one segment away when the expansion began, has receded to a distance equal to two of the original segments. Astronomer Z, who had been two segments away, is now at a distance equal to four of the original segments. Thus, even though the jungle gym's expansion rate is uniform throughout, astronomer X sees astronomer Z moving away at twice the velocity of astronomer Y.

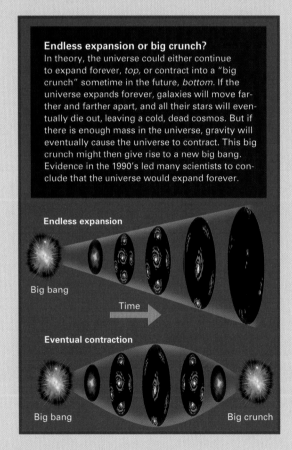

**Endless expansion or big crunch?**
In theory, the universe could either continue to expand forever, *top,* or contract into a "big crunch" sometime in the future, *bottom.* If the universe expands forever, galaxies will move farther and farther apart, and all their stars will eventually die out, leaving a cold, dead cosmos. But if there is enough mass in the universe, gravity will eventually cause the universe to contract. This big crunch might then give rise to a new big bang. Evidence in the 1990's led many scientists to conclude that the universe would expand forever.

Endless expansion

Big bang

Time

Eventual contraction

Big bang

Big crunch

An expanding universe suggested a universe with a beginning. In 1965, two American physicists, Arno Penzias and Robert Wilson of AT&T Bell Laboratories in New Jersey, accidentally discovered evidence for that beginning. While working with a large radio antenna, they detected dim microwave radiation coming from every part of the sky. Astronomers concluded that this feeble radiation was the leftover glow of the universe's explosive origin—the big bang.

## The role of gravity

Whether the cosmic expansion that began with the big bang will continue indefinitely is determined by the force of gravity. Gravity is a predictable force. For example, the gravitational pull of Earth on a baseball determines the ball's path once it leaves the bat, enabling a skilled fielder to judge where the ball is going to land. If there is sufficient mass in the universe to exert a strong enough gravitational pull, then, like a fly ball that goes up for a long time but finally comes back down, the universe will eventually stop expanding and begin to contract. But if there is not enough mass to stop the outward flight of the galaxies, the universe will expand forever.

In the 1990's, the known amount of mass in the universe led many scientists to believe that the universe would expand forever—but at a slower and slower rate. This prediction was based on the assumption that the early cosmos had expanded at a faster rate than it does today, and that in the future, the expansion would continue to decelerate. (Astronomers refer to the current rate of cosmic expansion as the Hubble constant.)

## Revelations from distant supernovae

Observations of exploding stars, called type Ia supernovae, in distant galaxies provided astronomers with a way to check their prediction on cosmic expansion. Type Ia supernovae result from the explosion of dense stars called white dwarfs at the end of their lives. When a white dwarf is part of a two-star system, its gravity can pull gas from the companion star. Eventually, the white dwarf reaches such a great mass that its core ignites as a thermonuclear bomb.

Astronomers locate these exploding stars by using special cameras on giant telescopes to take electronic images of the same galaxies a few weeks apart. Type Ia supernovae show up as bright spots when one of these images is "subtracted" from another by a computer. The luminosity of a type Ia supernova is determined by how long it stays bright. The longer a supernova can be seen, the more luminous it is. Once the luminosities of the supernovae are known, they can serve as standard

candles to measure the distances to galaxies 1,000 times farther away from the Earth than those in which Cepheid variables serve as the measuring sticks. The velocities with which these galaxies are moving away from us can be determined by measuring the red shifts of the supernovae or of the galaxies themselves.

Distant galaxies are far removed from us not only in space but also in time. That is because the light we see from them, traveling across space at the speed of 9.5 trillion kilometers (5.9 trillion miles) per year, has taken a long time to reach us. (The distance that light travels in a year is called a light-year.) Therefore, we see extremely distant galaxies as they appeared billions of years ago when the universe was young.

Two teams of astronomers in the 1990's were studying type Ia supernovae that erupted in galaxies 4 billion to 7 billion years ago (and, thus, at the time they exploded, were 4 billion to 7 billion light-years distant). One team was headed by Saul Perlmutter of the Lawrence Berkeley National Laboratory in Berkeley, California. The other team, which included me, was led by Brian Schmidt of Mount Stromlo and Siding Spring Observatories in Australia.

## An accelerating universe?

Both groups found that the supernovae were dimmer than would be expected if the universe has been expanding at a diminishing rate, or even if the expansion has been constant through time. For either of those two possibilities, we would expect a distant supernova to show a certain brightness at a given red shift. However, the fact that the supernovae were dimmer than expected at their red shifts indicated that these stars and their galaxies were farther away from us than expected. This implied that the expansion has actually been accelerating over time!

But how could this be happening? As far as physicists knew, gravity can only slow things down in an expanding universe. Therefore, our findings indicated that some other force was at work in the universe to make the expansion speed up over time. Although scientists aren't sure what this force might be, the answer might lie in an idea first proposed, and later rejected, by the German-born physicist Albert Einstein—the *cosmological constant*. The cosmological constant is a type of energy in the vacuum of space that causes the space to expand. This stretching of the space between galaxies makes the red shifts of the galaxies grow larger and larger as the distance between them increases.

Is such an acceleration really occurring? Although the evidence favors it, perhaps there are other explanations for our observations. There is always a possibility that both teams of astronomers could have made some error in their measurements or that intergalactic dust or some other factor diminished the brightness of the faraway supernovae. Another possibility is that the supernovae we examined may be somehow different from nearby (that is, more recent) supernovae, making them unreliable as standard candles. We are investigating these possible pitfalls. Scientists expect that additional research, some of it using the Next Generation Space Telescope, scheduled to be launched in 2007, will help us understand how the universe has been expanding from its earliest times.

If our conclusion is correct, it means that the expansion of the universe will not only continue forever but will do so at an accelerating rate. If this is the case, the far future will be one in which distant galaxies approach the speed of light as they move ever farther away from each other, and all the stars that make up galaxies will eventually burn out forever. A cold, black night that never ends may be the ultimate fate of the universe.

# How Does an Organism Develop from a Single Cell?

## By Edwin L. Ferguson

A new child born to a family brings great joy. The parents coo, and young brothers and sisters marvel at the appearance of the baby—so small but so much like themselves. Yet, just nine months before, the baby was a single fertilized egg. How did the genetic instructions contained within that one-celled egg coordinate the complex process of development from the single cell to the living, breathing infant? This question has long been one of the great mysteries of biology.

At the conception of a baby, the father's sperm fertilizes the mother's egg. Soon afterward, the fertilized egg begins to undergo cell division, becoming an embryo. By six days after fertilization, the embryo is a flat disk of cells that has implanted itself into the lining of the mother's *uterus* (womb). At 13 days after fertilization, when the embryo is less than 3/100 centimeter (1/100 inch) long, it undergoes the beginning of a remarkable process called gastrulation. During gastrulation, extensive cell movements transform the embryo from a single-layered disk of cells to an organism made up of three distinct cell layers. The significance of this process to the further development of the embryo prompted British developmental biologist Lewis Wolpert to say, "It is not birth, marriage, or death, but gastrulation, which is truly the most important time in your life."

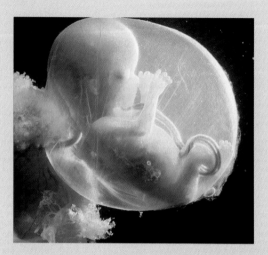

A tiny 17-week-old human fetus already has a recognizably human form. The major cellular processes of development occur before the ninth week after conception, by which time all the characteristic features of an infant have developed.

**The author:**

Edwin L. Ferguson is an associate professor in the Department of Molecular Genetics and Cell Biology at the University of Chicago.

The extensive cellular rearrangements that occur during gastrulation trigger the onset of a genetic program that ultimately leads to the development of the embryo's human form, including the appearance of organs. For example, the earliest stage of the brain and spinal cord, called the neural plate, forms 18 days after conception. At 24 days, an early, tubular heart takes shape, and the rudiments of the gut form. The limb buds, which will grow to become the baby's arms and legs, become visible at 26 days.

All of these events happen when the embryo is very small, less than 3/5 centimeter (1/4 inch) long. During the second month of pregnancy, the face of the embryo gradually becomes more human. By the end of the second month, when the embryo is about 3 centimeters (1 1/4 inch) long, all the major features of a human are present. During the remaining seven months of pregnancy, the embryo—now called a fetus—grows until it becomes a fully developed infant.

Although philosophers since ancient times had speculated about the mechanisms of embryonic development, it was not until the late 1800's that scientists began performing experiments to learn what actually occurs. These early studies revealed that cells within all embryos undergo a gradual process of specialization called differentiation.

Cells in a very young embryo are undifferentiated—that is, each cell has the ability to give rise to cells of many different types, such as blood, muscle, or nerve cells. Later during development, however, cells within the embryo become differentiated, which is to say each cell becomes a particular kind of cell that can give rise to only other cells of its type.

Differentiation is achieved partly through induction, a process in which one group of cells sends a chemical signal to an adjacent group of cells to cause, or induce, certain developmental changes within the adjacent cells. Differentiation also occurs through a process called pattern formation. During this process, cells "sense" their relative positions within an embryo and differentiate into cell types characteristic of those positions. For example, nerve cells at the front end of the

The development of an *embryo* (unborn organism) is controlled by a number of different processes operating at the cellular and molecular level.

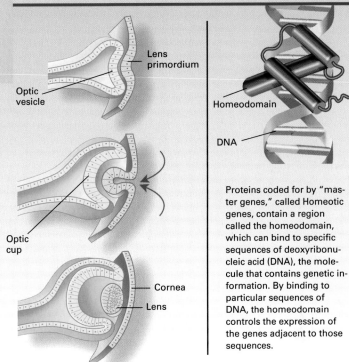

During the process of induction, groups of cells in an embryo cause developmental changes in other cells. For example, in a developing eye, the *optic vesicle* (an outgrowth of the brain) induces the formation of the *lens primordium* (the primitive lens) from underlying cells, *top*. The lens primordium, in turn, causes the optic vesicle to form a cup, *center*. The optic cup then causes the lens primordium to fold in on itself, forming the lens, and this induces the development of the cornea, *bottom*.

Lens primordium

Optic vesicle

Optic cup

Cornea

Lens

Homeodomain

DNA

Proteins coded for by "master genes," called Homeotic genes, contain a region called the homeodomain, which can bind to specific sequences of deoxyribonucleic acid (DNA), the molecule that contains genetic information. By binding to particular sequences of DNA, the homeodomain controls the expression of the genes adjacent to those sequences.

embryo become brain cells, while nerve cells toward the back become part of the spinal cord.

Although early researchers uncovered many principles of embryonic development, they knew little about the molecular mechanisms underlying these principles. In the second half of the 1900's, additional research opened the way to a greater understanding of embryonic development.

## Genes and proteins

In 1953, the American biologist James Watson and the English biologist Francis Crick determined the structure of deoxyribonucleic acid (DNA), the molecule that contains the genetic information within all organisms. A DNA molecule is similar to a long, twisting ladder, with the rungs and sides made up of individual building blocks called nucleotides, of which there are four kinds. Certain stretches of nucleotides in the DNA are called genes, and the exact order of nucleotides in a gene encodes the information necessary to make one particular protein within a cell.

The first step in protein production is a process called transcription, in which the coded information in a gene is copied into a molecule similar to DNA called messenger ribonucleic acid (mRNA). Next, in a process called translation, other molecules read the mRNA molecule and produce the

protein it specifies. The protein is then released to carry out its assigned function in the cell.

In the early 1960's, two French scientists, geneticist Francois Jacob and biochemist Jacques Monod, studied genetic mechanisms in bacteria under the assumption that similar mechanisms might operate in more complex organisms, including humans. They found that not all the genes are being transcribed at all times during life. At any given time, some genes are "turned on" while others are "turned off." The scientists also identified a regulatory gene in bacteria that codes for a protein whose only purpose is to control the transcription of other genes.

After these findings, the main question for developmental biologists became: How does each cell within an embryo regulate the transcription of its genes so that it develops into a complete organism with an orderly arrangement of many different kinds of cells?

Seeking to better understand this process, scientists used the power of genetics to study embryonic development in a simple animal—the fruit fly *Drosophila melanogaster*. In a series of experiments beginning in the late 1970's, geneticists Christiane Nusslein-Volhard and Eric Wieschaus, working at the European Molecular Biology Laboratories in Germany, identified more than 200 genes necessary for the embryonic development of the fruit fly. To do so, the researchers looked for *mutant* (genetically abnormal) fruit flies that had defects in embryonic development. By observing the type of defect caused by a particular abnormal gene, the scientists inferred how the normal version of that gene functions during development. The geneticists' work earned them the 1995 Nobel Prize for physiology or medicine.

By 1999, scientists had identified hundreds of genes in *Drosophila, Caenorhabditis elegans* (a roundworm), and *vertebrates* (animals with backbones, including humans) that play important roles in embryonic development. Initially, researchers encountered difficulty in determining the precise functions of vertebrate genes involved in development. However, in 1988, geneticists Mario Capecchi and Susan Mansour of the University of Utah, working with mice, invented a technique called "gene knockout" to discover the function of any gene for which they knew the DNA sequence.

Other scientists soon began using this technique. Investigators first knock out, or disable, a gene by inserting a foreign piece of DNA into it. They then insert the gene into undifferentiated mouse cells grown in culture dishes. In some of the cells, the disabled gene takes the place of the normal gene. These cells are, in turn, injected into embryonic mice, where they develop into all the different cell types. Researchers then breed these mice and observe the developmental defects in their offspring caused by the improperly functioning gene.

## Developmentally important genes

Researchers have found that many of the genes necessary for embryonic development code for proteins involved in the processes of induction and pattern formation. For example, some proteins act as chemical signals during induction. Other proteins help cells receive and respond to those signals.

Another class of genes important to development codes for transcription factors, proteins that regulate the transcription of other genes. These proteins thus play the same role in vertebrate development that the regulatory protein identified by Jacob and Monod performs in bacteria.

Perhaps the most fundamental knowledge that has come from these genetic studies is that many of the molecular mechanisms underlying embryonic development are common to all animal species. These results indicate that most developmental processes originated early in the evolution of multicellular life.

One spectacular example of shared developmental mechanisms across species involves a group of genes called the Homeotic, or Hox, genes. These genes code for transcription factors that are important to the process of pattern formation along the head-to-tail axis of the embryo.

Each species of animal has at least one cluster of Homeotic genes in its DNA. The transcription factors coded for by Homeotic genes contain a region called the homeodomain, which binds to specific DNA sequences to control the transcription of the genes next to these sequences. Each Homeotic gene is expressed in a particular portion of the embryo—some are expressed in the head, others in the middle of the body, and still others in the tail. The Homeotic transcription factors act as master regulators to control the transcription of many other genes. Their action ultimately results in the formation of particular structures along the head-to-tail axis of the body, such as a wing in a fly or a vertebra in a mouse.

## Remaining questions

Although scientists have identified many genes important to embryonic development, they were far from a full understanding of this complicated process in 1999. Many questions remained to be answered. For example, when a cell receives a chemical signal from another cell during induction, how many of the receiving cell's genes are turned on or off by the signal? And how can the action of a single Homeotic gene be enough to specify a complete structure within an organism?

In the future, scientists may be able to answer these questions. A revolution in our understanding of developmental processes—the identification of the complete DNA sequences of many organisms—was underway in the 1990's. In December 1998, researchers at Washington University in St. Louis, Missouri, and the Sanger Centre in Cambridge, England, published the first complete DNA sequence of an animal, the roundworm *C. elegans*. The *genome* (total amount of genetic information) of *C. elegans* is 97 million nucleotides in length and contains approximately 19,000 genes. The deciphering of the roundworm's genome allowed scientists, for the first time, to identify all the genes of a multicellular organism and all the proteins coded for by those genes.

The genomes of a number of other animals and plants were also being sequenced, or decoded, in 1999. The largest such effort was the sequencing of the human genome, which is approximately 3 billion nucleotides long. This effort was underway at several research centers around the world and was expected to be completed by 2003. The challenge for researchers of the 2000's will then be to understand how this genetic program is used by a growing human embryo to achieve its final form. And once we know that, we may have a better understanding of what it means to be human.

# What Is the Nature of Consciousness?

## By Bernard J. Baars

For more than 2,000 years, philosophers and religious thinkers have speculated about the nature of consciousness. Buddhism and Hinduism both developed with the intent of lifting the everyday state of consciousness to higher levels. Ideas about consciousness also played a major role in the development of Jewish, Christian, and Islamic ethics and the concept of free will. Many religious believers identify consciousness with the soul, which they say survives the body after death. Artists and writers have tried to communicate conscious experiences to other people for centuries.

Despite humanity's age-old quest to understand consciousness—particularly self-awareness and other high-level thought processes—it remains a profound puzzle. What does it mean for you to be aware of these printed words? What does it mean to talk to yourself—who's talking and who's listening? And how is it that you can have feelings that only you know about? In the 1990's, researchers sought a deeper understanding of consciousness, one of the great mysteries of modern science.

## The mind-body problem

Philosophies about consciousness have long revolved around the "mind-body problem," a term that refers to the relationship between conscious experience and the body, particularly the brain. The three basic philosophical positions on the mind-body problem are mentalism, physicalism, and dualism. If you believe you have free will to decide what you want to do, independent of any mechanisms in the body, you are taking the mentalist position, which states that mind is more fundamental than matter. If, on the other hand, you believe that your thoughts and actions are caused solely by the electrochemical activity of *neurons* (nerve cells) in your brain, you are taking the physicalist position. Finally, if you believe in some combination of these two views, you are a dualist.

Mystical religions, such as Hinduism and Buddhism, are mentalist, considering the mind to be the fundamental reality. Most scientists, however, are physicalists and insist that the mind will be explained strictly in terms of brain activity. A number of philosophers, notably Rene Descartes (1596-1650), have been dualists, asserting that mind and body are distinct yet related.

In everyday life, most people act as if they are continually moving from one position on the mind-body problem to another. For example, you act as a physicalist when you take an aspirin to make a headache go away, never worrying how it is that the conscious experience of a headache

Philosophers, religious thinkers, and scientists have all tried to explain the mystery of consciousness—how the physical brain can give rise to mental experiences, such as taking pleasure in a starry sky, the sound of a flute, or the smell of flowers.

can be helped by the chemicals in a pill. But you are more of a mentalist when you dial the phone number of an attractive person you would like to get better acquainted with, not pausing to wonder how it is that the physical act of dialing the telephone number was dictated by your thoughts and feelings. Students tend to think like dualists when they take personal credit for good test scores—"I chose to study hard for this test"—but then blame physical circumstances beyond their control when they do badly—"I had the flu."

All three standard positions on the mind-body problem present difficulties. Mentalists are unable to explain why people who have suffered a serious head injury, as from a motorcycle accident, lose consciousness. Physicalists need to explain what

**The author:**

Bernard J. Baars is a professor at the Wright Institute, a mental health education center and clinic in Berkeley, California.

use consciousness is to us if we are nothing but a collection of cells. Finally, dualists struggle in describing how physical and mental events interact with each other.

## Science studies consciousness

In the 1990's, scientists used an advanced medical imaging technology called positron emission tomography (PET) to study consciousness. This technology produces images of the brain that enable researchers to watch a living brain in action.

PET scans have revealed that when people talk silently to themselves—not making a sound or even moving their vocal muscles—much activity occurs in the parts of the brain that control speaking and listening. These parts, known as Broca's area and Wernicke's area, respectively, show up in the PET scans as bright spots. Such images prove that people carry on silent conversations with themselves during most of their waking hours. However, PET scans of this type of brain activity raise a question as to whether the bright spots on the scans truly represent conscious experiences. This is because many aspects of speech, including the way in which the brain usually retrieves the right word at the right time, are unconscious.

In studies during the 1990's, neurobiologist Nikos Logothetis (first at the Baylor College of Medicine in Houston and later at the Max Planck Institute for Biological Cybernetics in Germany) distinguished between conscious and unconscious brain activity by measuring electrical activity in the brains of macaque monkeys. He found that when a monkey gazed at an image, brain cells

were stimulated in a number of different sites along a neural pathway in the visual cortex, an area at the rear of the brain. The unconscious information from these brain cells was assembled into an object of conscious visual awareness in a part of the brain at the end of the neural pathway.

Biologist Francis Crick, corecipient of the 1962 Nobel Prize for discovering the structure of DNA (deoxyribonucleic acid), said that Logothetis's work indicated that consciousness arises in specific groups of brain cells. These cells process information collected from the complex interconnections of billions of other brain cells. Crick, who in the 1990's was conducting research on consciousness at the Salk Institute for Biological Studies in La Jolla, California, believes that scientists will eventually be able to explain how brain activity gives rise to conscious experiences.

But even if conscious activity is thoroughly mapped out in the brain, a host of other questions will remain. For example, what is the role of consciousness in the brain? Couldn't unconscious zombies do the same things that conscious people do? Or, to state the question even more fundamentally, why does consciousness exist at all?

In an attempt to describe the role of consciousness, several scientists (including myself) have proposed that conscious attention is like a spotlight in a theater. Just as a spotlight shines on actors performing on a stage, the brain's spotlight of attention shines on significant ideas and important sensory information, such as sounds and visual images, to produce conscious experiences. Most of the brain is unconscious, however, and consists of neural networks involved with lan-

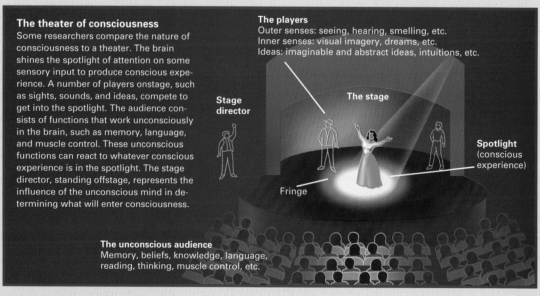

**The theater of consciousness**
Some researchers compare the nature of consciousness to a theater. The brain shines the spotlight of attention on some sensory input to produce conscious experience. A number of players onstage, such as sights, sounds, and ideas, compete to get into the spotlight. The audience consists of functions that work unconsciously in the brain, such as memory, language, and muscle control. These unconscious functions can react to whatever conscious experience is in the spotlight. The stage director, standing offstage, represents the influence of the unconscious mind in determining what will enter consciousness.

**The players**
Outer senses: seeing, hearing, smelling, etc.
Inner senses: visual imagery, dreams, etc.
Ideas: imaginable and abstract ideas, intuitions, etc.

**Stage director**

**The stage**

**Spotlight** (conscious experience)

**Fringe**

**The unconscious audience**
Memory, beliefs, knowledge, language, reading, thinking, muscle control, etc.

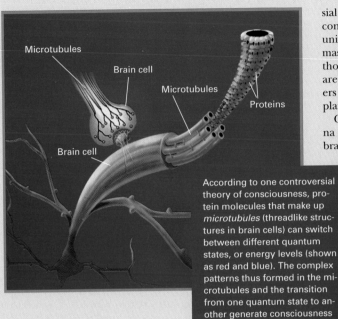

Microtubules

Brain cell

Microtubules

Proteins

Brain cell

According to one controversial theory of consciousness, protein molecules that make up *microtubules* (threadlike structures in brain cells) can switch between different quantum states, or energy levels (shown as red and blue). The complex patterns thus formed in the microtubules and the transition from one quantum state to another generate consciousness and free will.

guage, memory, muscle control, and other functions that we are usually unaware of. These networks are like the audience in the theater, sitting in the dark and observing events on the stage. Just as the actors and audience are able to respond to each other, the conscious and the unconscious can share information and influence each other.

The comparison of the mind to a spotlighted performance favors the views of physicalists, who believe that conscious experiences always involve electrochemical communication through neural networks in the brain. But some scientists and many philosophers argue that this position fails to address all the aspects of our personal experiences, such as perceptions of colors and sounds, and that such experiences remain a deep mystery.

Physicist Roger Penrose of Oxford University in England and physician Stuart Hameroff of the University of Arizona in Tucson propose a theory of consciousness that goes beyond chemical explanations. They speculate that consciousness may be based on *quantum mechanics* (the branch of physics that deals with the behavior of atoms and subatomic particles). According to this theory, protein molecules inside *microtubules* (threadlike structures in brain cells) switch between different quantum states, or energy levels. Penrose and Hameroff suggest that the complex patterns thus formed in the microtubules and the transitions from one quantum state to another generate consciousness and free will.

Philosopher David Chalmers of the University of California at Santa Cruz proposes a controver-

sial theory of the mind holding that consciousness is a basic quality of the universe—much like space, time, mass, and energy. He contends that although some aspects of consciousness are understandable scientifically, others may be forever impossible to explain with science.

Chalmers's list of mental phenomena that are understandable in terms of brain activity—which he calls the "easy problems" of consciousness—include the focus of attention, the difference between wakefulness and sleep, and the ability of humans to react to environmental stimuli. In contrast, Chalmers believes that subjective experiences, such as being dazzled by the red sky of a sunset, feeling overwhelmed by a Wagnerian opera, or deriving pleasure from smelling a bouquet of flowers, cannot be explained purely in terms of physical processes in the brain. Understanding such personal experiences, he says, is the "hard problem" of consciousness.

The current inability of scientists to fully understand consciousness leaves many important issues in society unresolved. For example, developmental biologists are unable to state when consciousness arises in a *fetus* (a baby prior to birth), an issue that complicates the ethics of abortion. Nor can it be determined with certainty when consciousness in a person has ceased. Physicians have no way of knowing beyond a reasonable doubt that an anesthetized patient in surgery is completely unconscious and out of pain. They do not even know for sure that individuals in a coma are unconscious. Nevertheless, life-and-death decisions are made on the assumption that physicians do know when consciousness begins and ends.

Some people assume that consciousness is limited to humans, or at least to highly evolved mammals. However, biologists believe that the neural activity giving rise to consciousness begins in the brain stem, a primitive part of the animal brain that is found even in fish. It is therefore quite possible that many kinds of animals possess some form of consciousness.

The ancient mystery of the mind may never be completely solved. Few human pursuits seem as complicated and difficult as the quest of the brain to understand itself. The many remaining questions about consciousness will be among the most intriguing for a new generation of scientists to tackle in the 2000's.

# Why Do We Age and Die?

## By David B. Finkelstein

In youth, people and other complex organisms are typically healthy and vigorous. But the youthful body is somewhat like a new automobile, which begins to show its age not long after first hitting the road. Within the body's cells, changes are already underway that will lead to physical decline and death. Scientists in the 1990's were seeking to learn why the aging process occurs and if anything can be done to stop it. This was one of the great remaining mysteries of science.

Many physical changes occur in the body during the normal process of aging, though the age at which a particular change occurs varies greatly from one individual to another. Some of the signs of physical decline, such as graying hair and wrinkling skin, are easy to see.

Various other aspects of human aging are less obvious, however, and have been discovered only through scientific studies. Researchers have found, for example, that after about the age of 40, many brain cells die, causing the brain to slowly shrink. In addition, the heart becomes slightly larger as its muscle walls thicken. The blood vessels of the

**The author:**
David B. Finkelstein is a science writer and an authority on the aging process.

heart also thicken and become less elastic, making it more difficult for the heart to pump blood. Lung capacity declines, and the kidneys become less efficient at extracting wastes from the blood.

There are other changes as well. The immune system becomes less able to fight off infections as certain types of white blood cells, which attack infectious microorganisms, stop functioning. *Osteoblasts* (bone-building cells) become outnumbered by *osteoclasts* (bone-destroying cells), resulting in a loss of bone mass. Joints wear out as cartilage deteriorates, a change that can lead to arthritis. Muscle mass normally declines.

Thus, for humans and other multicellular organisms, it appears that physical decline—ending in death—is inevitable. While improvements in public health and advances in medical science have raised the average life expectancy in the United States from 47 in 1900 to 76 in 1999, the maximum recorded life span for human beings—about 120 years—has remained unchanged.

But why are humans apparently limited in how long they can live? And what molecular mechanisms are at the root of the aging process? Researchers were pursuing several theories of aging in the late 1990's.

### Accumulation of errors

A number of investigators were exploring an idea called the error accumulation theory. This theory argues that aging is due to a lifetime buildup of unrepaired damage to *DNA* (deoxyribonucleic acid, the molecule that makes up genes), proteins, and other components of the body's cells. Researchers propose that much of this damage is caused by *free radicals* (highly reactive molecules that are a by-product of the digestion of food) and *glycosylation* (the reaction of glucose, or blood sugar, with proteins).

During digestion, the sugars, fatty acids, and *amino acids* (the building blocks of proteins) in food combine with oxygen to produce carbon dioxide, water, and a molecule called adenosine triphosphate, which provides energy for cellular activities. These chemical reactions occur in the mitochondria, tiny "power plants" within the cell. As part of this process, many oxygen molecules are transformed into free radicals, which can react chemically with DNA and other vital molecules. As a result, these molecules can be damaged and lose their ability to function.

In the process of glycosylation, glucose, a simple sugar formed in the body by the breakdown of complex carbohydrates in food, reacts with vari-

Everyone passes from youth to old age. Even in youth, the body is undergoing many cellular changes that will lead to physical decline and death. Scientists are seeking ways to slow or stop this process.

ous proteins in the body. In one such reaction, glucose reacts with molecules of *collagen* (an elastic protein found in arteries, tendons, ligaments, and the lungs), causing them to form cross-links with neighboring molecules. This process leads to a loss of elasticity in the tissue.

## Genetic changes

Another theory of aging argues that physical decline results mostly from the accumulation of genetic changes, such as *mutations* (random alterations) in DNA and the rearrangement of *chromosomes* (the structures that carry the genes). Support for this theory comes in part from studies of people who suffer from an inherited disease called Werner syndrome. This disorder causes people to develop signs of aging in their 20's and die of age-related illnesses by the time they are 50. In 1996, researchers at the University of Washington found that Werner syndrome is caused by a defective gene. The gene carries coded instructions for producing an *enzyme* (a protein that speeds up a biochemical reaction) called a helicase. Although the scientists were not sure what the role of the Werner helicase was, most helicases are involved in DNA *replication* (duplication) during cell division.

In 1997, researchers led by biologist Leonard Guarente of the Massachusetts Institute of Technology showed that a version of the helicase gene in baker's yeast, when mutated, causes premature aging in yeast cells. Normally during a yeast cell's life, a small segment of DNA may break off from *ribosomal DNA* (genetic material important to the manufacture of proteins) and form a circle. This happens in a tiny cellular structure called the nucleolus. The breakaway circle is capable of self-duplication every time the yeast cell divides. The resulting daughter DNA circles stay within the nucleolus of the original yeast cell—the mother cell. Ultimately, the mother cell becomes clogged with these DNA circles and protein production is halted. This causes the yeast cell to die.

The researchers discovered that when the helicase gene in yeast is mutated, this process begins earlier than normal. The scientists said their findings indicated that some human cells may begin aging because of damage to the DNA in their nucleoli—though not necessarily the same kind of damage seen in yeast.

Another theory of aging being investigated by scientists concentrates on telomeres, DNA "caps" at the ends of chromosomes. Telomeres are needed for cell division and—just as the plastic tips on shoelaces keep the laces from fraying—they help prevent the chromosome ends from being damaged. In normal DNA replication, the very ends of a chromosome do not reproduce, so each time a

cell divides, each chromosome in the cell loses a bit of its telomeres. The telomeres become shorter and shorter until the cell can no longer divide or function.

However, certain cells in the body have an enzyme called telomerase that enables them to replicate their chromosome's telomeres during cell division. These cells include *germ cells* (eggs and sperm) and *stem cells* (young cells in the bone marrow that give rise to blood cells). In 1998, researchers at Geron Corporation in Menlo Park, California, said they had extended the life span of human skin cells in laboratory culture dishes by inserting in them a gene that codes for telomerase. The gene caused the chromosomes in the old cells to regrow their telomeres to lengths characteristic of young cells.

Although the role of telomerase in aging remained controversial, this experiment indicated that a part of the aging process, at least in some cells, can be slowed or reversed. But can a whole

---

### What happens to the body during aging?

A number of physical changes occur during the aging process. The ages when these processes may occur vary greatly from one individual to another.

**Hair.** Hair turns gray and thins out or is lost.

**Brain.** Some brain cells are lost and others become damaged with age. The brain shrinks beginning in middle age.

**Sight.** Eyes may have difficulty focusing on close objects beginning in the 40's. Eyes have greater difficulty seeing in dim light beginning in the 50's. The ability to distinguish details may begin to decline in the 70's.

**Hearing.** The ability to hear higher frequencies declines with age.

**Immune system.** The immune system becomes less able to fight off infectious diseases as certain types of white blood cells stop functioning.

**Heart.** The heart grows slightly larger with age. Cardiovascular problems become more common.

**Lungs.** The maximum breathing capacity of the lungs may decline by 40 percent between the ages of 20 and 70.

**Kidneys and urinary tract.** The kidneys gradually become less efficient at extracting wastes from the blood. Bladder capacity declines, and urination becomes more frequent.

**Body fat.** Fat is redistributed from just under the skin to deeper parts of the body. Women tend to store fat in the hips and thighs, while men store it in the abdomen.

**Skin.** Wrinkles develop in the skin as the skin thins and underlying fat shrinks.

**Muscles.** Muscle mass may decline by more than 20 percent in both men and women between the ages of 30 and 70.

**Glands.** The levels of sex hormones and certain other hormones decline, or the body becomes less responsive to them.

**Bones and joints.** Bones become brittle as they lose their ability to hold calcium. Joints wear out as cartilage deteriorates.

Source: Baltimore Longitudinal Study of Aging, National Institutes of Health.

## Possible causes of aging

Scientists in 1999 were investigating a number of suspected causes of aging, including the possible involvement of factors called free radicals, telomeres, and DNA circles.

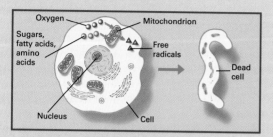

Free radicals are highly reactive molecules formed in the *mitochondria* (structures inside cells that convert energy in food to energy cells can use) when food molecules react with oxygen molecules. Although free radicals attack invading bacteria, they can also attack and destroy the body's own cells.

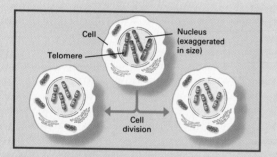

Telomeres are caplike structures at the ends of *chromosomes* (the structures that carry the genes). Telomeres prevent chromosome tips from being degraded and play an essential role in cell division. Telomeres become shorter and shorter with each cell division until the cell can no longer function or divide.

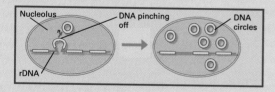

A yeast cell begins to age when circles of *DNA* (the molecule that contains genetic information) pinch off of a specialized genetic material called rDNA in the *nucleolus* (a component in the cell that plays an important role in making proteins). As the DNA circle multiplies, the nucleolus becomes clogged and stops functioning. Experts believe that similar processes may be involved in human aging.

organism be made to age more slowly? In a number of animal studies (not involving telomeres), researchers have demonstrated that this is possible. For example, in 1994, biologists Rajindar Sohal and William Orr of Southern Methodist University in Dallas increased the life span of the fruit fly *Drosophila melanogaster* by 34 percent. They did this by inserting into fly cells extra copies of genes responsible for breaking down free radicals before they can do their damage.

In addition, researchers have found a number of genes in the fruit fly and the roundworm *Caenorhabditis elegans* that, when mutated, increase the life span of the animals. These mutations have various effects, including helping the animal's body fight free radicals.

## Longer human life spans?

As of 1999, no life-lengthening genes or genetic mutations had yet been identified in humans or other *vertebrates* (animals with backbones). A reduction of food intake is the only method that researchers have found can slow aging in mammals. Pathologist Roy Walford of the University of California at Los Angeles kept rats and mice on diets containing 30 to 60 percent fewer calories than they would normally eat. The animals lived as much as 40 percent longer than rats and mice that were allowed to eat all they wanted. In addition, the onset of age-related diseases was delayed, and the animals remained energetic for a longer period. Although scientists do not know why restricting calories prolongs life, they speculate that one of the reasons might be that eating less food results in fewer free radicals being produced in the body.

Many investigators predicted in 1999 that research would eventually uncover a number of genes that regulate human longevity. If so, scientists might then be able to alter those genes in some way to extend life span. Scientists said that life span might also be extended in the future with drugs that modify the functions of aging-related genes.

While research into the aging process continues, scientists advise that the best way to lead a long life is to maintain a healthy lifestyle. This starts with a nutritious, well-balanced diet, limited in fats and sweets and containing many fruits and vegetables. In addition, researchers say that regular exercise will help keep one's body in shape at any age. Finally, they caution that alcohol should be consumed only in moderation—if at all—and that everyone should avoid tobacco, illicit drugs, and intense sunlight. Following this healthy regimen should help people live long enough to see scientists unravel many more of the molecular mysteries of aging and death.

# What Has Caused Mass Extinctions?

## By Donald L. Wolberg

Throughout the history of life on Earth, many species of plants and animals have come into existence, and nearly all have since disappeared. The complete, worldwide disappearance of a species is called extinction. Scientists believe that extinction has claimed more than 95 percent of the 5 billion to 50 billion species that have ever existed.

Extinctions have usually occurred at a gradual rate. That is, every century a handful of species have died out from such causes as an inability to compete with other species for food or living space. However, during several interludes in the past 600 million years, vast numbers of species around the world have become extinct in relatively short periods. The causes of these mass extinctions has long been a great mystery of science.

### Six mass extinctions

Most *paleontologists* (scientists who study prehistoric fossils) believe that six mass extinctions have occurred—during the Cambrian Period (which spanned the time from about 570 million to 505 million years ago), the Ordovician Period (505 million to 438 million years ago), the Devonian Period (408 million to 360 million years ago), the Permian Period (286 million to 245 million years ago), the Triassic Period (245 million to 208 million years ago), and the Cretaceous Period (144 million to 65 million years ago). Paleontologists disagree on the precise timing of these periods.

The Cambrian mass extinction actually consisted of three separate extinction episodes over a span of about 20 million years, between 525 million and 505 million years ago. Much of the planet's life, which at that time was all in the sea, died out. Species that became extinct included many types of trilobites (soft-shelled animals related to crabs), brachiopods (wormlike animals with shells), and coral-reef organisms.

Most theories of what caused the Cambrian extinctions have focused on possible cooling of the ocean or even of the entire globe. For example, several geologists, including Allison Palmer of the Geological Society of America, propose that cold, oxygen-poor water from the ocean depths spread upward onto the continental shelves, replacing warm, oxygen-rich water. This upwelling of cold water might have been produced by *tectonic activity* (the folding and shifting of the Earth's crust) beneath the sea. Species unable to tolerate the harsher conditions would have died out.

The Ordovician extinction episode occurred from 450 million to 438 million years ago, when life was still confined to the sea. Many of the same

Although dinosaurs are probably the most famous victims of mass extinction, large extinction events have claimed the lives of many other kinds of animals and plants during the past 600 million years.

groups of animals affected by the Cambrian extinctions were again hit in the Ordovician.

There is strong evidence for a glacial epoch, or ice age, during this period of time. Ice-scoured rocks in Africa indicate that huge ice sheets moved across the land. As the glaciers advanced around the world, possibly as a result of a fluctuation in the energy output by the sun, the ice would have reflected an increasing amount of heat from the sun back into space. Such heat loss would have chilled all of the land and sea, leading to the death of many organisms. In addition, the buildup of glaciers would have slowly transferred water from the sea to ice on land, causing ocean levels to drop and reducing the available living space for marine creatures.

The Devonian extinction event dates sometime between 370 million and 360 million years ago, after plants and animals had finally colonized the land. During this event, as many as 70 percent of the existing species of marine *invertebrates* (animals without backbones) along with several kinds of fish became extinct. However, land organisms were little affected.

The cause of the Devonian extinctions is a matter of much debate. In 1991, paleontologists Dig-

**The author:**
Donald L. Wolberg is the Chief Executive Officer and President of Natural History Development Company, Inc., in Voorhees, New Jersey.

## Major mass extinctions

| Geological period or epoch | Approximate time of extinction event | Major groups of species affected |
|---|---|---|
| Cambrian Period (570 million-505 million years ago) | Three episodes from 525 million-505 million years ago | Much sea life, including many trilobites (shelled animals related to crabs), brachiopods (wormlike animals with shells), and reef-building organisms |
| Ordovician Period (505 million-438 million years ago) | 450 million-438 million years ago | Numerous brachiopods, bryozoans (mosslike animals), trilobites, graptolites (worm-like creatures), and reef-building organisms |
| Devonian Period (408 million-360 million years ago) | 370 million-360 million years ago | Jawless fish, many placoderms (early jawed fish), ammonites (coiled-shell animals), and reef-building organisms |
| Permian Period (286 million-245 million years ago) | 255 million-245 million years ago | As much as 90 percent of all species, including pelycosaurs (sail-back reptiles), remaining trilobites, and many fish, corals, brachiopods, ammonites, and crinoids (sea lillies) |
| Triassic Period (245 million-208 million years ago) | 218 million-213 million years ago | Many land and sea organisms, including some reptiles and various ammonites, brachiopods, and other shelled animals |
| Cretaceous Period (144 million-65 million years ago) | 65 million years ago | Dinosaurs, pterosaurs (flying reptiles), marine reptiles, remaining ammonites, some mammals, clams, and plants |
| Holocene Epoch (11,500 years ago to present) | Present | Wide variety of plant and animal species, including many unknown to science |

by McLaren and Helmut Geldsetzer of the Geological Survey of Canada reported that a huge meteorite may have struck the Earth at the end of the Devonian. The evidence for such a catastrophe is an unusually large amount of the element iridium in rock dating from this time. Iridium is rare in most of the Earth's crust, but it is present in relatively high concentrations in meteorites.

The impact of a large meteorite would have cast up an enormous amount of dust into the atmosphere. Winds would have distributed the dust around the world, blocking sunlight for many months and causing surface temperatures to plummet. As a result, many plants and animals would have died out within just a few years.

However, a contrary opinion was offered by Thomas Dutro, a paleontologist with the United States Geological Survey. Dutro concluded, based on fossil evidence, that the Devonian extinctions

did not happen as quickly as implied by the meteorite theory. Rather, he said, species disappeared gradually over time. Some scientists have cited sedimentary deposits left by ancient glaciers in South America as evidence that an ice age may have caused the Devonian extinctions.

Between about 255 million and 245 million years ago, at the end of the Permian Period, life on Earth experienced the most devastating mass extinction ever. As many as 90 percent of all species became extinct. In the oceans, many kinds of fish and large numbers of invertebrates, including all remaining trilobites, died out. Land animals were also decimated.

Most paleontologists believe that the cause of the Permian extinctions was a worldwide cooling of the climate. Geologists have discovered large glacial deposits dating from this time. Another possible cause was massive volcanic eruptions. Scientists have dated vast lava flows discovered in Siberia from the late Permian Period. Those eruptions would have spewed thousands of cubic kilometers of ash into the atmosphere, blocking sunlight around the world and lowering global temperatures.

After the Permian extinctions, only a small number of species remained. Throughout the following Triassic, Jurassic, and Cretaceous periods, many new species, including the dinosaurs, evolved to fill the ecological *niches* (opportunities for existence) left vacant by the animals that had become extinct. This illustrates a common theme that scientists have found in extinction episodes: Mass extinctions have typically led to new spurts in evolution.

The next mass extinction was the Triassic episode, which occurred from approximately 218 million to 213 million years ago. This event is poorly understood. The fossil record indicates that the Triassic catastrophe overtook both land and sea organisms, with the land extinctions occurring several million years earlier than the marine extinctions. Explanations for these extinctions range from global cooling to impacts of asteroids or comets.

The most famous extinction event occurred about 65 million years ago, when the Cretaceous Period and the Age of Dinosaurs ended. Besides dinosaurs, all flying reptiles and marine reptiles became extinct. In addition, various other species did not survive. However, many organisms, including alligators, turtles, frogs, sharks, birds, and mammals, did survive the devastation. These animals had little in common with one another, and scientists do not know why they survived.

**Possible causes of mass extinctions**
Scientists have proposed many causes for mass extinctions. Chemicals and ash released by volcanoes, *left,* could have altered climate and ocean chemistry. The spread of glaciers around the globe, *below,* might have cooled the climate by reflecting heat back into space. And a collision between Earth and a massive asteroid or comet, *right,* would have wreaked ecological havoc around the world.

In the early 1980's, geologist Walter Alvarez and his father Louis, a physicist, proposed an extraterrestrial cause for the Cretaceous extinctions. Their theory, based on the presence of a large layer of iridium in 65-million-year-old rock, was that a large meteorite crashed into the planet, drastically disrupting the global climate. Evidence for this theory lies just off the coast of Mexico's Yucatan Peninsula, where there is a submerged crater about 300 kilometers (190 miles) wide. Rock samples indicate that the crater dates from the end of the Cretaceous. Many scientists believe that the crater was created by the impact of an asteroid about 16 kilometers (10 miles) in diameter—large enough to trigger a global catastrophe.

Despite this evidence, a number of scientists challenge the extraterrestrial explanation of the Cretaceous extinctions. These scientists note that dinosaurs and many other organisms had been in decline for hundreds of thousands of years prior to the end of the Cretaceous. Paleontologist J. David Archibald of San Diego State University believes that prolonged volcanic activity and drastic habitat changes caused by geological upheaval, including the rise of the Rocky Mountains, played roles in the extinctions.

In examining the long history of extinction, some researchers, including paleontologists David Raup and John Sepkoski of the University of Chicago, have suggested that a recurring astronomical event, such as cyclic fluctuations in the sun's energy output or the periodic bombardment of the Earth by comets, might explain all large extinction episodes. Raup and Sepkoski base their theory on an analysis of the fossil record, which they say shows that large extinction events have occurred roughly every 26 million years. Other scientists, however, question whether there really is such a regular period to extinction events and maintain that each mass extinction had its own unique cause.

## Today's mass extinction

Although the causes of past mass extinctions remain a matter of much debate, there is little dispute regarding what is causing hundreds of species of animals and plants to be threatened with extinction today: human activities. The spread of human populations and the destruction of natural habitats are the major factors now responsible for threatening the survival of species. Rain forests in South America, Africa, and Asia are among the main areas affected by habitat destruction. Biologists fear that millions of rainforest species, especially large numbers of insects and plants, are being wiped out before they have even been discovered. Some scientists believe that the current round of extinctions may constitute one of the greatest extinction episodes of all time.

While many scientists worked to save endangered species in 1999, research continued into the causes of past mass extinctions. Both of these endeavors highlighted just how vulnerable life is to changes in the environment. Scientists believe that human beings would be equally vulnerable to such changes as drastic warming or cooling of the global climate. Therefore, people have ample reason to contemplate their own future and hope that they don't end up like the dinosaurs.

Contributors report on the year's most significant developments in their respective fields. The articles in this section are arranged alphabetically.

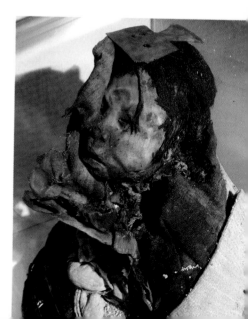

Page 250          Page 186

Page 214

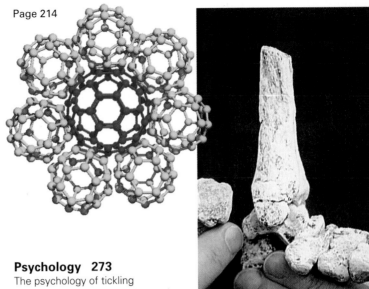

Page 181

Page 205

A report describing how scientists unmasked a fungal culprit that hindered world sales of U.S. wheat in 1996 and 1997 was published in August 1998 by the United States Agricultural Research Service, a division of the U.S. Department of Agriculture (USDA). The report described the molecular and structural differences between two species of fungi—one that causes serious damage to wheat and one that looks much the same but is harmless.

The harmful fungus is called *Tilletia indica.* Discovered in Arizona and California in 1996, it causes a wheat disease known as Karnal bunt. Wheat afflicted with Karnal bunt may become blackened and foul smelling. The other fungus, *Tilletia walkeri,* resembles *T. indica* and was discovered in wheat samples in a number of areas in the Southeastern United States the same year.

The ability to distinguish between the two species of fungus is important, because many countries *quarantine* (refuse to purchase) wheat from regions where Karnal bunt has been found. At first, however, it was not apparent that the two fungi were separate species.

In 1996, wheat from the Southeast came under international quarantine. However, USDA scientists had doubts about the supposedly infected wheat samples from this part of the country. Although the samples contained *spores* (seedlike bodies) that appeared to be from *T. indica,* the scientists could find no sign of infection in the wheat.

This finding led the researchers to suspect that the spores were from a different fungus that lived on some plant other than wheat. For example, a weed growing in wheat fields could carry a fungus whose spores are harvested with the wheat.

One candidate for such a weed was ryegrass, which is common in the Southeast. Investigators found that ryegrass in this region was indeed infected with a fungus whose spores looked like those of *T. indica.* However, the only available molecular test could not distinguish between spores of *T. indica* and ones that might represent an "imposter" fungus.

In early 1997, USDA *mycologists* (fungus experts) Lisa A. Castlebury and Mary E. Palm, along with plant pathologist Lori M. Carris of Washington State University in Pullman, used a powerful instrument called a scanning electron microscope (SEM) to study the fungi. The SEM revealed the ryegrass fungus to be a new species, which the researchers named *T. walkeri.* The researchers could make out various structural differences between the two fungal species, including differences in the thickness and spacing of tiny spines on the outer coats of the spores.

The results allowed restrictions to be lifted on wheat shipments from the Southeast in 1997. In 1998 and 1999, federal officials were using the new detailed descriptions of fungal spores to distinguish between *Tilletia* species.

**Better crop rotation.** One of the most successful crop production systems in American agriculture, the yearly *rotation* (alternation) of corn and soybeans, can be improved upon, according to a March 1999 report. Agriculture researchers, led by *agronomist* (specialist in managing farmland) Paul Porter of the University of Minnesota's Southwest Research and Outreach Center in Lamberton, said that a rotational system incorporating additional grains and legumes would boost agricultural yields and benefit the environment. Grains are plants in the grass family, such as corn and oats. Legumes are plants in the pea family, such as soybeans and alfalfa.

The rotation of corn and soybeans in a field is done to add nutrients to the soil. The nutrients are provided by microbes in legume roots, which take nitrogen from the air and convert it into a form plants can use. Crop rotation also breaks up the life cycle of insect pests that feed on one crop or the other.

In research begun in 1989, Porter's team compared the two-year corn-soybean rotation system to a four-year rotation in which corn, soybeans, and oats are each planted for one year, alfalfa is grown with oats the third year, and alfalfa is grown by itself the fourth. Without the aid of *herbicides* and *insecticides* (substances that kill weeds and insect pests, respectively), the four-year rotation produced greater crop yields than the two-year rotation.

There were several reasons for the greater yields. One reason is that alfalfa is a deep-rooted crop that uses nutrients and moisture from deeper levels of the soil than other crops can reach. In addition, soil erosion is reduced with the

A sheep automatically sheds its fleece after being injected with a solution containing epidermal growth factor, a protein that causes wool fibers to break. The solution, called Bioclip, was developed by scientists at the Commonwealth Scientific and Industrial Research Organisation, Australia's largest scientific research agency. Bioclip, introduced in October 1998, provided farmers with a substitute for shearing.

longer rotational cycle, because alfalfa and oats are planted in rows that are spaced closer together than corn or soybean rows. The greater ground cover helps hold soil and nutrients in place. Finally, insect problems are reduced, because the use of four different crops is more effective than two crops in breaking up insect life cycles.

**Livestock weight loss.** Scientists at Pennsylvania State University in State College announced in November 1998 that they had proved the effectiveness of a method for preventing weight loss in livestock animals that are being transported to auction. During long trips in warm weather, some animals can lose as much as 10 percent of their body weight through perspiration. In an effort to reduce such weight losses, some livestock breeders feed their animals a liquid similar to Gatorade and other "sports drinks" used by athletes.

Lowell Wilson, a professor of animal science at Penn State's College of Agricultural Sciences, studied hogs, lambs, and calves to determine if an electrolyte-restoring liquid given before and during

transportation helps animals maintain body weight. Electrolytes are molecules necessary for keeping blood and other body fluids in proper chemical balance.

After providing one group of animals with electrolyte-containing drinking water and another group with regular water, Wilson and his assistant Darron Smith simulated conditions experienced by livestock going to auction. To do this, the researchers transported the animals 80 kilometers (50 miles), held them in pens for four hours, and then transported them another 80 kilometers. During the trip, the animals were videotaped to record their behavior.

The scientists found that the electrolyte-fed animals ate normally, whereas the animals that received only plain water alternately fasted and binged. As a result, the animals in the electrolyte group maintained a more consistent weight throughout the transportation process. The researchers said this finding might encourage more livestock breeders to feed their animals electrolytes when taking them to market.

[Steve Cain]

A multinational team of scientists in April 1999 announced the discovery of a new prehuman species, *Australopithecus garhi.* The team was led by anthropologists Berhane Asfaw of the Rift Valley Research Service in Addis Ababa, Ethiopia, and Tim White of the University of California at Berkeley.

*A. garhi* belongs to a group of early *hominids* (members of the human lineage) known as australopithecines, that lived in Africa about 4.4 million to 1.2 million years ago. Australopithecines were relatively small creatures, weighing about 30 to 45 kilograms (65 to 100 pounds). More like apes than modern humans, they had small brain cases and large, projecting faces. Unlike apes and like people, australopithecines walked on two feet, but they had an apelike ability to move through trees using their long, powerful arms.

The researchers discovered the *A. garhi* fossils, including a skull believed to be from an adult male and fragments from other individuals, in a region of Ethiopia called the Middle Awash. Using scientific dating methods, the investigators determined that the *A. garhi* fossils are approximately 2.5 million years old. Because of when and where the species lived, it may fill a gap in the hominid lineage.

Most anthropologists agree that the earliest species of the genus *Homo,* to which modern humans belong, was *H. habilis,* which first appeared in East Africa about 2 million years ago. The australopithecine species that many researchers believe was ancestral to *H. habilis* is *A. afarensis,* which lived in the same region about 3.9 million to 2.8 million years ago.

Learning what occurred during the gap between these two species is critical to understanding hominid evolution, but the fossil record from that time has been confusing. Previously discovered australopithecines that existed between 2.8 million and 2 million years ago either lived in other regions of Africa or belonged to a side branch of the family tree that was an evolutionary dead end —the "robust" australopithecines.

The word *garhi* means *surprise* in the language of the local Afar people, and

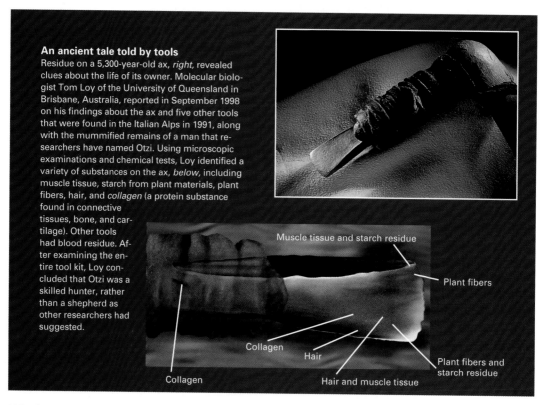

**An ancient tale told by tools**
Residue on a 5,300-year-old ax, *right,* revealed clues about the life of its owner. Molecular biologist Tom Loy of the University of Queensland in Brisbane, Australia, reported in September 1998 on his findings about the ax and five other tools that were found in the Italian Alps in 1991, along with the mummified remains of a man that researchers have named Otzi. Using microscopic examinations and chemical tests, Loy identified a variety of substances on the ax, *below,* including muscle tissue, starch from plant materials, plant fibers, hair, and *collagen* (a protein substance found in connective tissues, bone, and cartilage). Other tools had blood residue. After examining the entire tool kit, Loy concluded that Otzi was a skilled hunter, rather than a shepherd as other researchers had suggested.

Muscle tissue and starch residue

Plant fibers

Collagen

Hair

Plant fibers and starch residue

Collagen

Hair and muscle tissue

**Most complete australopithecine**
A skull still embedded in rock, *right,* is part of the best-preserved australopithecine skeleton ever found, according to an October 1998 report by anthropologists at the University of Witwatersrand in Johannesburg, South Africa. Australopithecines were *hominids* (members of the human lineage) that lived 4.4 million to 1.2 million years ago. The researchers expected that an analysis of the remains, such as the foot bones, *inset,* would help them better understand australopithecine anatomy and the evolution of the human lineage.

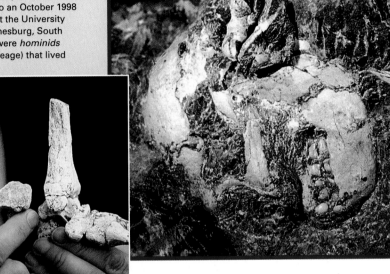

the scientists chose that name because the fossils exhibit some surprising characteristics. In many respects, such as the shape of the premolars and the size ratio of the canine teeth to the molars, *A. garhi* is similar to early *Homo.* However, unlike *Homo, A. garhi* has enormous molars, even larger than average for robust australopithecines, which are distinguished by their huge molars. This unique combination of features convinced Asfaw and White's team to suggest *A. garhi* as a good candidate for the ancestor of *Homo.*

Also in April 1999, another dramatic find from the same site in the Middle Awash was reported by Tim White and other scientists from Belgium, the United States, and Ethiopia. The scientists recovered 2.5-million-year-old bones of animals that appeared to have been killed by meat-eating hominids.

The researchers said that cut marks and numerous fractures on the bones indicated that the animals had been butchered with stone tools and that the bones had been broken open to expose the nutrient-rich marrow. The investiga-

tors, however, did not find stone tools at the site and could not clearly link the animals to the *A. garhi* fossils. Nonetheless, many scientists noted that the discovery fit predictions that the evolution of the first *Homo* species was somehow related to a change from a vegetarian to a meat-based diet.

**Most complete australopithecine.** The discovery of the most complete australopithecine skull ever found was reported in October 1998 by paleoanthropologist Ronald J. Clarke of the University of Witwatersrand in Johannesburg, South Africa. The specimen also included nearly complete bones of an australopithecine foot, arm, and lower leg.

The new specimen, called StW-573, was preserved in South Africa's Sterkfontein Cave, a site that has yielded numerous australopithecine fossils since 1936. The discovery of StW-573, however, was somewhat of a detective story.

Clarke first found some of the foot bones in 1994 while searching for animal remains in a storage shed containing rock removed from Sterkfontein. Three years later, he happened upon

# The First Americans

In February 1999, a team of six scientists began studying the skeleton known as Kennewick Man. At about 9,300 years old, it is one of only a few skeletons found in the Americas that is more than 8,000 years old. Anthropologists and archaeologists believe, therefore, that it has enormous potential for contributing to our understanding of how the Americas were peopled.

Kennewick Man, found in 1996 in Washington state, is also famous because of legal disputes over the remains. The Native American Graves Protection and Repatriation Act of 1990 requires that excavated human remains and cultural objects be given to culturally affiliated tribes. When the federal government planned to turn Kennewick Man over to local tribes, a group of scientists sued. Eventually the scientists won permission to study the skeleton at Seattle's Burke Museum as long as they used nondestructive research methods.

The story is further complicated because researchers in 1996 initially described some features of the Kennewick skull, such as the small cheekbones, as Caucasoid. The word *Caucasoid* comes from a classification scheme for human races introduced in the 1700's and refers to a racial category that includes European, North African, and Middle Eastern populations.

Many anthropologists believe that this classification system is outdated. Most so-called racial classifications depend on obvious characteristics such as skin color. However, skin color varies enormously, even within one supposed race, and relying on one characteristic obscures the significance of variability in other traits. Moreover, populations are always changing, and the identifying physical traits of a particular population may be different from those of earlier generations. Most researchers, therefore, are more interested in overall human variation rather than in fitting people into fixed groups defined by arbitrarily selected characteristics.

Nonetheless, some anthropologists use this classification scheme because they believe certain features can generally be classified as most resembling those of native Europeans, Africans, or Asians. Most researchers who use this classification, however, would not make sweeping assumptions based on a skeleton's "racial identity." For example, the reportedly Caucasoid features of Kennewick Man would not lead scientists to draw conclusions about the color of his skin. Also, few anthropologists would conclude that Kennewick Man's ancestors were Europeans. Instead, the Caucasoid features suggest a need to reevaluate the physical diversity among populations in Asia—geographically the most probable origin of American migrations.

More importantly, anthropologists are interested in Kennewick Man's unusual features because he provides an example of diversity among ancient American populations. Other examples include two skeletons from Nevada, Wizards Beach Man and the Spirit Cave Mummy, both dating to more than 9,000 years ago. While Wizards Beach Man's features look more like those of Native Americans, the Spirit Cave Mummy looks different, with a narrower face, high skull, and smaller cheekbones.

Another ancient skeleton is from Buhl, Idaho. The Shoshone-Bannock Tribes, who claimed ownership of the remains, permitted a portion of the skeleton to be used for laboratory tests. In July 1998, a research team, led by archaeologist

**A face from the past**
A cast of the skull of Nevada's Spirit Cave Mummy, 9,400-year-old remains discovered in 1940, provides clues for how the man may have looked. Forensic facial expert Sharon Long, a research associate at the Nevada State Museum in Carson City, created an approximation of the Spirit Cave Mummy's face, revealing that the man's features were unlike those of most Native Americans today.

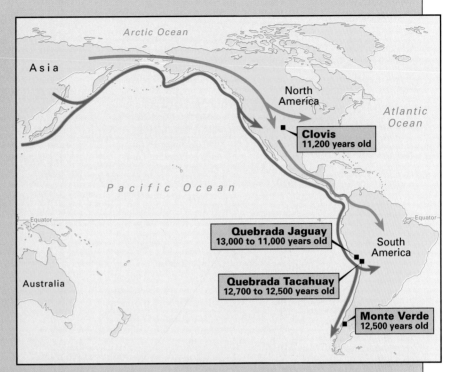

**Ocean voyagers?**
The standard explanation of migration to the Americas, *red,* holds that a population group crossed a land bridge between northeast Asia and North America about 15,000 years ago and, about 3,000 years later, began to move through North America. Some scientists now believe, however, that people migrated from several Asian regions by boat to various points along the west coast of the Americas, *blue.* Sites in South America dating from more than 12,000 years ago seem to support this theory of early multiple migrations.

Arctic Ocean

Asia

North America

Atlantic Ocean

**Clovis**
11,200 years old

Pacific Ocean

Equator

**Quebrada Jaguay**
13,000 to 11,000 years old

South America

**Quebrada Tacahuay**
12,700 to 12,500 years old

Australia

**Monte Verde**
12,500 years old

Thomas J. Green of the Arkansas Archaeological Survey, reported that the Buhl skeleton is more than 10,000 years old. According to preliminary analysis of the skull, the features resemble those of Native American and East Asian populations.

Many researchers claim that this diversity among ancient remains suggests that the peopling of the Americas may be much more complex than the generally accepted explanation. The standard theory holds that the original inhabitants of the Americas came from northeast Asia. By about 15,000 years ago, these people are believed to have been in Beringia, a land bridge between Siberia and Alaska that was exposed during the last Ice Age, when sea levels were lower than today.

This group of people, according to the standard explanation, subsequently moved down into North America about 11,000 to 12,000 years ago, when an ice-free corridor opened up between North American glacial masses in the region that is now western Canada. Eventually, the descendants of these people spread throughout North America and into South America. The earliest well-dated sites that seem to support this theory belong to the so-called Clovis culture of big-game hunters, which arose about 11,200 years ago in North America.

As researchers gather evidence of the physical diversity among ancient Americans, this land-migration theory appears less probable. Some an-

thropologists propose an alternative migration route by sea from Asia to the west coast of the Americas. The sea-route hypothesis, which does not depend on the opening of an ice-free corridor, allows for more than one crossing to the Americas—that is, different regional populations may have migrated at different times.

Some very early archaeological sites in North and South America appear to support this alternative theory. A settlement known as Monte Verde in south-central Chile has a well-established occupation date of 12,500 years ago, much older and much further south than the Clovis sites. Also, two separate studies reported in September 1998 identified maritime settlements in Peru that may be up to 13,000 years old.

Consequently, many researchers in 1999 were calling for a revision of the traditional theory of how the Americas were settled, and many hoped that research on Kennewick Man would provide additional insights into early populations. Whatever the outcome of that study, the Kennewick skeleton and other specimens being examined were sure to remain controversial. Some scientists did not believe that any of these remains are as distinct as many claim. And to many Native Americans, the study of migrations to the Americas was a nonissue because their sacred oral traditions tell them where they came from.     [Kathryn Cruz-Uribe] additional hominid bones, from the same area of the cave, that had been stored at the University of Witwatersrand's Sterkfontein repository.

Clarke then asked two Sterkfontein fossil specialists, Stephen Motsumi and Nkwane Molefe, to investigate the original site. They located the skull and other skeletal remains by matching a bone fragment from the stored collection with a fragment still encased in rock.

Excavation at Sterkfontein is very difficult because the fossils are embedded in *breccia,* a hard rock made up of a jumble of sand, silt, and stones that are cemented together. The researchers removed a block of breccia containing StW-573 from the cave, but by 1999 they had excavated only a small portion of what they hoped would be a nearly complete skeleton. Although the scientists could determine that the skull was clearly that of an adult australopithecine, they decided to finish excavations before identifying the exact species.

Based on the preliminary research on StW-573 foot bones, some anthropologists suspected that the specimen is a member of the species *A. afarensis.* If StW-573 is indeed *A. afarensis,* it would mean that populations of this species covered a much wider geographical region than researchers had suspected.

Several anthropologists theorized that StW-573 might instead be *A. africanus,* a species previously found at Sterkfontein and other South African sites. *A. africanus* lived from about 3 million to 2.5 million years ago. Clarke reported that, based on the initial dating, StW-573 is embedded in breccia that is probably more than 3 million years old, perhaps as much as 3.5 million years old. Thus, if the new fossil belongs to the species *A. africanus,* it would be the oldest known fossil of this species.

In any case, scientists agreed that the most exciting aspect of the new find was the relative completeness of the skeleton. Anthropologists expected that StW-573 would help them better understand the anatomy of australopithecines.

**Could Neanderthals talk?** In February 1999, a team of anthropologists from the University of California at Berkeley and the University of the Pacific in San Francisco challenged research claiming that Neanderthals were able to speak. In the latest chapter of a debate about Ne-

anderthals, Berkeley graduate student David DeGusta and his colleagues questioned the assumption that a hominid's ability to speak was directly linked to the size of the *hypoglossal canal,* a bony passage in the skull that houses the nerve connecting the brain and the tongue.

Neanderthals, who lived in Europe and the Middle East, arose sometime between 230,000 and 200,000 years ago and died out between 35,000 and 27,000 years ago. There is much disagreement among anthropologists over whether Neanderthals, who belonged to the genus *Homo,* were physically and behaviorally similar to the anatomically modern humans who succeeded them. Many researchers believe that Neanderthals did not have the capacity for certain modern behaviors, such as speaking.

The ability to use spoken language is clearly a hallmark of modern humans, but it is a capability that is very difficult to study in the fossil record. In 1997, a team of scientists led by anthropologist Richard Kay of Duke University in Durham, North Carolina, determined that the size of the hypoglossal canal in modern humans was relatively large compared with the average size in chimpanzees and gorillas. The scientists concluded that the size of the canal was related to the human capacity for speech. After examining two fossilized Neanderthal skulls, they determined that the relatively large size of the canals suggested that Neanderthals also had the ability to speak.

DeGusta's team challenged those conclusions. The researchers measured the canals in the skulls of 104 modern humans, 75 modern nonhuman primates, 4 australopithecines, and 2 Neanderthals. They found a wide range of canal sizes among modern humans, from 4.4 to 36.5 square millimeters (0.007 to 0.057 square inch) in cross section. When the scientists measured the canals of the other specimens, factoring in relative differences based on the size of the mouth, they determined that the other species had a comparable range of sizes. They concluded, therefore, that the hypoglossal canal "is not a reliable indicator of speech." Thus, the debate over the Neanderthal capacity for speech remained unresolved.

[Kathryn Cruz-Uribe]
See also ARCHAEOLOGY.

The excavation of a rich tomb within Mexico's Pyramid of the Moon was reported in January 1999 by archaeologist Saburo Sugiyama of Arizona State University in Tempe. The tomb dates to around A.D. 100, making it one of the oldest burial sites ever found in the ancient city of Teotihuacan.

The enormous site of Teotihuacan, which lies north of Mexico City, was occupied from about 100 B.C. to the A.D. 700's. Hundreds of thousands of people lived in Teotihuacan at its peak around A.D. 500, and its ruler controlled vast regions of present-day Mexico and Guatemala. The city is famous for its two massive structures, the Pyramid of the Sun and the Pyramid of the Moon, which sit at either end of a temple-lined street.

The tomb discovered by Sugiyama and his colleagues was apparently enclosed in a small building, which was covered by later stages of construction that eventually became the Pyramid of the Moon. The archaeologists were astounded by the number of valuable objects in the tomb, including tools made of *obsidian* (volcanic glass), conch shells, figurines carved from a stone that resembles jade, and mirrors made from a stone called pyrite. The bones of birds and of two jaguars were also in the tomb. Sugiyama reported that one of the jaguars may have been buried alive in the tomb as part of a ritual sacrifice.

Sugiyama did not think the man buried in the tomb was an elite individual but rather, perhaps, a servant. Indeed, the man appears to have been bound. Sugiyama speculated that he had been sacrificed at the time of a royal burial or as part of a building dedication ritual.

**Mysterious circle.** In the heart of Miami, Florida, among hotels and skyscrapers, excavations have revealed an ancient circle 11.5 meters (38 feet) in diameter outlined with 200 holes cut into limestone bedrock. Archaeologist Robert Carr of the Miami-Dade Historic Preservation Division announced the discovery of the circle near the mouth of the Miami River in December 1998.

Carr and his colleagues uncovered the circle during an archaeological investigation in an area that had been designated for real estate development.

A stucco portrait of a Maya ruler was found in the ruins of a palace in the ancient city of Palenque in southern Mexico. The discovery of several such portraits and a limestone bench or throne was reported in April 1999 by a Mexican and U.S. research team led by archaeologist Alfonso Morales of the University of Texas at Austin. The researchers theorized that the palace, which was dated to about A.D. 760, was one of several elaborate construction projects intended to impress the ruler's subjects during the years when Maya civilization began to collapse.

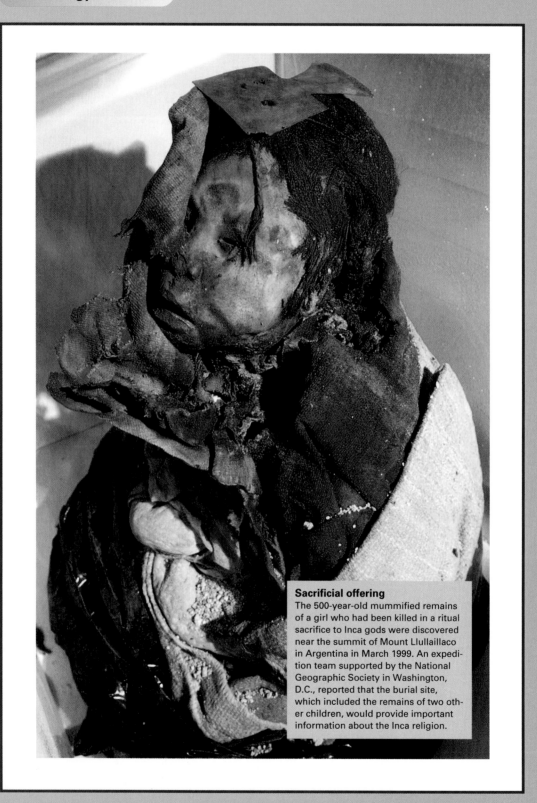

**Sacrificial offering**
The 500-year-old mummified remains of a girl who had been killed in a ritual sacrifice to Inca gods were discovered near the summit of Mount Llullaillaco in Argentina in March 1999. An expedition team supported by the National Geographic Society in Washington, D.C., reported that the burial site, which included the remains of two other children, would provide important information about the Inca religion.

Florida's historic preservation laws require such investigations before new construction can begin.

Most of the excavations at the site revealed debris from a modern landfill. However, when the researchers dug down to bedrock, they exposed several holes. After recognizing a pattern, the team expanded their excavation until they had revealed the entire circle. Inside the circle, Carr's team uncovered 24 irregularly shaped basins carved into the limestone, but the archaeologists were unable to determine the function of those structures. Also found at the site were fragments of stone tools, including axes made of basalt from Georgia. The researchers concluded, therefore, that the community had far-reaching trade contacts. A 1.5-meter (5-foot) shark had been buried in the exact center of the circle, perhaps as part of a ceremony.

Because this kind of archaeological feature had never before been found in Florida, the discovery caused speculation to run rampant. Some observers felt that the holes in the circle formed a celestial calendar. Others added to that argument the theory that the circle had been inspired, and perhaps constructed, by the Maya of southern Mexico, who were specialists in astronomy.

Carr and his colleagues, however, argued that the site, which they called the Miami River Circle, was the foundation of a council chamber, or the house of a leader, of the Tequesta Indians. The holes at the site would have held wooden posts that supported the structure.

Although the archaeologists do not know when the circle was created, they found pottery fragments resembling the Tequesta style from about 1200 to 1400. Other pottery fragments around the site were believed to be 2,000 years old. (The Tequesta were devastated by war and disease shortly after the arrival of Europeans in the early 1500's.)

Although many questions remained in 1999, Carr's team believed that the Miami River Circle would contribute to a better understanding of prehistoric Florida Indians. Negotiations were underway in 1999 for the preservation of the circle, perhaps by cutting it up in blocks to be reassembled nearby.

**Early footwear.** New research at a 40-year-old excavation site at Arnold Research Cave in Calloway County, Missouri, revealed that some footwear found in the cave was more than 8,300 years old. In July 1998, archaeologist Michael J. O'Brien of the University of Missouri at Columbia reported his findings on 35 sandals and moccasins. Those items were among dozens of objects made from plant fibers, wood, and other materials that had been preserved in the dry, dusty soil of the cave.

O'Brien took advantage of a relatively new *radiocarbon dating* technique called accelerator mass spectrometry (AMS). (Radiocarbon dating involves measuring the amount of radioactive carbon-14 remaining in materials that were once alive to determine how long ago they died and stopped absorbing the element.) The researchers needed only tiny pieces of the sandals and moccasins for the AMS dating and, therefore, removing a sample did not significantly damage the artifacts.

The footwear dates revealed a sequence of changing styles. The most recent styles were moccasins, about 750 to 1,000 years old, made of deerskin and lined with grass. Those were preceded by woven sandals. One sandal, which was between 1,000 and 1,245 years old, was made from a plant called rattlesnake master and was constructed with a round toe and a cupped heel.

A number of other sandals, which were also made from plant materials, were 4,390 to 5,575 years old. O'Brien and his colleagues characterized those sandals as "slip-ons." They noted that the most complex sandals, with loops for tying them on and interior padding, were dated as early as 8,325 years ago. The ancient footwear appears to have been discarded when the heels wore out, though sometimes efforts were made to repair the footwear.

**Pacific coast settlements.** In 1998 and 1999, three archaeological research teams working independently at sites in Peru and Ecuador reported on evidence of human occupations along the Pacific Ocean coast from as early as 13,000 years ago. At all three locales, the archaeological evidence points to a specialized *maritime* (ocean-based) economy that depended on fish and other coastal resources.

The oldest site is probably Quebrada Jaguay 280, near the extreme southern coast of Peru. Archaeologist Daniel Sandweiss of the University of Maine in Orono, along with several colleagues, announced in September 1998 that radiocarbon dates of 11,000 to 13,000 years ago had been obtained from the site, which lies in a desert about 2 kilometers (1.2 miles) from the Pacific coast. At the time of the initial human occupation, however, when ocean levels were much lower because of the Ice Age, the site would have been as much as 8 kilometers (5 miles) from the coast.

Excavations at Quebrada Jaguay 280 yielded large amounts of seashells and bones from small saltwater fish. This evidence and fragments of fishnets led the researchers to conclude that the inhabitants of the site were skilled ocean fishermen. Stone artifacts found at the site included several pieces of obsidian, which the researchers traced to a source about 130 kilometers (80 miles) inland.

Also at the site were a series of holes, which Sandweiss's team interpreted as postholes for a *pit house* about 5 meters (16 feet) in diameter. Such dwellings had sunken floors, often with a fire pit located in the center.

At another Peruvian coastal site, called Quebrada Tacahuay, a team of geologists and archaeologists investigated evidence of a different maritime culture. The lead scientist, geologist David Keefer of the United States Geological Survey, reported in September 1998 that settlement began there between 12,500 and 12,700 years ago.

The excavations at Quebrada Tacahuay yielded abundant remains of small fish that were probably caught with nets. However, the most common element of the people's diet was sea birds.

The geologists also recorded 19 separate deposits of rock and mud laid down by floods. The floods were evidently caused by episodes of *El Niño,* a periodic warming of equatorial Pacific Ocean waters that affects global weather patterns and often brings torrential rainfall to this part of South America. The evidence suggested that El Niño events occurred at least as far back as the time when Quebrada Tacahuay was first occupied, but may have lessened or halted altogether between 8,900 and 5,700 years ago. After that time, severe flooding resumed, and about 3,500 years ago, flood-water debris apparently forced

the inhabitants to abandon the site.

Farther up the Pacific coast, a collection of 30 maritime sites were also being excavated near the mouth of the Las Vegas River in Ecuador. Archaeologist Karen Stothert of the University of Texas at San Antonio and her colleagues reported in April 1999 that radiocarbon tests showed that the sites were inhabited as early as 10,800 years ago.

The remains of ocean-going boats and rafts and the bones of large fish led the research team to conclude that the Las Vegan people fished offshore. Stothert also speculated that the Las Vegans may have migrated to coastal Ecuador from the north coast of Peru.

The researchers also noted changes in the Las Vegan culture. By 9,000 years ago, the people had largely switched from fishing and hunting to farming. By 7,000 to 8,000 years ago, they were burying their dead in tombs and cemeteries.

**Ancient Japan.** New excavations at an important prehistoric site near the city of Aomori on the northern coast of Japan's main island were announced in April 1999 by archaeologist Yasuhiro Okada of the Aomori Prefectural Board of Education. The new evidence at the site was expected to change the general understanding of the Jomon period, the earliest major culture in Japan.

The Jomon culture existed at various sites in Japan from about 10,000 B.C. to 300 B.C. Although archaeological evidence shows that the Jomon were primarily hunters and gatherers, scientists have also found seeds of domestic plants, indicating that some farming was being done. At other sites, archaeologists have excavated Jomon pit houses and ceremonial or assembly buildings.

The excavations near Aomori demonstrate a level of sophistication greater than researchers had suspected. One building site from about 4000 B.C. contained evidence of wooden pillars that were at least 1 meter (3.3 feet) across. Constructing a building with such large columns would have required a high level of social organization and technology.

Although the community of several hundred people at Aomori survived mostly on hunting and gathering, they also practiced a limited form of agriculture, growing chestnuts and millet. The presence of jade from southern Japan and obsidian from the northern Japanese island of Hokkaido revealed that the Aomori community had extensive trade contacts. Excavations also revealed that, for reasons not yet understood, the Jomon buried adults and children in separate cemeteries.

**The first writing?** Scholars have long theorized that writing originated about 5,000 years ago in the Near East, probably in ancient Sumeria, a civilization that thrived in what is now Iraq. In December 1998, Gunter Dreyer of the German Archaeological Institute in Cairo, Egypt, challenged that theory with new findings that set off a hot debate.

In excavations at the 5,300-year-old tomb of an Egyptian king known as Scorpion, Dreyer uncovered clay tablets marked with pictographic symbols, such as mountains, plants, and animals. He found more than 300 examples of such inscriptions either on the tablets, which were about the size of postage stamps, or on clay jars.

Although the markings are symbols, Dreyer considers them to be true writing that directly corresponded to spoken language. In other words, each sym-bol represented not an object or idea but rather a sound in spoken Egyptian. For example, one tablet is inscribed with an early Egyptian symbol for a throne and the depiction of a stork. Dreyer translated the inscription as *Ba-* (throne) *set* (stork), or *Baset,* the name of an ancient Egyptian city. In studying the tablets, Dreyer and his colleagues deciphered what they believed were early inventories, mainly accounts of linen and other goods delivered as taxes or as tribute to Scorpion.

Dreyer asserted that the inscriptions were a form of writing that originated in Egypt, not one that was introduced from Sumeria. Not all scholars agreed. *Linguist* (language specialist) Holly Pittman of the University of Pennsylvania in Philadelphia and other researchers reviewed and challenged Dreyer's theory at a symposium on the origins of writing at the university in March 1999.

Many scholars at the symposium defended the Sumerian-origin theory. Sumerian writing, they maintained, began as pictorial symbols about 5,500 years ago, and by 5,000 years ago it had

**Candidates for the earliest writing**
Small clay tablets found in a 5,300-year-old Egyptian tomb bear inscriptions that may be the earliest examples of writing, according to researchers at the German Archaeological Institute in Egypt. The scientists, claiming that the impressions represent a fully developed writing system, challenged the traditional theory that writing had originated about 5,000 years ago in Sumeria (part of present-day Iraq).

**Tomb in an ancient pyramid**
Recent excavations inside the Pyramid of the Moon, *right,* at the ancient city of Teotihuacan in southern Mexico revealed a richly furnished tomb dating from about A.D. 100. Archaeologist Saburo Sugiyama of Arizona State University in Tempe reported the discovery in January 1999. Teotihuacan was the center of a vast civilization that existed from about 100 B.C. to the A.D. 700's.

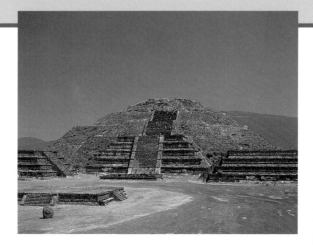

Artifacts inside the tomb, *right,* included stone figurines and blades made from *obsidian* (volcanic glass). A researcher excavates the skeleton of a man found in the tomb, *below.* Sugiyama concluded that the man, who had been bound, was probably a servant who had been sacrificed in a religious ritual. The tomb also contained the skeletons of birds and jaguars.

evolved into a standardized system for recording spoken language with slash-like marks in clay, a type of writing known as cuneiform. Therefore, they said, Sumerian writing in some form would have predated the symbols found in Scorpion's tomb. The defenders of the Sumerian origin argued that the idea of writing was probably transmitted from Sumeria to Egypt.

Other researchers noted that the early inscriptions in both Egypt and Sumeria were tax records and property inventories that reflected the need of these early civilizations to keep track of their expanding economies. Writing thus may have developed independently in the two regions because of similar cultural needs. These experts tended to support the theories of archaeologist Denise Schmandt-Besserat of the University of Texas at Austin, who began studying Sumerian writing in the 1970's.

Schmandt-Besserat argues that before there was actual writing in Sumeria, accountants made "tokens" of molded clay that were shaped to represent certain commodities, such as livestock or grain.

Over time, as more goods had to be inventoried, the use of tokens became impractical. Standardized symbols, according to Schmandt-Besserat, were then substituted for the tokens. These symbols were impressed on clay tablets, and eventually cuneiform writing was born.

The origins of writing remained in dispute in 1999. While most specialists thought the evidence from Sumeria was stronger than that from Egypt, further analysis of the inscriptions from Scorpion's tomb was needed. Furthermore, reports of excavations in the Indus Valley of Pakistan suggested that writing may also have developed there as much as 5,300 years ago. Archaeologists Mark Kenoyer of the University of Wisconsin at Madison and Richard Meadows of Harvard University in Cambridge, Massachusetts, reported their Indus Valley findings to the Pennsylvania symposium. They suggested that writing arose independently in the Indus and Sumerian civilizations.          [Thomas R. Hester]

See also ANTHROPOLOGY. In the Special Reports section, see DEEP INTO THE PAST.

## Astronomy

A group of American geophysicists reported in December 1998 that they had made the first precise estimate of the amount of water ice at the north pole of Mars. The team, led by Maria T. Zuber of the Massachusetts Institute of Technology in Cambridge, made the measurement with an instrument called the Mars Orbiter Laser Altimeter (MOLA) aboard the Mars Global Surveyor spacecraft.

Mars Global Surveyor began orbiting the red planet in September 1997. At the same time, MOLA began repeatedly firing a laser beam at the Martian surface. By measuring how long it took for each laser pulse to reach the surface and be reflected back to the spacecraft, and by tracking Surveyor's orbit, MOLA scientists determined the varying elevations of the Martian landscape. With that data, they mapped the surface features of Mars's northern hemisphere.

The map indicated that the permanent ice cap at the planet's north pole—which earlier studies had shown is composed mostly of water in the form of ice—rises some 3 kilometers (2 miles)

above the surrounding terrain and is about 1,200 kilometers (750 miles) across. Knowing the height of the ice cap, as well as its area as determined from photo images, the scientists were able to calculate the volume of the ice, which turned out to be about half that of the Greenland ice cap on Earth.

Other evidence indicated that the Martian ice cap is smaller than it once was. Scientists know that ice caps, because of their enormous weight, press down on the land surface beneath them, causing a depression. The great ice sheets that spread across much of the Earth during the last Ice Age, which ended about 11,500 years ago, left a depression on northern continental areas that is still evident today. Likewise, Zuber's team found that the ice cap at the Martian north pole is surrounded by a hemispherical depression 5 kilometers (3 miles) deep.

The land in the Martian depression slopes gently toward the pole, and dried-up channels that ran with water several billion years ago appear to be inclined in the same direction. These topographic features indicate that in the past, the ice cap was several times larger than it is now. Although the MOLA data did not indicate when the cap declined in size nor how rapidly the change occurred, they were consistent with scientists' general view that Mars has lost much of its initial store of water.

**Water vapor on Titan.** U.S. and European scientists announced in October 1998 that they had detected water vapor in the atmosphere of Titan, the largest of Saturn's moons. The group was led by astronomer Athena Coustenis of the Paris Observatory in Meudon, France.

Coustenis and her colleagues observed Titan in 1997, using an Earth-orbiting telescope called the Infrared Space Observatory (ISO), launched in 1995 by the European Space Agency. With ISO, the scientists studied the universe in the infrared part of the electromagnetic spectrum. Infrared light has lower energy and longer wavelengths than visible light. It represents the energy radiated by warm bodies—including human beings—but even cold objects like Titan emit small amounts of infrared. Astronomers are interested in infrared light because many molecules in the atmospheres of other planets, and in gas clouds where planetary systems are forming, emit or absorb infrared light at particular wavelengths.

ISO used a telescope cooled by liquid helium. This extremely cold fluid prevented the instrument from generating infrared rays that would have interfered with the detection of small amounts of infrared energy coming from cold bodies such as Titan. ISO's helium coolant gradually became depleted, and the observatory's mission ended in 1998.

Coustenis and her team studied the pattern of light coming from Titan and detected telltale energy peaks at certain infrared wavelengths characteristic of water vapor. By comparing the strengths of the peaks with laboratory measurements of the infrared spectrum of water vapor, Coustenis and her colleagues inferred that the amount of water vapor in Titan's atmosphere is about 8 parts per billion. That is much less than the amount of water vapor in Earth's dry upper atmosphere, which averages a few thousand parts per billion.

The team's findings supported earlier data provided by the Voyager spacecraft, which flew past Saturn and its moons in 1980 and 1981, revealing that Titan is not Earthlike in terms of supporting life. Voyager findings showed that Titan's atmosphere is composed of nitrogen with lesser amounts of methane. The water vapor probably plays an important part in the sunlight-driven chemistry of Titan's atmosphere. Scientists speculated that the vapor may contribute oxygen to convert some of the methane into carbon monoxide and carbon dioxide.

In addition, because Titan is more than 1 billion kilometers (almost 1 billion miles) from the sun, its surface temperature is only $-178$ °C ($-288$ °F). Thus, water could exist only as ice. Titan's lower atmosphere is so cold that any water vapor introduced into it would immediately turn to snow. Coustenis's group determined from computer modeling of the moon's atmosphere and from the appearance of the infrared spectrum that the water vapor detected by ISO lies some 400 kilometers (250 miles) above Titan's surface. At this height, a thick haze of gases traps any ultraviolet light from the sun, so the temperature there is about $-73$ °C ($-99$ °F), enough to allow the amount of water vapor seen by ISO.

Because of the tremendous cold of Titan's surface, Coustenis's team speculated that the moon is probably not the source of the water vapor in its upper atmosphere. A more plausible origin, they believe, is the small *meteoroids* (bits of ice and rock traveling through the solar system) that find their way from comets into the atmospheres, or onto the surfaces, of all bodies in the solar system.

**Comet Hale-Bopp.** New findings about the 1997 close encounter of Comet Hale-Bopp with Earth were announced in March 1999 by planetary scientists at the California Institute of Technology (Caltech) in Pasadena and other universities. The researchers' report cast doubt on the idea that Earth's oceans originated largely from comets.

The team, led by Geoffrey A. Blake of Caltech, observed Hale-Bopp from a radio-telescope facility in California's Owens Valley. The team used multiple antennas linked electronically to serve as a single radio "eye" that provided a much clearer view of celestial objects than was possible with single receivers.

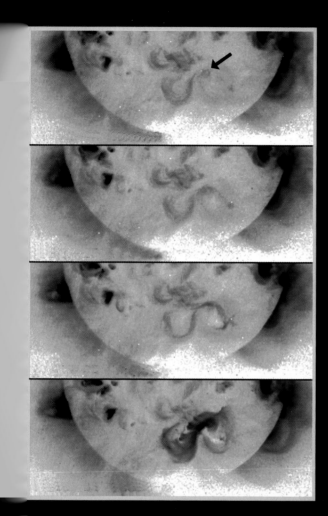

**The shape of sun storms**
An S-shaped pattern (arrow) that sometimes appears on the sun's surface is usually followed by a large, violent discharge called a coronal mass ejection that reaches Earth in two to four days. This finding, based on images taken with an X-ray camera aboard a Japanese-U.S.-British satellite called Yohkoh, was reported in March 1999 by a team of U.S. planetary scientists. The group, led by astronomer Richard C. Canfield of Montana State University in Bozeman, said they hoped that the new information would help scientists predict days ahead of time when a solar storm is likely to occur. Such storms disrupt satellite transmissions and can damage electrical power grids on Earth.

Blake and his colleagues turned the array toward Hale-Bopp in March 1997, when the comet passed close to the sun and near enough to Earth to make it one of the brightest comets in more than 100 years. The array of telescopes allowed the team to examine the radio energy from individual jets of dust and gas issuing from the rocky and icy core of the comet as it was heated by the sun.

Variations in the radio signals revealed that the cometary jets contained water. More importantly, the scientists were able to distinguish two types of water: normal water (composed of one oxygen atom and two hydrogen atoms) and "heavy water," which contains deuterium, rather than hydrogen. Deuterium is an *isotope* (variant form) of hydrogen. While the nucleus of a hydrogen atom contains just a proton, a deuterium atom contains both a proton and a neutron.

Deuterium is present in all natural sources of hydrogen, including seawater, which contains about 150 parts per million of heavy water—water containing at least one deuterium atom. Because heavy water is a stable, permanent part of Earth's oceans, scientists believe that the cosmic materials from which the Earth formed must contain a similar abundance of heavy water.

In Comet Hale-Bopp's individual jets, Blake and his colleagues found heavy water to be 20 times more abundant relative to normal water than it is in Earth's oceans. In earlier observations, scientists (using different techniques) had noted a proportion of heavy water in the dust and gas of two other comets, Halley and Hyakutake, roughly twice that in Earth's oceans. Thus, Blake's team concluded that while comets may well have seeded the Earth with some of its water, they could not have been the principal source of our planet's oceans.

**A new solar system.** Two teams of U.S. astronomers announced in April 1999 that they had independently discovered another solar system. Located 44 *light-years* from Earth, the system consists of three large planets orbiting a star called Upsilon Andromedae. (A light-year is the distance light travels in one year, about 9.5 trillion kilometers [5.9 trillion miles]).

Although Upsilon Andromedae is visible to the naked eye, its three planets cannot be seen with even a large telescope. The astronomers—one team at San Francisco State University and the other at the Harvard-Smithsonian Center for Astrophysics in Cambridge, Massachusetts, and the National Center for Atmospheric Research in Boulder, Colorado—inferred the planets' existence by indirect means. They measured variations in the light from Upsilon Andromedae, caused by slight wobbles in the star as it is pulled by its planets' gravity.

The astronomers noted that none of the three planets is likely to be habitable. All three are probably large gaseous bodies similar to Jupiter and Saturn, with no solid surface. They also are probably either too close or too far from their sun to support life. However, unlike the more than one dozen single planets orbiting other stars that have been detected since 1995, the discovery of the Upsilon Andromedae system provides clear evidence that our multiplanet solar system is not a unique feature in the universe.

**Surprises in interstellar dust.** New findings about the dust in *interstellar*

*space* (the regions between stars) were reported in October 1998 by two groups of astronomers in Germany and New Zealand. The researchers said their data showed that most of the mass in interstellar dust is contained in larger grains than astronomers had suspected. Because stars originate from interstellar dust clouds, this discovery challenged current theories about star formation.

Astronomers have long known that the space between the stars is filled with a haze of tiny, solid particles. These interstellar dust grains affect the light from distant stars, causing the stars to appear dimmer—and thus farther away —than they really are. Measurements had shown the grains to be less than 1 *micrometer* (millionth of a meter) in diameter, and astronomers thought that grains larger than a micrometer were rare or nonexistent in interstellar space.

Astronomers led by Eberhard Grun at the Max Planck Institute in Heidelberg, Germany, analyzed dust grains collected by the Galileo spacecraft, in orbit around Jupiter since 1996, and the Ulysses spacecraft, launched in 1990 to or-

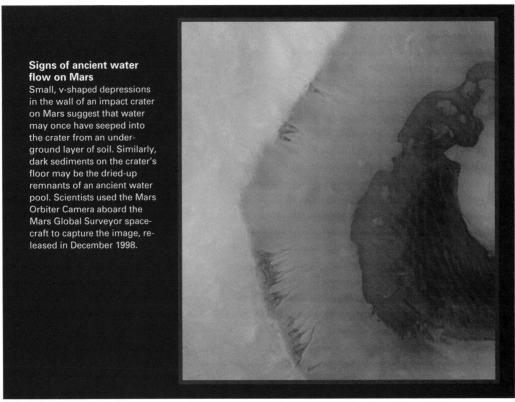

**Signs of ancient water flow on Mars**
Small, v-shaped depressions in the wall of an impact crater on Mars suggest that water may once have seeped into the crater from an underground layer of soil. Similarly, dark sediments on the crater's floor may be the dried-up remnants of an ancient water pool. Scientists used the Mars Orbiter Camera aboard the Mars Global Surveyor spacecraft to capture the image, released in December 1998.

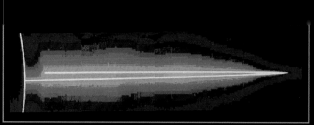

**The rings of Jupiter**
An image taken by the Galileo space-craft reveals that Jupiter's main ring and halo, *above,* are made of dust particles. Scientists at Cornell University in Ithaca, New York, and the National Optical Astronomy Observatories in Tucson, Arizona, announced in September 1998 that the rings are formed of debris flung into space when fragments of comets and meteoroids hit Jupiter's four small inner moons. Using images taken by Galileo, in orbit around Jupiter since 1995, the scientists found that the main ring and halo were formed from the moons Metis and Adrastea, *right,* while the so-called gossamer rings—which scientists once thought were only one ring—were formed from Amalthea and Thebe.

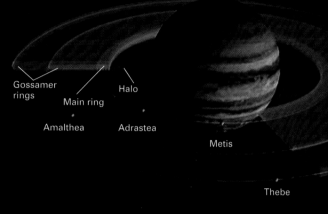

Gossamer rings

Main ring

Halo

Amalthea

Adrastea

Metis

Thebe

bit and study the sun. The dust-collecting instruments aboard the two spacecraft were to provide data on the size, number, and composition of particles that orbit the sun between the planets. However, the instruments also gathered particles from the local interstellar medium—a thin cloud of dust that surrounds the sun and planets—that were passing through the solar system. Grun's team found that some of the particles were up to several micrometers in size. (That the particles were from outside the solar system was inferred from their speed and direction of motion.)

The other group, led by astronomer Jack Baggaley at the University of Canterbury in New Zealand, used a radar telescope to measure *ionization tracks* caused by interstellar particles entering the Earth's upper atmosphere. Ionization tracks are strings of ionized, or electrically charged, air molecules created when particles from space strike the molecules and dislodge electrons. The size of tracks observed by Baggaley's group indicated that they were made by particles even larger than those

studied by Grun's team of researchers.

The new findings forced astronomers to consider that large particles may be common in interstellar clouds of gas and dust. But if that is so, they wondered, then why is it that stars—which form from these clouds—do not contain a greater proportion of the chemical elements, such as silicon, iron, and magnesium, that the dust particles are made of? Answering that question will be a challenge for astronomers who study the process of star formation.

**Magnetars—exotic neutron stars.**
On Aug. 27, 1998, an enormous blast of energy from outer space saturated every X-ray and gamma-ray detector operated by researchers throughout the world. This burst, astronomers believe, came from a source called SGR1900+14 and fit all of their expectations of how a magnetar—a hypothetical type of star first proposed in 1992—would behave.

Gamma rays are the most intense form of electromagnetic radiation, possessing wavelengths less than a thousandth that of visible light. Astronomers have been puzzled for years by occasional bursts of gamma rays coming from sources in the remnants of *supernovae* (exploding stars).

The association of these bursts with supernovae remnants suggested that the sources—dubbed soft gamma-ray repeaters (SGR's) because they emitted relatively low-energy gamma rays—might be neutron stars. A neutron star is a very small (about 20 kilometers [12 miles] in diameter), very dense object formed by the collapsed core of a star that blew off its outer layers in a supernova explosion. It is made almost entirely of neutrons, the electrically neutral particles that, with protons, make up an atomic nucleus.

A neutron star rotates very rapidly, in periods of a few *milliseconds* (thousandths of a second), because the original rotation of a star is magnified when the star falls in on itself. A neutron star also has a very strong magnetic field—more than 1 trillion times stronger than that of the Earth—because when a star collapses during the supernova process, its original magnetic field is compressed and amplified in the neutron star.

Astronomers realized in the mid-1990's that the properties of SGR's could be explained if the objects were neutron stars with magnetic fields about 100 times stronger than that of a normal neutron star. These hypothetical objects were dubbed magnetars.

The magnetic forces of a magnetar would create extreme stresses on the star's surface. These stresses would cause the surface to suddenly crack and release enormous amounts of energy, mostly in the form of low-energy gamma rays. The losses of energy would cause the star's magnetic field to gradually weaken, and as a result, the star's rotation would slow.

The first confirmation that magnetars exist came in May 1998, when a team of astronomers led by Chryssa Kouveliotou of the Marshall Space Flight Center in Huntsville, Alabama, used data from the orbiting Compton Gamma Ray Observatory (GRO) to analyze a burst of soft gamma rays coming from a source within a supernova remnant. The analysis revealed that the object—called SGR-1806-20—had a relatively slow rotational

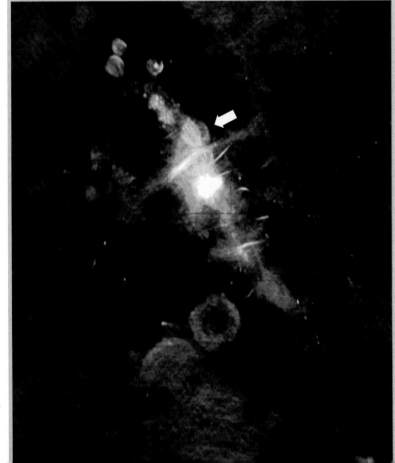

Thick clouds of gas and dust swirl around a massive object at the center of the Milky Way Galaxy that scientists think is a black hole. The remnant of a new *supernova* (exploding star) glows just above a bright band marking the path of an intense magnetic field (arrow). Astronomers at the Naval Research Laboratory in Washington, D.C., used an array of 27 ground-based radio telescopes in Socorro, New Mexico, to produce the image— the clearest view of the core of the Milky Way ever obtained. The group presented its findings in January 1999.

**A ring of infant stars**
Newborn stars circle the center of galaxy NGC 4314 in an image taken with the Hubble Space Telescope and released in June 1998. Astronomers noted that the ring was the only place in the galaxy, located about 40 million *light-years* from Earth, in which stars were being born. (A light-year is the distance that light travels in one year, about 9.5 trillion kilometers [5.9 trillion miles].) The purple clouds surrounding the circle indicate the presence of hydrogen gas, which is glowing from the energy being emitted by the hot, young star clusters.

period of 7.47 seconds and that its rate of rotation was declining, characteristics that fit exactly with those expected of a magnetar. Similarly, SGR1900+14, observed in August 1998, was found to be rotating once every 5.16 seconds, with definite signs of slowing.

Astronomers estimate that there may be as many as 30 million magnetars in the Milky Way. Therefore, in 1999 they were examining the properties of other gamma-ray and X-ray sources in supernova remnants in an effort to identify additional magnetars.

**Observing a gamma-ray source.**
Another type of gamma-ray source made the news in 1999. On January 23, astronomers in New Mexico, led by astrophysicist Carl W. Akerlof of the University of Michigan in Ann Arbor, made the first visual observation of a high-energy gamma-ray burst, the most powerful source of energy in the universe.

Astronomers had detected occasional flashes of gamma rays coming from seemingly random locations in the sky since the 1960's. The phenomenon was first discovered by U.S. defense satellites

equipped to detect gamma rays emitted by nuclear bomb explosions on Earth. However, because the bursts typically last only about 10 seconds, astronomers did not have enough time to locate their sources.

Beginning in 1996, when the Italian and Dutch space agencies launched a combined gamma-ray and X-ray observatory called BeppoSAX, a few sources were finally pinpointed. After a gamma-ray burst occurred, an X-ray camera aboard BeppoSAX detected the more easily traced X rays that immediately follow each burst. In this way, the astronomers were able to identify the point in space from which the burst originated.

BeppoSAX revealed to astronomers that gamma-ray bursts originate in very distant galaxies, billions of light-years away. Thus, they may be caused by a process that occurred primarily during the early evolution of the universe. (Objects seen far away in space are also seen as they were far back in time, because it took billions of years for the light from those objects to reach us.) But little was learned about how the bursts actually occurred. All astronomers could see of each burst was an afterglow, the remnant radiation at X-ray and radio wavelengths that follows a gamma-ray burst. But with the Jan. 23, 1999, gamma-ray burst, astronomers finally got a look at the source.

That burst was detected by a U.S. orbiting spacecraft called the Compton Gamma Ray Observatory. As soon as the observatory picked up the burst, computers in contact with the spacecraft alerted an automated optical telescope in Los Alamos, New Mexico. Within 22 seconds, the telescope, operated by Akerlof's team, began obtaining images of the region of sky in which the burst, named GRB990123, was occurring. The burst reached its peak brightness 5 seconds later and was briefly bright enough to be seen with even a small telescope.

Within a day, two other teams of astronomers used larger telescopes to observe the source of the burst. Shrinivas R. Kulkarni of Caltech and his colleagues located a faint, faraway galaxy that housed the gamma-ray source. And a team led by Daniel Kelson of the Car-

negie Institution of Washington (D.C.) determined that the galaxy was about 9 billion light-years from the Milky Way, or more than halfway to the edge of the universe. The distance and brightness of the burst revealed that at its peak, GRB990123 was emitting at least 10,000 times as much energy as any other object ever observed.

Although the cosmic events that produce gamma-ray bursts are not known, astronomers speculate that they may be caused by the collision and merger of neutron stars or black holes. Another possibility is a theorized type of supernova explosion known as a *hypernova*, in which a massive star is vaporized instantly and converted entirely into energy.

Hypernovas were first proposed as a possible source of gamma-ray bursts by Polish-born American astronomer Bohdan Paczynski in 1997. In April 1999, a team of astronomers at Northwestern University in Evanston, Illinois, and at the University of Illinois in Urbana-Champaign, reported that they had observed the first evidence of two hypernovae in a galaxy called M101—or the

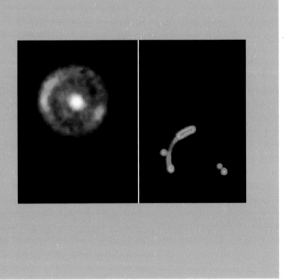

**The power of gravity**
Images from the Hubble Space Telescope and from the MERLIN radio telescope in England show light from a distant galaxy distorted by a phenomenon known as gravitational lensing. In the Hubble image (left), the galaxy's light is spread into a fiery circle known as an Einstein ring. The bright spot at the center of the ring is an intervening galaxy whose gravity is bending the more distant galaxy's light around it. The Merlin image (right) was made from radio waves emitted by two sources in the same distant galaxy. Because the alignment of the radio sources with the nearer galaxy was slightly off-center, the gravitational warping of the radio waves created just a partial ring plus three dots. The two images were released in September 1998.

Pinwheel Galaxy—located 25 million light-years from Earth. The team estimated that the two hypernovae were about 100 times more powerful than average supernovae.

**An accelerating universe?** By the end of 1998, two international teams of astronomers had gathered compelling evidence that the universe is expanding at an ever-increasing rate—and that it will continue to expand forever. These findings turned the world of *cosmology* (the study of the universe as a whole) upside down.

Both groups had been trying to determine the rate at which the expansion of the universe is slowing. (Many astronomers had long assumed that the universe's expansion is slowing because of the gravitational pull that all of the galaxies in the universe exert on each other.) One group was led by astrophysicist Saul Perlmutter at the University of California at Berkeley and the other by astronomer Brian Schmidt of Australia's Mount Stromlo and Siding Spring Observatories in Canberra. Both groups calculated the rate of the universe's expansion by measuring the distance from Earth to certain bright supernovae and calculating the speed at which the galaxies these stars reside in were moving away from the Earth. Perlmutter's group studied the most distant supernovae, while Schmidt's group collected data on those in nearby galaxies. By comparing the speeds of the galaxies with their distances, the astronomers concluded that the expansion of the universe, rather than slowing down, has been accelerating.

The discovery in October 1998 by Perlmutter's group of the oldest and farthest supernova ever seen fully supported this finding. However, many more supernovae will need to be studied before astronomers can confirm the accelerated expansion of the universe and begin to understand what sort of energy may be causing it.

[Jonathan I. Lunine and Theodore P. Snow]

In the Special Reports section, see HOW THE MOON WAS BORN. In the Special Section, see WHAT IS THE ULTIMATE FATE OF THE UNIVERSE?

## Atmospheric Science

The strongest El Niño in recorded history persisted midway through 1998, but Hurricane Mitch was the weather event of the year. An extremely powerful storm that lasted for more than a week in October and November 1998, Mitch became one of the deadliest and most destructive hurricanes on record. In May 1999, a series of severe tornadoes struck portions of Arkansas, Kansas, Oklahoma, and Texas, leveling entire neighborhoods and killing more than 50 people.

**Hurricane Mitch,** an intense, slow-moving storm, first appeared as a *tropical depression* (low-pressure zone) in the Caribbean Sea north of Panama on Oct. 22, 1998. The depression soon became a monster hurricane. By the time it headed out to sea again on November 5, Mitch had carved a swath of death and destruction across Central America, moved across the Gulf of Mexico as a tropical storm, and battered southern Florida. According to the National Oceanic and Atmospheric Administration (NOAA), Mitch was the second most deadly hurricane ever to hit the Western Hemisphere, killing more than 11,000 people.

The deadliest was the "Great Hurricane of 1780," which killed an estimated 22,000 people in the West Indies.

On Oct. 26, 1998, Mitch achieved Category 5 status, the highest ranking on the Saffir-Simpson scale of hurricane strength. The storm had sustained winds of 290 kilometers (180 miles) per hour, with gusts exceeding 320 kilometers (200 miles) per hour. By achieving this intensity, Mitch reached fourth place on the list of the strongest Atlantic hurricanes on record, behind Gilbert (1988), the Florida Keys Hurricane (1935), and Allen (1980). In addition, Mitch set or tied several other hurricane records. Most notably, the storm remained at Category 5 strength for 33 hours—the longest continuous period for a Category 5 storm since 1979's Hurricane David (36 consecutive hours). Although direct comparisons with some of these earlier storms are complicated by changes in storm-monitoring procedures, Mitch was undoubtedly one of the strongest hurricanes ever recorded in the Atlantic Basin, which includes the Gulf of Mexico and the Caribbean Sea.

**Mitch's path of destruction.** After initially moving north toward Jamaica and the Cayman Islands, Mitch turned westward, and by October 27 it was approaching the northern coast of Honduras. Wave heights on the Honduran coast reportedly reached 13 meters (44 feet). Drifting slowly southward, Mitch hung offshore for two days, pouring heavy rains on Honduras and Nicaragua. When it finally did make landfall, two days later, Mitch began a slow journey through the mountainous interior of Honduras, reaching Guatemala on October 31.

As it drifted west across Central America, Mitch drew on water vapor from both the Caribbean Sea to its east and the Pacific Ocean to its south. The combined effects of the abundant moisture, the cooling of air rising over the mountainous region, and Mitch's slow movement resulted in rainfalls of up to 50 centimeters (20 inches) per day in some regions. Total rainfall in some localities during the storm was as much as 190.5 centimeters (77 inches).

Mitch expended much of its strength traveling over the mountains of Central America. It then turned northward, crossing southern Mexico as a weak depression and entered the Bay of Campeche on November 2. However, the warm waters of the Gulf of Mexico and favorable conditions aloft enabled Mitch to reform. It regained tropical-storm status and moved northeastward. Mitch weakened slightly as it crossed Mexico's Yucatan Peninsula, but when it entered the Gulf it attained tropical-storm status for the third time. Mitch headed northeastward to pound Florida's Key West with strong winds and rain on November 4 and 5. It dropped 15 to 20 centimeters (6 to 8 inches) of rain on southern Florida and spawned several tornadoes. Finally, on November 5, the storm headed out into the Atlantic Ocean and was again downgraded to a tropical depression.

Mitch's high winds and the massive ocean waves they generated ravaged the Caribbean coastal regions of Belize, Guatemala, and Honduras. However, it was the tremendous rainfall in Central America that produced a true disaster. Flooding and mudslides destroyed nearly all the roads and bridges in Honduras and

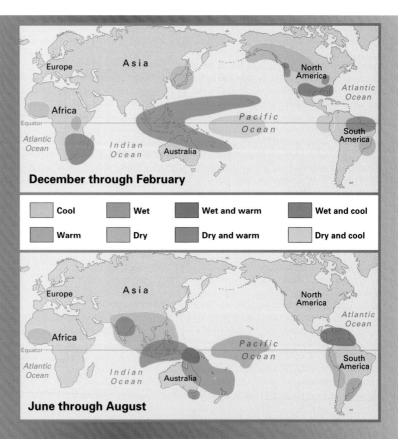

**La Niña: The flip side of El Niño**

Maps based on data from the National Oceanic and Atmospheric Administration show colored areas where the phenomenon known as La Niña typically causes abnormal weather conditions from December through February, *top*, and June through August, *bottom*. La Niña, a periodic cooling of the waters of the equatorial Pacific Ocean, alters normal weather patterns around the globe, but it works in reverse of its more common counterpart, El Niño, which is a periodic warming of equatorial Pacific waters. As predicted in the maps, the La Niña of 1998-1999 caused unusually cool and wet conditions that triggered record winter snowfalls in the Northwestern United States and abnormally dry conditions in Florida in the spring of 1999, which led to uncontrolled wildfires.

**December through February**

Cool  Wet  Wet and warm  Wet and cool

Warm  Dry  Dry and warm  Dry and cool

**June through August**

devastated parts of Nicaragua, Guatemala, Belize, and El Salvador. Whole villages and their populations were swept away by torrents of water and mud that came racing down mountainsides. Hundreds of thousands of homes were destroyed.

In addition to Mitch's death toll of about 11,000, some 20,000 or more people were missing. Accurate figures were difficult to arrive at because many victims were swept out to sea or buried beneath deep layers of mud. Mitch also left more than 3 million people homeless or otherwise severely affected.

**"Tornado Alley" twisters.** On May 3 and 4, 1999, a string of powerful thunderstorms spawned 59 tornadoes across portions of Arkansas, Kansas, Oklahoma, and Texas in a region of the United States known as "Tornado Alley." Oklahoma was the most severely hit, with 38 deaths across the state. In Kansas, 5 people were killed and nearly 150 were injured. One death was reported in Texas.

In the city of Moore, Oklahoma, a suburb of Oklahoma City, more than 200 houses were completely flattened. The level of destruction indicated that the twister that struck Moore was nearly 1 kilometer (0.6 mile) wide and generated winds of about 400 kilometers (250 miles) per hour, which qualified it as an "F5" tornado, a category that signifies "incredible" levels of wind damage on the Fujita Scale of tornado intensity. Early damage estimates from the storms exceeded $1 billion.

Scientists credited modern storm-prediction technology with helping save many lives. Using advanced weather radar and other resources, meteorologists at the Storm Prediction Center at the University of Oklahoma in Norman were able to track the deadly storm system from the time it developed. Local television and radio stations broadcast continuous reports that warned people to seek shelter.

**"El Niño of the century."** The strong El Niño of 1997-1998 caused scientists to reassess the history of the phenomenon. El Niño is a warming of the tropical Pacific Ocean that affects weather patterns worldwide. It appears that the 1980's and 1990's were a period of unusually persistent El Niño activity that also included two extremely strong events. The event of 1982-1983 had been termed the "El Niño of the century" based on weather

records and historical accounts. The 1982-1983 event took climate watchers by surprise and was almost over before significant measurements were collected. Its occurrence—and the great damage it caused worldwide—led to the establishment of modern global ocean monitoring and prediction systems. However, the 1997-1998 event was even larger, and so in 1999 scientists declared that it had unseated its predecessor as the most extreme El Niño of the 1900's—and perhaps in recorded history.

**Learning from El Niño.** The El Niño of 1997-1998 was the most intensely studied and documented El Niño ever. In 1999, researchers around the world were analyzing the immense amount of data collected from early 1997, when the first signs of a developing El Niño were recognized, through mid-1998, when conditions in the equatorial Pacific had given way to La Niña conditions. (La Niña, a periodic cooling of waters in the equatorial Pacific Ocean, works in reverse of the more common El Niño.)

The data included ocean temperature readings, wind speeds, wave heights in the Pacific, and rainfall totals from around the world. Also included were statistics on El Niño's impacts—floods in Africa, droughts in Indonesia, and scorching summer heat in the southern Great Plains of the United States. Weather experts estimated that by the time this El Niño had faded from the equatorial Pacific, it had caused thousands of deaths and billions of dollars in damage.

The appearance and behavior of the 1997-1998 El Niño caught many observers by surprise. The warnings that went out in the late spring of 1997 were based on observations of an El Niño already underway, not on predictions from scientists' favorite tools, computer *models* (simulations). According to oceanographer Antonio Busalacchi of NASA's Goddard Space Flight Center in Greenbelt, Maryland, this most recent El Niño started earlier, peaked at higher intensity, and dropped off more quickly than scientists had expected. Looking back, it became clear that most of the computer models that scientists had developed to predict El Niños, which were based largely on the last intense El Niño (1982-1983), performed poorly. Most failed to accurately predict the intensity of the 1997-1998 event, and some missed it alto-

**A deadly hurricane strikes Central America**
In a color-enhanced satellite image of Hurricane Mitch, the eye of the powerful storm, some 560 kilometers (350 miles) wide, hovers off the northern coast of Honduras on Oct. 26, 1998. Mitch came ashore two days later, carving a trail of devastation across Central America (map). The storm brought heavy rains that triggered flash flooding and mudslides, which swept away entire villages. The second-deadliest Atlantic storm ever, Mitch was blamed for approximately 11,000 deaths and billions of dollars' worth of damage.

gether. Indeed, the behavior of the latest El Niño raised questions about how much scientists really understood about the phenomenon.

The new observations enabled scientists to improve their models of El Niño. Perhaps the best existing model for forecasting El Niños was the one created by Mark Cane and Steven Zebiak, weather researchers at Columbia University's Lamont-Doherty Earth Observatory in Palisades, New York. In late 1997 and early 1998, tests of this model using weather data from early 1997 indicated that an El Niño would form. However, the model expected the phenomenon to be less intense and begin much later than the one that had actually developed. Cane and Zebiak realized that the wind measurements they had used in their model, which were based on readings taken by ships crossing the tropical Pacific, contained too little detailed information about what was occurring in some parts of the world. In addition, they found that by adding data about ocean tidal heights, the computer's El Niño more closely resembled the actual one. (Water expands

slightly when warmed. As a result, ocean tides tend to run a bit higher when a layer of warm water lies beneath the ocean surface. This is why tide levels are a reliable indication of undersea warming.)

However, much remained to be learned from the data collected during 1997 and 1998. Given its magnitude, this El Niño was likely to serve as the benchmark for climate system analysis for many years to come, providing data that scientists can use to both refine their understanding of the El Niño phenomenon and prepare computer models to predict the next event.

**Global warming: Natural or not?** In 1998 and 1999, the reason behind a trend of rising average global temperatures continued to be a hotly debated question among climate researchers around the world. The debate centered on the cause of the warming. Was it due to human activity or was it simply an instance of natural climatic variation?

Scientists were hampered in their efforts to resolve this question by the complex nature of Earth's climate system. That system includes not only the atmosphere but also the interactions of the atmosphere with the oceans and land surfaces and with the living things both on the land and in the oceans. In addition, reliable atmospheric temperature measurements go back only about 120 years. Temperatures from earlier times must be inferred from other evidence, such as tree rings, ice-core samples at the poles, and sediments on the ocean bottom. These factors limit the reliability of even the best computer models of the global climate system. They also make it difficult to estimate how much (if any) of the recent global warming trend has been due to human activities, including the burning of fossil fuels, such as coal, gasoline, and natural gas.

The burning of fossil fuels releases carbon dioxide into Earth's atmosphere. In the atmosphere, carbon dioxide traps infrared energy (heat) radiated by the Earth, which helps keep the planet warm. Most scientists believe that global warming caused by rising carbon dioxide levels increases the amount of water vapor in the atmosphere. Because water vapor is the main substance in the atmos-

phere that traps solar heat, this cycle leads to even more heat becoming trapped near Earth's surface. Although other factors appear to offset some of this warming, such as *aerosols* (tiny particles of matter suspended in the air) that produce a sun-blocking haze, the net effect appears to be a warming trend.

While most scientists accept the idea that an increase in carbon dioxide in Earth's atmosphere could raise global temperatures, many are unsure whether this effect can actually be seen in temperature data. Physical evidence shows that large climatic changes have occurred periodically in the Earth's past, even when humanity could not have played a part.

Other scientists considered the debate irrelevant. The real issue, they argued, is the rapid increase in global temperatures that climatologists have recorded since 1976. They said these measurements could be a signal that a period of significant global warming is underway.

[John T. Snow]

See also OCEANOGRAPHY. In the Special Reports section, see TURBULENCE: HIDDEN THREAT IN THE SKIES.

## Biology

The first "census" of the world's bacteria, published in August 1998, suggested that there are many more of the microorganisms than scientists had previously thought. Researchers at the University of Georgia in Athens reported that Earth is home to more than 5 million trillion trillion bacteria—that's a five with 30 zeroes after it. The vast majority of these microbes, an estimated 92 to 94 percent, live in ocean sediments or deep underground in soil and rock.

The scientists based their estimate on a review of scientific literature that had previously estimated the number of bacteria in individual habitats. Among these habitats were the ocean, soil, and air, as well as the surfaces of leaves and the bodies of animals.

**Biggest bacterium.** The discovery of a giant bacterium was announced in April 1999 by researchers at the Max Planck Institute for Marine Microbiology in Bremen, Germany. The bacterium, found in ocean sediment off the southwestern coast of Africa, can grow to a diameter of 0.75 millimeter (0.03 inch)— about the size of a period on this page.

To put this in perspective, the researchers said if the size of an ordinary bacterium were represented by a mouse, the new bacterium would be the size of a blue whale.

The scientists said the bacterium, *Thiomargarita namibiensis*, forms strands of individuals that resemble a pearl necklace. Most of the bacterium consists of a *vacuole* (storage chamber) holding large amounts of nitrogen-containing compounds that the bacterium uses in chemical reactions to obtain energy.

**How bacteria "talk."** The molecular mechanism by which some bacteria communicate with each other was described in February 1999 by molecular biologist Bonnie L. Bassler of Princeton University in Princeton, New Jersey. Her description was expected to help pharmaceutical companies design more effective drugs.

A crowd of bacteria often behaves differently than an individual bacterium. For example, bacteria that produce dangerous *toxins* (poisons) in humans and other animals do so only when they are present in large numbers, a phenomenon known as quorum sensing. The reason for this is that if the microbes were to release the toxin when there were only a few bacteria present, they would trigger an immune response from the host organism that would quickly wipe them out. Scientists knew that such bacteria communicate with each other to determine when to release the toxin, but they did not know how this communication is regulated.

To gain insights into a form of bacterial communication similar to the process that occurs in toxic microbes, Bassler studied two species of marine bacteria that are *bioluminescent* (capable of emitting light). These bacteria release chemical signals that are received by other bioluminescent bacteria and cause them to emit a blue glow when the microbes are present at a site in large numbers. Scientists presumed this glow is somehow important to the survival of the microbe group.

Bassler identified the *receptors* (structures on a cell that make contact with substances in the environment) that enable the bacteria to measure the quanti-

ty of the chemical signals being released by the group. When the bacteria detect a sufficient number of signals with the receptors—indicating a large population—the microbes start to glow.

Researchers said this finding could have medical applications because it might be used to hinder communication between harmful bacteria that have become resistant to antibiotics. If pharmaceutical companies could find drugs that block the action of bacterial receptors similar to the ones Bassler identified, those drugs might be able to prevent bacteria from releasing their disease-causing toxins.

**Octopus lights.** Suckers lining the arms of the deepwater octopus *Stauroteuthis syrtensis* are bioluminescent, flashing on and off like the lights of Broadway, marine biologists reported in March 1999. Although bioluminescence is common in many deepwater organisms and biologists had previously observed bioluminescent circles around the mouths of certain octopuses, the fact that some octopus suckers can glow came as a surprise to scientists.

**First photo of a rare species**
A wild saola stares into the camera in the first photograph ever taken of the elusive antelope, which was discovered in 1993 along the Vietnam-Laos border. The photograph, published in December 1998, was taken by researchers from Fauna and Flora International, a conservation organization based in Cambridge, England. A drawing of the animal, *inset,* reveals its unusual appearance.

**Pollen transfer on bird tongues**
A double-collared sunbird sips nectar from flowers of *Microloma sagittatum,* a South African plant. Botanist Anton Pauw of the University of Cape Town reported in August 1998 that when the bird drinks nectar from an *M. sagittatum* flower, pollen sacs become attached to the bird's forked tongue, *above.* The sacs are then transferred to the next flower the bird visits, where the pollen fertilizes seeds in the flower.

The biologists, led by Edith A. Widder of the Harbor Branch Oceanographic Institution in Fort Pierce, Florida, discovered that *S. syrtensis* has suckers that emit a blue-green light that can be detected at considerable distances underwater. The octopus, which lives off the East Coast of the United States, can make its suckers glow continuously for up to five minutes or flash on and off at intervals of one to two seconds.

Unlike the suckers of most octopuses, the suckers of *S. syrtensis* are not used for grasping prey. Instead, Widder believes, the glowing suckers serve to attract prey. Many ocean creatures tend to be drawn to light, partly because bioluminescent microbes in the fecal matter of fish and other marine organisms point the way to a rich source of nutrients. When the octopus's prey, tiny crustaceans called copepods, approach the animal, they may become trapped in mucus that is secreted near the octopus's mouth.

Widder speculated that this unusual feeding method may have evolved in *S. syrtensis* because its prey is much smaller than that of other octopuses, which typically catch fish and other larger organisms.

**Living fossils.** The discovery of a community of coelacanths in the Celebes Sea near the Indonesian island of Manado Tua was reported in September 1998 by biologist Mark V. Erdmann of the University of California at Berkeley. Coelacanths are large fish that were previously thought to exist only in one small group in the Comoros Islands off the eastern coast of Africa. Until the discovery of the Comoros Islands group in 1938, scientists had thought that coelacanths became extinct more than 80 million years ago. The discovery of the second group, almost 10,000 kilometers (6,200 miles) away from the first, raised the possibility that coelacanths may also exist elsewhere in the world's oceans.

Coelacanths are very distinctive fish. They grow to about 1.5 meters (5 feet) in length and have muscular, limblike fins, on which they rest on the ocean bottom. In addition, they give birth to live young rather than lay eggs, as most fish do.

# Life in Extremes

Drilling through sandstone more than 3 kilometers (2 miles) below the sea floor off the western coast of Australia, in one of the most hostile environments on Earth, geologists from the University of Queensland in Brisbane discovered the smallest organisms ever found. Philippa Uwins and her colleagues reported in March 1999 that thread-like bacteria—some as small as 20 *nanometers* (billionths of a meter) in diameter—were thriving at a temperature just over 150 °C (300 °F). Previously, scientists had thought that life could not exist in temperatures above 113 °C (235 °F).

The microorganisms, which the team called *nanobes* because of their small size, were in the same size range as the "fossilized bacteria" that researchers from the National Aeronautics and Space Administration reported finding in a meteorite from Mars in 1996. However, while the possible Martian bacteria were egg shaped, the nanobes more closely resembled filaments of fungi. Many scientists had argued that the Martian specimens could not be bacteria, in part because such tiny specimens had never been found on Earth. They speculated that chemical processes having nothing to do with life may have produced the markings in the rock.

Similarly, some experts were skeptical that Uwins had found living organisms. They argued that such a small organism cannot possibly contain all the biological machinery necessary for growth and division, the criteria for life. The smallest certified bacteria ever found on Earth—mycoplasma, which cause a type of pneumonia—are generally 150 to 200 nanometers in diameter. Nevertheless, the Australian discovery provided new support for the idea that the structures in the Martian meteorite may truly represent life.

Only nine months earlier, researchers from Montana State University in Bozeman had reported their discovery of microbes in an equally unlikely place at the opposite temperature extreme—the McMurdo dry valleys of Antarctica, 1,290 kilometers (800 miles) from the South Pole. The dry valleys are some of the most arid places on Earth, receiving no rainfall and little snow throughout the year. In summer, sunlight in the region is just strong enough to form pockets of liquid water in the valleys' frozen lakes, though the water temperature never rises above 1 °C (34 °F). In winter, the lakes freeze solid. The bacteria found by microbial ecologist John C. Priscu and his colleagues thrive in the 150 days of summer, then hibernate for the long, cold winter. These

Clumps of bacteria hibernate through a long winter in bubbles trapped in ice in Antarctica. The image was taken by microbial ecologists at Montana State University in Bozeman and released in June 1998. Such microbes, called *extremophiles* because of the environments in which they live, are changing scientists' views about the conditions under which life can exist.

bacteria include blue-green algae, *nitrogen-fixing* bacteria (bacteria that convert nitrogen into a form that can be used by other organisms), and microorganisms called heterotrophs that decompose dead algae.

Uwins's and Priscu's discoveries, however, were not completely unexpected. Since the 1970's, microbiologists had been finding life in environments thought too hostile for any organism—the boiling springs of Yellowstone National Park, deep underground in oil fields and solid basalt rock, in underground caves, and in high-temperature water vents on the ocean floor. The microorganisms found in such places, collectively called *extremophiles,* are changing scientists' ideas about where life can exist, how it began, and how common it may be throughout the universe.

Extremophiles include a broad range of species. At one end of the spectrum are tube worms that live with their tails immersed in the boiling water of deep-ocean vents. At the opposite end of the spectrum are strange microbes more closely related to plants and animals than they are to common *terrestrial* (earth-living) bacteria. These organisms are called *Archaea* (ancient ones), a name that emphasizes their strong resemblance to the first, extraordinarily simple microorganisms that came into being on Earth more than 4 billion years ago. Some researchers, such as microbiologist William B. Whitman of the University of Georgia in Athens, estimate that more than 90 percent of the microbes on Earth may be Archaea thriving in such hostile environments, particularly buried deep below the Earth's surface.

Many scientists now believe that life on Earth may have started at an ocean vent rather than in a shallow tidal pool, as the British naturalist Charles Darwin had suggested. The hot undersea plumes may have supplied the first microbes with the sulfur compounds, methane, and other naturally occurring chemicals that would have been their first source of energy. If that is the case, the odds increase greatly that life could also have arisen on Mars, Venus, and the moons of Jupiter. Although such planetary bodies receive little sunlight, they generate great amounts of internal heat because of gravitational forces exerted on them by other celestial objects. The sulfur-spouting volcanoes of Io and what may be ice-covered oceans on Europa—Jupiter's two innermost moons—could harbor the biochemical cousins of the Archaea or even species unlike anything scientists now know.

Beyond their great scientific interest, the microbes discovered in extreme conditions on Earth have great economic importance as well. As early as the 1880's, visitors to Yellowstone National Park observed that articles of clothing dropped into the hot springs emerged totally clean. Decades later, scientists who analyzed the springs found that microbes living in the scalding water produce enzymes that eat up stains in clothing. In the 1980's and 1990's, enzymes with other uses—such as in making perfume, brewing beer, and replicating genetic material for DNA fingerprinting—were found in the Yellowstone springs. In all cases, the key feature of the enzymes was their ability to carry out reactions at high temperatures in which previously used enzymes would be destroyed.

In 1997, the National Park Service gave permission to biotechnology companies to exploit microorganisms found in Yellowstone. By 1999, one firm, Diversa Corporation of San Diego, reported that it had identified more than 700 unique enzymes from Yellowstone and other hot springs— more than the total number of enzymes then in use by industry. (A federal judge suspended the Diversa contract in April 1999 until an environmental-impact study could be completed at Yellowstone.) Other companies have collected microbes from deep within caves throughout the world to produce chemicals that show promise as sources of antibiotics and anticancer drugs.

The more microorganisms thriving in extreme environments that scientists discover, the more they realize the vast numbers of such creatures that may remain to be found. As they continue to explore the varied forms that life can take, researchers hope to find ways to enrich life on Earth as well as to learn whether life beyond Earth is truly possible.                    [Thomas H. Maugh II]

Erdmann's wife first noticed a coelacanth in September 1997 in a fish market on Manado Tua. Erdmann then began interviewing local fishermen about coelacanths. Finally, in July 1998, he found a crew of fishermen that had captured one of the primitive fish.

The Indonesian coelacanths look much like the African fish, except that they are brown in color, while those from Africa are usually steel blue. Despite this similarity, genetic tests revealed in March 1999 that the two populations of coelacanths vary enough in their genetic makeup to be classified as different species.

**Blinded sea-floor shrimp.** British researchers reported in March 1999 that scientists visiting deep-sea hydrothermal vents may be harming some of the very organisms they are there to study. Hydrothermal vents are volcanic fissures on the ocean floor that spew hot, mineral-rich water into the cold waters around them. Microorganisms thrive on these minerals, and the microbes, in turn, support many other organisms, including shrimp.

The scientists concluded that some of these shrimp may have become blinded by the floodlights of research submersibles. The eyes of shrimp and other organisms living at the vents are extremely sensitive in order to enable the animals to see in the virtually lightless vent surroundings. However, scientists use submersibles with bright floodlights to study the vent communities.

Biologist Peter J. Herring and his colleagues at the University of Leicester in England collected two species of shrimp from vents in the Atlantic Ocean previously visited by other researchers. They found that the shrimp's eyes were severely damaged, some showing the complete destruction of eye pigment.

Although the team could not prove that the shrimp were blinded by scientists' floodlights, many researchers acknowledged that the findings seemed to be more than coincidental. Because of this, some deep-sea scientists took steps in 1999 to establish ocean sanctuaries that would be off-limits to most hydrothermal-vent research.

**Drifting iguanas.** Animals can move from island to island, and perhaps even from continent to continent, by floating on natural tree rafts, according to find-

Fisherman Om Lameh Sonathan poses with a coelacanth shortly after catching the big fish in Indonesian waters in July 1998. The fish was the first evidence that coelacanths exist in other locations than the eastern coast of Africa, where they had been discovered in 1938. Until the 1938 discovery, scientists had thought that coelacanths became extinct 80 million years ago. Biologist Mark Erdmann of the University of California at Berkeley reported the discovery of the second population in September 1998.

ings reported in October 1998. *Herpetologist* (reptile expert) Ellen J. Censky of the Connecticut State Museum of Natural History in Storrs said a group of at least 15 iguanas had apparently traveled 322 kilometers (200 miles) from the Caribbean island of Guadeloupe to the island of Anguilla in 1995 on trees uprooted by a hurricane.

Researchers have long speculated about how animals reach isolated islands or travel from one continent to another. Among the several possibilities that have been suggested by scientists is rafting on natural debris. The Anguilla discovery marked the first time such rafting had been documented.

Fishermen on Anguilla noticed the iguanas landing on the island shore in a large clump of floating trees in October 1995. Native iguanas on Anguilla are brown and have a plain tail. The newly arrived group, on the other hand, were blue-green and had rings around their tails. Censky checked prevailing currents in the area and concluded that the iguanas, which normally live in trees on Guadeloupe, were swept out to sea by

one of two hurricanes that passed through the region in September 1995.

Although she tracked down 15 of the iguanas, she said she suspected there were more. The new arrivals appeared to be weak, dehydrated, and hungry, and some were injured. However, at least one of the females has since given birth, suggesting that the animals would establish a self-sustaining population on the island.

**Bird daddies.** Fatherhood is a major role for the male wattled jacana, a tropical Central American bird, though he may not be the genetic father of all the young he raises. That finding was reported in December 1998 by a team of researchers led by biologist Stephen T. Emlen of Cornell University in Ithaca, New York. Emlen and his colleagues found that more than 40 percent of the broods tended by the male jacanas they studied contained chicks sired by a different father.

The scientists studied the long-legged wading birds over a six-year period on the Chagres River in Panama, observing more than 1,400 matings between the large, aggressive—and promiscuous—females and the much smaller, submissive males. They obtained their results by performing genetic analyses on tissue samples from 465 adult and juvenile jacanas.

The researchers noted that after the male wattled jacana mates, he is left to incubate the eggs and raise the chicks for three months. Meanwhile, the female jacana mates with other males. These other males too will end up incubating clutches of the female's eggs, eventually forming a harem of male birds tending chicks that may or may not be their own offspring.

Why the male jacanas tolerate raising young that are the offspring of other birds is uncertain. However, the scientists speculated that each male seems to understand that the female will provide no care at all to the chicks, and if he abandons them, they will all die—including the few that are his own offspring. (At least some of the young he cares for are likely to carry his genes.) The researchers also noted that because female wattled jacanas mate with as many males as possible, every male gets a chance to leave some offspring. This may provide an advantage over other bird mating systems, in which many males never get a chance to mate.

**Bat sonar.** The sonar of bats—a system for detecting and locating objects by the reflection of sound waves—is much more efficient than scientists had previously suspected, according to researchers at Brown University in Providence, Rhode Island. The investigators, led by neuroscientist James A. Simmons, reported in October 1998 that bat sonar is better than the best sonar equipment produced by humans.

Bats use their sonar to locate night-flying insects, such as moths. While flying, a bat emits a continuous series of high-frequency sound waves, which bounce off of objects and are reflected back to the bat as echoes. Echoes reflected by a flying insect give the bat information about the movement of the insect, as well as the insect's direction and distance.

The scientists trained captive bats to respond to pulses of sound with large gaps between the pulses, giving the bats a reward of mealworms for their performance. Then, the researchers gradually shortened the gaps between the pulses and observed whether the bats still responded to the sounds.

Using this method, the scientists concluded that the bats are able to discriminate between echoes arriving as little as 2 microseconds (2 millionths of a second) apart. This is an ability roughly three times better than scientists had previously believed possible. A two-microsecond gap between echoes means that a bat can tell the difference between two positions that differ by only the width of a pencil line. Simmons noted that the best electronic sonar used by the United States Navy can detect echoes no less than 6 to 12 microseconds apart. The researchers hoped that a better understanding of how bats discriminate between echoes will help electronics experts produce more effective sonar systems for the military.

**Popular poplars.** Researchers at the University of Georgia in Athens announced in October 1998 that they had developed trees that can remove toxic mercury from polluted soil. However, many environmentalists expressed concern that this solution to cleaning up mercury contamination might be just as bad or even worse than the problem.

Many mercury compounds, created by industrial processes such as the manufacture of electric switches and thermometers, are very toxic. If accidentally consumed or inhaled, they can damage brain cells. Some kinds of bacteria have the ability to absorb toxic mercury compounds and convert them into less harmful substances. However, cleaning up a polluted site with these microorganisms would require a very long time.

Botanist Clayton L. Rugh and his colleagues used genetic engineering techniques to insert a bacterial gene for breaking down mercury into the cells of a fast-growing tree, the yellow poplar. The gene enables the tree to convert the toxic mercury, picked up through the tree's roots, into a less harmful form, which evaporates into the air from the tree's leaves.

Although this may clean up polluted soil, the release of mercury into the air concerned many environmentalists. They said the mercury vapors could pose a serious threat to anyone who lived near a site being treated.

[Thomas H. Maugh II]

Here are 16 important new science books suitable for the general reader. They have been selected from books published in 1998 and 1999.

**Astronomy.** *Strangers in the Night: A Brief History of Life on Other Worlds* by David E. Fisher and Marshall Jon Fisher, a father and son writing team, recounts the history of scientific research on extraterrestrial life. The Fishers proceed from the work of the American astronomer Percival Lowell, who thought he saw canals on Mars, through the planetary missions of the U.S. National Aeronautics and Space Administration (NASA) in the 1960's and 1970's, to the SETI (Search for Extraterrestrial Intelligence) project of the 1990's. (Counterpoint, 1998, 348 pp. illus. $25)

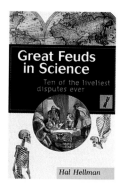

Great Feuds in Science
Ten of the liveliest disputes ever
Hal Hellman

**Biology.** *Time, Love, Memory: A Great Biologist and His Quest for the Origins of Behavior* by Jonathan Weiner, Pulitzer Prize-winning author of *The Beak of the Finch,* introduces readers to Seymour Benzer, an American biologist. Benzer's work has revealed that genes shape not just how animals look but also how they behave. For more than 30 years, Benzer and his students worked with fruit flies, identifying genes that govern biological rhythms, learning ability, and sexual behavior. While the application of these results to humans is only in its infancy, the linkage of genetics with the workings of the body and the mind is one of the great dramas of modern biology. (Knopf, 1999, 320 pp. illus. $27.50)

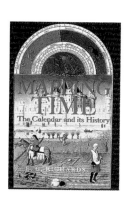

MAPPING TIME
The Calendar and its History
E. G. RICHARDS

*Throwim Way Leg* by Tim F. Flannery, an Australian biologist who studies exotic animals in the wilds of New Guinea, describes 20 years of Flannery's field work. The title refers to a phrase in New Guinea pidgin English meaning "to begin a journey." Flannery discovered new species, battled pythons, and endured tropical diseases. Most significantly, he hobnobbed with remote cultures that as late as the 1930's knew nothing of the outside world. Flannery offers insights into the last area on Earth to remain largely untouched by modernization. (Grove/Atlantic, 1998, 336 pp. illus. $25)

**Chemistry.** *Molecules At An Exhibition: Portraits of Intriguing Materials in Everyday Life* by John Emsley, a chemist, examines more than 80 of the elements commonly found in our homes and offices. Each element is the subject of a brief es-

say, which includes information on the element's usefulness, properties, and history and on the compounds it forms. Emsley includes discussions on such substances as methyl mercaptan, the primary ingredient of bad breath, and sodium azide, the explosive used to inflate auto air bags, as well as molecules as familiar as water. (Oxford University Press, 1998, 250 pp. $25)

**General Science.** *What Remains to Be Discovered* by John Maddox, a long-time editor of the magazine *Nature,* surveys a wide range of scientific topics that are likely to occupy the frontline of research in the 2000's. The problems are large ones—the unification of the theories of relativity and quantum mechanics, the nature of consciousness, and the origin of galaxies and of life itself—but the prospects are exciting. Maddox not only explains the science but also argues for the importance of each of the fields in which breakthroughs are thought to be imminent. (Free Press, 1998, 384 pp. $26)

**History of Science.** *Great Feuds in Science* by veteran science writer Hal Hellman centers on 10 passionate debates by some of the greatest figures of science. Hellman includes brief histories of Galileo, Isaac Newton, Charles Darwin, and other noted scientists, who argued about such thorny issues as the structure of the universe, the age of the Earth, and the origin of our species. (John Wiley & Sons, 1998, 240 pp. $24.95)

*Mapping Time: The Calendar and its History* by E. G. Richards, a former biophysics teacher at the University of London, explores the development of the most important calendars of the world. From the calendar of the Egyptians through those of the Aztecs and Maya to the Gregorian calendar of the present day, Richards explains the problems of attempting to impose a human framework on the flow of time. For computer buffs, Richards includes instructions on how to convert from one calendar system to another. (Oxford University Press, 1998, 460 pp. illus. $27.50)

**Mathematics.** *Towing Icebergs, Falling Dominoes, and Other Adventures in Applied Mathematics* by Robert B. Banks, a former professor of engineering, shows how simple calculations can transform numbers into insights about the world around us. He considers such questions as: How does the energy of an atom bomb compare with that of a falling meteorite? and Is there a maximum height a human being can attain? Banks explores such topics at an introductory calculus level. (Princeton University Press, 1998, 328 pp. illus. $29.95)

*The Man Who Loved Only Numbers* by Paul Hoffman, a science writer and the former editor in chief of *Discover* magazine, recounts the life of Paul Erdos, one of the most brilliant—and most eccentric—mathematicians of the 1900's. Erdos lived out of a suitcase, relying on the hospitality of his colleagues to obtain housing, transportation, and the amenities of daily life. He spoke in a technical jargon of his own invention, calling children "epsilons," the mathematical symbol for a small quantity. Hoffman offers insights into the man and his contributions to mathematics. (Hyperion, 1998, 289 pp. illus. $22.95)

**Medicine.** *Blood: An Epic History of Medicine and Commerce* by Douglas Starr, a professor of journalism at Boston University, discusses the invention and development of transfusions. Starr shows how, spurred on by two world wars, science learned to use blood efficiently and safely, despite thorny economical, medical, and social problems that still affect its use. Scarcely 100 years ago, the transfusion of blood was a rare and risky medical procedure. Today, transfusions are routine procedures, and blood has become the keystone of our medical system, more precious a commodity than oil. (Knopf, 1998, 441 pp. illus. $27.50)

**Physics.** *Nothingness: The Science of Empty Space* by physicist Henning Genz, explores such questions as Does nature, as the ancient Greeks claimed, "abhor a vacuum"? Or can there be space, perhaps between atoms, in which there is absolutely nothing? Genz reveals that even the emptiest space is filled with energy-bearing fields and with particles that flash in and out of existence like sparks in a fireworks display. (Perseus Books, 1999, 340 pp. illus. $30.00)

*Physics in the Twentieth Century* by journalist Curt Suplee describes the remarkable advance of modern physics and illustrates the process with more than 200 pictures. In the early 1900's, the atom was unexplored territory, the extent of our Milky Way Galaxy was unknown, and the only computers were people who added columns of numbers. As we enter the 2000's, mammoth particle accelerators split the atom into ever tinier particles, space telescopes see almost to the edge of the universe, and supercomputers reveal the workings of natural processes ranging from the weather to the explosive death of stars. (Harry N. Abrams, 1999, 223 pp. illus. $49.50)

**Psychology.** *Visual Intelligence: How We Create What We See* by psychologist Donald D. Hoffman explains how we really see not with our eyes but with our brains. While the eye records patterns of light and darkness, it is the brain that turns these patterns into a perception of physical reality. Hoffman reveals the brain's secrets by analyzing optical illusions (many included in the book), color and depth, and the nature of reality and perception. (W.W. Norton & Company, 1998, 294 pp. illus. $29.95)

**Space exploration.** *This New Ocean: The Story of the First Space Age* by William E. Burrows, a science journalist and a professor at New York University, reviews the history of rocketry and space travel. Burrows traces space flight from the earliest experimental rockets to the 1960's race to the moon and the development of reusable spacecraft such as the U.S. space shuttle in the 1970's and 1980's. (Random House, 1998, 723 pp. illus. $34.95)

**Technology.** *Glass* by William S. Ellis, a former writer and editor for *National Geographic* magazine, takes readers on a tour of the many places where glass has been made, used, and appreciated. Ellis shows that glass has become indispensable to modern society. Glass is a major component of computer screens, light bulbs, home insulation, auto bodies, optical fibers, and telescope mirrors, just to name a few of its many uses. (Avon Books, 1998, 306 pp. illus. $25)

*Visions of Technology,* edited by Pulitzer Prize-winning author Richard Rhodes, explores the controversy over whether the many advances in technology that occurred during the 1900's have served mostly to improve human life or to enslave people. Rhodes illustrates the debate with quotations by notable figures such as inventor Orville Wright, industrialist Henry Ford, and authors Kurt Vonnegut and George Orwell. (Simon and Schuster, 1999, 400 pp. illus. $30)

[Laurence A. Marshall]

A team of Russian and American scientists announced in January 1999 that they had apparently created chemical element number 114—the heaviest element yet produced. Atoms of the new element had 114 *protons* (positively charged subatomic particles) in their *nucleus* (central region). The work was done at Russia's Joint Institute for Nuclear Research near Moscow.

The scientists, whose achievement needed to be confirmed by additional research, reported that nuclei of the unnamed, radioactive element remained intact for 30 seconds before decaying, or splitting up, into the nuclei of lighter elements. In contrast, several other heavy elements that scientists had created in recent years lasted for only fractions of a second.

The relatively long life of element 114 confirmed predictions made by chemical theorists that such a superheavy element would lie within an "island of stability," surviving long enough to enable researchers to study chemical compounds formed by it. The discovery also suggested that other stable superheavy elements would eventually be found by researchers.

Physicist Yuri Oganessian of the Joint Institute for Nuclear Research led the group that created element 114. American scientists in the group were from the Lawrence Livermore National Laboratory in Livermore, California.

The researchers made their discovery with a cyclotron, a type of *particle accelerator*, a device that uses intense magnetic fields to accelerate atoms or subatomic particles to extremely high speeds and energies. With this apparatus, the scientists bombarded atoms of a radioactive isotope called plutonium-244 with atoms of another radioactive isotope called calcium-48. Isotopes are different forms of an element, with nuclei containing the same number of protons but different numbers of *neutrons* (subatomic particles with no charge). The collisions caused atoms of the two isotopes to fuse into larger atoms.

When the nuclei of these large atoms disintegrated, the scientists concluded (based on the nature of the decay products) that the atomic fragments most

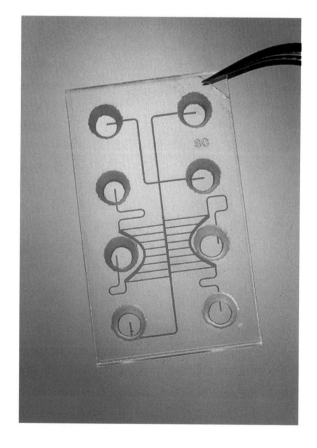

A pair of forceps holds a miniature glass chip that has tiny channels and reaction chambers etched into it. The chip, developed by Caliper Technologies of Palo Alto, California, was one of several such devices being tested by researchers in 1999 at microscopic-chemistry laboratories. The researchers reported that the etched microchips were capable of analyzing chemical solutions with greater accuracy and in much less time than is possible using beakers, test tubes, and other large equipment of conventional chemical laboratories.

likely came from element 114. However, because the decay products were somewhat difficult to identify, the scientists could not prove that element 114 had been created. Nevertheless, most chemists agreed that the group's conclusion was correct. Meanwhile, investigators at the Lawrence Berkeley National Laboratory in Berkeley, California, and the Institute for Heavy Ion Research in Darmstadt, Germany, said they too would try to produce the new element. The German researchers had been the first to create elements 107 through 112. (The heaviest naturally occurring element is number 94, plutonium.)

**Molecular machines.** The dream of molecule-sized machines that could be used to assemble complex chemicals or store data in computers came a step closer to reality in January 1999. Researchers at New York University (NYU) in New York City reported that they had constructed a simple machine from deoxyribonucleic acid (DNA), the double-stranded molecule that carries the genetic blueprint for all organisms. The leader of the research team, NYU chemist Nadrian C. Seeman, said the tiny device was able to rotate its two armlike projections by as much as 6 nanometers (6 billionths of a meter, or 0.2 millionths of an inch) in response to chemical changes in a solution. Previous molecular devices developed by researchers had only been capable of sideways motion of about 1 nanometer.

Seeman's group chose DNA as the building material for its machine because each DNA strand contains chemical groups called bases, which naturally form links to the bases of other strands. This makes it possible to assemble lengths of DNA into a wide variety of sizes and shapes.

DNA strands are normally not stiff enough to be capable of doing mechanical work. The researchers solved this problem by preparing an unusually rigid form of DNA called double-crossover DNA (DX DNA), which contains two double-strands of DNA instead of one. They then connected these rigid, armlike DX fragments to each other with a normal DNA molecule that acted as a bridge.

When the scientists introduced positively charged *ions* (electrically charged atoms) of cobalt into a water solution

containing copies of the assembly, the ions caused the bridge DNA of each assembly to twist into a different shape. In the process, the bridge DNA rotated one of the DX DNA arms to the opposite side of the bridge.

The scientists said their research might lead to the development of *nanorobots* (molecular-scale robotic devices) capable of fitting together simple chemical building blocks to create complex molecules. Many such molecules are impossible to make in the large reaction vessels that have long been used for the manufacture of chemicals.

Engineers also envision DNA machines as possible memory storage units for tremendously powerful computers. Each tiny machine, depending on the way it was twisted, would represent either a 0 or a 1 in the digital language of computers, in which all data is converted into strings of 0's and 1's.

**Nanotube arrays.** An improved method for depositing parallel and uniform arrays of *nanotubes* (microscopic, hollow cylinders of carbon atoms) on sheets of glass was announced in November 1998 by scientists at the State University of New York (SUNY) in Buffalo and Sandia National Laboratories in Albuquerque, New Mexico. The researchers, led by SUNY physicist and chemist Zhifang Ren, said their achievement might pave the way for new types of low-cost flat-panel displays for televisions and computers.

Although other scientists had previously deposited arrays of nanotubes on flat surfaces, their methods required high temperatures, typically above 700 °C (1,292 °F). This meant that the nanotubes could only be deposited on expensive, heat-resistant *substrates* (surfaces) made of mesoporous silica, a silicon and oxygen compound. The high cost of the silica substrates ruled out large-scale applications of the arrays.

With their new method, the researchers were able to deposit nanotubes at temperatures lower than 666 °C (1,231 °F), the temperature at which glass begins to soften. The technique thus made it possible to lay down the nanotubes on substrates of ordinary, relatively inexpensive glass.

To create the aligned arrays of nanotubes, the researchers heated a mixture of acetylene—a gas containing atoms of carbon and hydrogen—and ammonia by feeding it into a chamber filled with a *plasma* (a hot, electrically charged gas). The heat caused the acetylene to decompose, leaving clusters of closely packed nanotubes standing upright on a nickel-coated glass surface, which was also in the chamber.

The process used by the scientists was similar to procedures used by previous researchers, except that those investigators had used nitrogen instead of ammonia, and mesoporous silica instead of glass. Ren's group speculated that the ammonia and nickel acted as *catalysts* (compounds that speed up reactions), causing the acetylene to decompose into nanotubes under slightly milder conditions than normal.

The scientists said their nanotube arrays could be induced to emit streams of *electrons* (negatively charged subatomic particles) from their ends. They said these electrons, in turn, could be used to bombard a flat screen coated with *phosphors* (light-emitting chemicals that are widely used in television screens) to create images.

Nanotube arrays on glass, the researchers noted, could be made cheaply enough to become the key components of flat-panel TV displays, which people could hang on their walls like paintings. The arrays might also be fabricated into laptop computer displays that would be clearer and easier to view from any angle than the liquid-crystal display screens with which laptops of the 1990's were equipped.

Yet another potential application of the arrays, according to the investigators, would be to improve the quality of images produced by scanning electron microscopes, instruments in which a probe scans the surface of a specimen with a beam of electrons. In these microscopes, electron-emitting nanotubes would be grown on the tip of the probe.

**Porous polymers.** Chemists at Pennsylvania State University in State College reported in February 1999 that they had prepared a group of unusual polymers, materials made of long-chain molecules. The polymers, made of carbon and hydrogen compounds called hydrocarbons, contained countless pores, all of a

Seven fullerene molecules (yellow)—each a hollow ball made of 36 carbon atoms—surround a single molecule of buckminsterfullerene (blue), which consists of 60 carbon atoms. Researchers at the University of California at Berkeley reported in June 1998 that they had isolated the 36-carbon fullerene, the first hollow carbon molecule to have fewer atoms than buckminsterfullerene, from a carbon-rich gas. The scientists said the 36-atom molecule was chemically more reactive and easier to make into useful materials than other fullerenes.

uniform, microscopic size. Previous polymer research had been unable to create such tiny pores. The new materials had many potential applications, including the purification of drugs.

To make the porous polymers, the Penn State team, under the direction of chemist Thomas Mallouk, started with extremely tiny spheres of the mineral silica. Using a powerful laboratory press, the scientists compressed millions of these spheres to form a dense solid in which all the spheres were closely packed in an orderly arrangement.

The researchers used this solid as a mold, forcing liquid *monomers* (chemical building blocks of polymers) into the microscopic spaces between the spheres. Next, they chemically treated this monomer-saturated solid to transform the monomers into polymers. In the final step, the scientists dissolved the silica with an acid.

What remained after this treatment was a polymer material containing a pore where each of the silica spheres had been. There were also many channels connecting the pores. In different experiments, the scientists were able to vary the diameter of the pores by making the polymers out of different mixtures of monomers.

The pore sizes of the new polymers, together with the ability to tailor the pore diameters, could make the polymers useful tools in industry. For example, the researchers noted that the polymers could perhaps be used by the pharmaceutical industry to separate mixtures of *chiral molecules*. These are molecules that exist in both right- and left-handed forms—molecular shapes that are mirror images of each other. Typically, only one form of a chiral molecule is useful as a drug. However, drugs produced in industrial reaction vessels often have equal amounts of both the right- and left-handed forms, and separating the two kinds of molecules from each other is time consuming and expensive. Mallouk said it might be possible to tailor the polymer pores to capture only the desired right- or left-handed molecules as the drug mixture streams through the polymer channels.          [Gordon Graff]

## Computers and Electronics

In early 1999, the retail price of some personal computers (PC's) dropped below $500 for the first time—in some cases considerably below—resulting in a dramatic alteration of the computer marketplace. The April 1999 introduction of the $499.95 Webzter and the $299.95 Webzter Jr. PC's from Microworkz Computer Corporation of Seattle, Washington, provided consumers with powerful computers that were capable of connecting to the Internet and running state-of-the-art applications *software* (programs), but for a price less than half that of comparable computers. The Webzter included a CD-ROM drive and other features not included with the Webzter Jr. Neither PC included a monitor, which had to be purchased separately. Consumer response to the Webzter Jr. was so strong that the Microworkz site on the World Wide Web received more than 7 million visits in the first week that the Webzter line became available.

**PC for free.** An even smaller price tag—nothing—was announced by Free-PC, Inc., of Pasadena, California, in late 1998. The computers to be given away were brand-name machines that came with a monitor, CD-ROM drive, and other features, as well as free Internet access. In return for the free computer, participants in the program agreed to use the computer at least 10 hours each month and give the company detailed personal information, including income, shopping and spending habits, hobbies, birthdays, and other data. Free-PC would use this information to develop a customized advertising and marketing profile of each consumer. A portion of the computer screen was devoted to advertising space. The ads were to be updated automatically each time the user accessed the Internet, but they remained on the screen whenever the computer was running. Free-PC planned to derive its revenues from sales of this advertising space. The company planned to distribute its first batch of 10,000 computers beginning in June 1999.

**iMac invigorates Apple.** Ironically, the bestselling PC through mid-1999 was one of the more expensive models, the iMac from Apple Computer Corporation of Cupertino, California. The iMac was the latest in Apple's Macintosh line of personal computers. Like the original Macintosh PC, which debuted in 1984, the iMac featured an all-in-one design, with the monitor and other components built into a single case. Apple shipped more than 1 million of the new computers to retailers in the eight months following its debut in August 1998. The first iMacs came only in blue, but in early 1999 Apple introduced four new colors for the line. Demand for the iMac resulted in shortages of some of the more popular colors and surpluses of the less popular ones. A typical iMac cost $1,199.

The popularity of the iMac and the PowerMac G3, a Macintosh laptop model aimed at business users, helped solidify Apple's return to profitability less than two years after many business analysts had predicted the company's demise.

**Music controversy.** The Internet and technologies associated with it offered a variety of opportunities for new businesses in 1998 and 1999. Among the most popular was the sale and distribution of music over the Internet. While Internet transmission of sound files had been popular since the introduction of the World Wide Web in the early 1990's, it was the growth in popularity of a new standard for storing and retrieving those files that raised the stakes in the Internet music business.

Storing audio and video information can result in very large data files. The larger the file, the more space it takes to store the file on a computer's hard disk and the longer it takes to transmit and receive it over the Internet. The new file standard was an evolution of existing MPEG (Moving Pictures Experts Group) standards for compressing audio, video, or multimedia files into smaller files that made it more practical to store, transmit, and receive high-quality audio, making it a perfect medium for recorded music. The new format was known as MP3 (for MPEG 1, Audio Layer 3).

Even though software compression techniques built into MP3 enabled a song to be stored in a fraction of the space required by other formats, the audio quality of MP3 approached that of a compact disc (CD). The decrease in file size without much loss in sound quality

In November 1998, Diamond Multimedia Systems, Inc., of San Jose, California, began marketing the Diamond Rio PMP300, *right,* a portable music player. The Rio played music that had been saved as computer files in a compressed format called MP3. Users download MP3 files from the Internet to a personal computer and then transfer them to the Rio's memory chips. Because illegal MP3 copies of copyrighted music recordings were widely available for free on the Internet, many people in the music industry worried that legitimate music sales would be hurt by MP3 players.

worried many people in the recording industry when MP3 versions of songs began to appear—often without the consent of the artist or record company—on Web sites throughout the world. Some industry groups filed copyright-infringement lawsuits against distributors of unauthorized MP3 music.

The controversy surrounding MP3 intensified in September 1998, when Diamond Multimedia Systems, Inc., of San Jose, California, announced its intention to market the Rio PMP300, a portable MP3 player. The Rio could store 30 minutes of MP3-formatted music downloaded from the Internet by a PC and then transferred to the Rio. The suggested retail price for the device was $199.95.

However, the release of the Rio was delayed when a federal judge issued a temporary restraining order in response to a complaint by the Recording Industry Association of America (RIAA) and the Alliance of Artists and Recording Companies (AARC). These groups alleged that the Rio violated the federal Home Audio Recording Act of 1992, which protects record companies from the illegal duplication of copyrighted material and prohibits the sale of certain types of recording devices in the United States.

When the restraining order expired in November 1998, Diamond began to market the Rio, and the RIAA and AARC moved forward with a lawsuit. In April 1999, Diamond announced plans to cooperate with the recording industry by incorporating technical measures designed to limit the ability of the Rio to play illegally copied MP3 files.

**Internet growth continues.** The willingness of computer manufacturers like Free-PC to give away hardware in return for advertising exposure illustrated the explosive growth of Internet advertising and its increasing importance as a complementary medium to more traditional advertising arenas like television, radio, and print. By some estimates, Internet advertising expenditures more than doubled in 1998, passing the $1.5-billion dollar mark. This made the Internet a larger advertising marketplace than outdoor advertisements such as billboards.

The growth in advertising was only one indication of the phenomenal popularity of the Internet and the World Wide Web, the portion of the Internet able to present animation, video, sound, and other multimedia information. It was estimated that in 1999 more than 150 million people around the world had access to the Internet.

Actual Internet traffic—both the number of people using the Internet and the volume of data being exchanged—was increasing even more rapidly. Some estimates placed annual Internet traffic growth in excess of 1,000 percent. Much of that traffic was focused on commerce, as Internet-based companies offered various products for sale on their Web sites. Among the most popular companies were Amazon.com, which specialized in book sales but in early 1999 began selling music, home video, and other products; eBay.com, an Internet auction service; Buy.com, which sold a variety of products; and Priceline.com, which specialized in discount airline tickets.

**Electronic stock market.** The rapid expansion of various business aspects of the Internet—advertising, retail sales, business-to-business sales, services, and other transactions—all of which together came to be referred to as e-commerce (electronic commerce), resulted in wild speculation in the stocks of Internet companies. Even though most of these companies had never made a profit, shares of many Internet stocks soared. Although many business analysts warned that Internet stocks were almost hysterically overpriced and predicted a collapse in their value, investors did not seem concerned. Internet stocks fueled much of the unprecedented growth in the stock market during 1998 and early 1999.

In fact, the buying and selling of stocks had itself moved to the Internet by the end of 1998. By mid-1999, according to Piper Jaffray Inc., a Minneapolis, Minnesota, investment firm, nearly one-third of all stock and mutual-fund trades were being made over the Internet, with approximately 7 million online brokerage accounts established at the end of 1998.

However, e-trading had its share of problems. Several online brokerage firms experienced system crashes that caused investors to lose money when their trades were not executed properly. In early 1999, two class-action lawsuits were filed in California against E*Trade, an online investment brokerage based in Palo Alto, California. The suits accused E*Trade of launching an aggressive ad campaign to attract clients even though the company

# Zapping the Y2K Bug

By 1999, a seemingly harmless programming short-cut left over from the early days of the computer age had snowballed into an apparent crisis that threatened to disrupt computer systems the world over. The possibility had many people very worried, because computers had become so vital to modern society. It was difficult to imagine any institution, private or public, that did not rely on computers to manage information. Vital government records, military secrets, airline schedules, financial documents—all were in computer files. The question was, would the world's computer systems be able to avoid an "information meltdown" when the calendar turned over to the year 2000?

The problem occurred because many important computer systems still relied on *hardware* (electronic components) and *software* (programs) dating from as far back as the mid-1960's that were not designed with the year 2000 in mind and so would not recognize it. The looming crisis became known as the Millennium Bug, the Year 2000 Problem, or simply Y2K (computer jargon for "Year 2000").

Y2K stemmed from a time when even the most advanced computers boasted just a fraction of the memory and processing power found in the average desktop computer of 1999. Software engineers, therefore, wrote programs that were as small and efficient as possible. One widely used space-saving technique was to denote the year with just two digits instead of four. For example, the year 1979 was represented as "79." In the early days of computing, this shortcut resulted in significant savings of memory and processing time. Although programmers realized that this method would cause major problems if a computer had to deal with years that did not begin with "19," many institutions chose to ignore the problem, assuming that the programs would not still be in use 20 or 30 years later. Unfortunately, that was an incorrect supposition.

For various reasons—mainly cost—many computer system administrators continued to put off plans to make their systems *Y2K compliant*—updating them with the ability to handle four-digit years. Beginning in about 1994, as the magnitude of the problem became clear, industry and government leaders realized that the issue could be ignored no longer. Technicians realized that reliably fixing the problem meant checking and updating literally every line of software code and every microchip that could not recognize a four-digit year. This was an enormously costly and time-consuming task, but in many cases it simply had to be done.

Keeping track of dates is essential to the proper operation of most computer systems and to the accuracy of information that they manage. Bookkeeping programs, for example, not only calculate financial balances and other figures but also track such date-related matters as payroll records and payments due. If left unremedied, an outdated system might execute improper instructions or simply become confused and shut down, with potentially disastrous results. For example, a city's power distribution system could malfunction, leaving whole communities without electricity. On a smaller scale, an older accounting program used at a bank might misread a mortgage payment due on Feb. 3, 2000, as being due on Feb. 3, 1900 and mark the account as overdue. If the error goes undetected, the bank might begin proceedings to foreclose on the property based on the computer's report that the homeowner had made no payments for 100 years.

Making Y2K fixes on the software being used to run many of the large mainframe computers still in wide use in the late 1990's was complicated by the fact that many of these programs had been written in computer languages that were no longer used. As a result, many programmers had to be lured out of retirement to help fix the problem.

Fixing hardware problems was even more difficult because it often involved replacing a special type of computer processor known as an embedded chip. Embedded chips are tiny dedicated computers that have the instructions for performing their tasks "hard-wired" into their circuits. In 1999, these chips were built into devices ranging from microwave ovens to nuclear power plant controls. Because noncompliant chips could not be reprogrammed, they had to be physically replaced. Fortunately, though there were billions of embedded chips in use throughout the world, most of them did not need to be Y2K compliant in order to function properly, because their functions did not involve dates. Nevertheless, the large number of outdated embedded chips requiring replacement made it impossible to guarantee that every one would be replaced by Jan. 1, 2000. As a result, most experts anticipated power outages, failures of telecommunications systems, and other service interruptions. Of particular concern were electronic devices used in medical equipment, emergency-response systems, and other essential services.

As news of the Y2K problem spread, owners of personal computers (PC's) worried about whether their machines would be affected. Fortunately, all Macintosh computers and most PC's built after 1997 were ready for the coming of 2000. Most software companies posted statements on the Internet telling whether a particular version of software was Y2K compliant or not.

The United States led the world in heading off a

Fears of malfunctions in computers not equipped to recognize the year 2000 were of particular concern to the airline industry in 1999. Although many industry experts thought it was unlikely that the problem would jeopardize flight safety, many predicted at least some difficulties in computerized reservation and ticketing systems.

Y2K crisis. In 1997, federal officials began a review of every U.S. government computer system, as well as those of vital industries. By mid-1999 the most crucial of the government's computer systems were well on their way to being ready for the turn of the century. Some industries had begun addressing the problem early on and were also prepared. By 1999, the banking, securities, and airline industries in the United States had all expressed confidence that they would be ready for New Year's Day 2000.

Making all the necessary changes for Y2K compliance was expensive. Analysts anticipated that the costs would be too high for some businesses—and even some countries—to bear. In October 1998, the Gartner Group, an information-technology consulting firm, estimated that the worldwide costs associated with becoming Y2K compliant would be about $1 trillion to $2 trillion, with U.S. costs being $150 billion to $225 billion.

While the United States and other wealthy nations were making great strides in addressing Y2K in 1999, many poorer nations had not even begun. In Asia and much of South America, for example, an economic crisis in 1998 had strained financial resources. Many observers expected the impact of Y2K to be largest in poor and economically besieged nations.

By mid-1999, speculation about what would happen when the New Year arrived ranged from mild concern to serious anxiety. Some observers predicted temporary fuel shortages and service inter-ruptions, while others foresaw far more serious and long-lasting disruptions. Those pessimists anticipated significant problems for even the most well-prepared governments and businesses.

The likelihood of severe Y2K problems in developing nations, where a great percentage of manufacturing work for international firms takes place, led some financial experts to predict a global *recession* (economic slowdown) as a result of Y2K. Analysts also warned that financial worries would cause many people to withdraw assets from banks and investment firms, perhaps causing an economic crash. Concern over the availability of fuel—and the mechanisms for managing its distribution—led many analysts to expect problems in the transportation of goods, which would in turn interrupt the distribution of food and merchandise.

Among the public, concern about the Millennium Bug caused some people to stockpile food in preparation for a complete breakdown of civilization. Although few experts foresaw the large-scale collapse of civil order, virtually none expected a painless transition to the New Year. One point that almost everyone could agree on, however, was how utterly dependent the world had become on computers during the 1980's and 1990's. Many commentators had taken notice of this dependence over the years, but it took the Y2K problem to drive home just how far it had gone. However, the full extent of the problem would not be known until 12:01 a.m., Jan. 1, 2000.          [Keith Ferrell] knew its systems could not handle heavy trading volumes during peak hours.

**Internet viruses cause concern.** Many Internet users in 1998 and 1999 became concerned about the spread of computer viruses. Computer viruses are destructive programs that copy themselves and spread from computer to computer, either over the Internet, by exchanges of floppy disks, or through other file-sharing methods. While all viruses represent the unauthorized introduction of new data into a computer system, some were designed merely as pranks and caused little or no damage to the systems they infected or to the information in them. Others, however, posed a real threat to computer systems.

One virus, named Melissa, caused minimal damage but spread more rapidly than any previous computer virus. Melissa began infecting systems worldwide on March 26, 1999. It sent a list of pornographic Web sites to each of the e-mail addresses listed in the victim's address book and inserted a quote from a popular animated television series into some word processing documents. It then e-mailed itself to other recipients using the victim's e-mail account. Melissa was allegedly created by a 30-year-old New Jersey man, who was arrested within a week of Melissa's appearance.

Chernobyl, a far more damaging virus, was activated on April 26, 1999, the 13th anniversary of an accident at the Chernobyl nuclear power plant in Ukraine, which was then part of the now-defunct Soviet Union. Allegedly created by a Taiwanese *hacker* (computer programmer who purposely develops harmful programs and illegally breaks into computer systems), Chernobyl destroyed the data on hard disks and rendered some computers inoperable. While virus-protection software—updatable over the Internet to deal with new viruses—protected many computers in the United States, hundreds of thousands of PC's in Asia were affected by the Chernobyl virus.

**A new way to watch TV.** The evolution of digital technologies such as file compression and digital storage devices also began to influence another popular entertainment medium—television. In March 1999 TiVo, Inc., of Sunnyvale, California, introduced TiVo Personal TV. A similar service, ReplayTV, from Replay Networks, Inc., of Mountain View, Cali-fornia, made its debut the following month. Both services offered a new way to watch and record television programs, because they were used along with a set-top box that contained computer chips and hard-disk storage.

The devices recorded video and audio information *digitally* (as strings of 1's and 0's) onto a hard disk for later viewing. The digital format enabled a viewer to pause, rewind, or replay a program even while it was being recorded. The devices could also be customized by the viewer to automatically record particular programs or even certain types of programming, such as horror movies.

ReplayTV offered three versions, ranging in price from $699 to $1,499, with storage capacities ranging from 10 to 28 hours. TiVo came in a 14-hour version priced at $499 and a 30-hour version for $999, excluding a fee for the service.

**Microsoft suit.** The United States government's suit against the Microsoft Corporation of Redmond, Washington, continued throughout 1998 and the first half of 1999. The case, brought by the U.S. Department of Justice, alleged that Microsoft, the world's largest software company, had violated federal antitrust laws by incorporating its Internet Explorer *Web browser* (software used for viewing sites on the World Wide Web) into its Windows *operating system* (the master program that controls all of a computer's basic functions). Because Windows was far and away the most popular operating system in the world, the government argued that Microsoft was exploiting the popularity of Windows to unfairly promote the sale and use of Explorer over other browsers, including its chief rival, Navigator, developed by Netscape Communications Corporation of Mountain View, California. As of mid-1999 there was no clear resolution of the case in sight, but many legal observers felt that Microsoft's attorneys had failed to present a persuasive defense of the company's practices.

As the Microsoft case dragged on, two developments emerged that could make the outcome of the suit almost meaningless. One was the purchase of Netscape by online service provider America Online, Inc., (AOL) of Dulles, Virginia, the world's largest online service. The other was the growing popularity of an operating system called Linux, which some industry analysts believed could one day

Visitors to an electronics store in Japan in February 1999 try out the new iMac computer, the latest offering from the Apple Computer Corporation of Cupertino, California. "Grape" was the most popular new color for the iMac, and many retailers in both Japan and the United States quickly sold out of purple models. On the other hand, "tangerine" and "blueberry" iMacs did not sell nearly as well.

threaten the dominance of Windows.

In November 1998, AOL acquired Netscape for $4.2 billion. Because of AOL's dominance of the online services market and its vast financial resources, some business analysts expected the purchase to revitalize Netscape's challenge to Microsoft in the browser market. As of February 1999, AOL had more than 16 million members plus another 2 million members of the AOL-owned CompuServe online service.

Linux (pronounced *LIHN ucks*) was created in 1991 by Linus Torvalds, a student at the University of Helsinki in Finland. Based on Unix, a widely used operating system initially developed for mainframe computers, Linux was developed by Torvalds in cooperation with programmers all over the world, working together over the Internet. Also using the Internet, Torvalds made the source code for Linux available to other programmers at no charge, encouraging them to use it as they wished. Such an approach to the distribution of software codes is called open-source programming. In contrast, programmers who wished to de-

velop applications for many other operating systems had to obtain an expensive source-code license or sign restrictive agreements that often limited what they could do with the source code.

Since its introduction, Linux had achieved great popularity in the computer community, particularly for running *network servers,* central computers on a network that store and distribute information to other, less powerful machines on the network. Linux offered all of the functionality of costly commercial operating systems and was thought to be in some ways more dependable. In addition, Linux ran on many different types of computers, unlike other operating systems, which were generally usable with only one type of machine.

In an attempt to make Linux more appealing to the general public, a growing number of software companies in 1998 and 1999 were developing a *graphical user interface* (a system that employs groups of icons manipulated by a mouse to operate the computer) for Linux, similar to the ones used by the Windows and Macintosh operating systems.     [Keith Ferrell]

Biologists from the United States Fish and Wildlife Service (USFWS) encountered serious problems in 1998 and 1999 in their efforts to reestablish wild populations of wolves. In November 1998, the biologists captured two lone male Mexican gray wolves, the last of 11 wolves that were released in March in Apache National Forest in Arizona and Gila National Forest in New Mexico. The other wolves, which had all been raised in captivity, were previously recaptured after straying too close to human settlements or were the victims of fatal shootings. Except for one wolf that was shot by a camper, authorities did not know who killed the animals.

Despite the dangers, biologists rereleased the two males in December 1998 after pairing them with new mates. Four additional wolves were released in March 1999, but one of these was soon killed by an automobile.

The Mexican gray wolf had been eradicated from the Southwestern United States by the 1970's after decades of trapping and poisoning, primarily by ranchers who regarded the animals as a threat to livestock. By 1999, fewer than 200 Mexican gray wolves remained. All of these animals, except for the few released in 1998 and 1999, were captives in zoos or breeding facilities in the United States or Mexico. Officials hoped to eventually establish a population of 100 Mexican gray wolves in the wild.

**Red wolf program halted.** In October 1998, USFWS biologists called off their effort to restore red wolves to Great Smoky Mountains National Park in Tennessee and North Carolina. Biologists had released 37 red wolves in the park between 1992 and 1996, but most of these animals died from various causes or were recaptured after straying onto private land. In addition, all but 2 of 30 pups produced by the wolves died from disease, parasites, or starvation or were killed by predators. These losses spurred biologists to remove the remaining four wolves in late 1998.

As with the gray wolf, shooting, trapping, and poisoning helped push the red wolf to the brink of extinction. Another factor that threatened the red wolf as a species was a process called *genetic swamping*, in which red wolves bred with coyotes to produce coyotelike *hybrids* (offspring of two different species).

By the 1960's, the last wild population of pure red wolves survived in the wetlands of southwestern Louisiana and southeastern Texas. Those animals were captured in the 1970's to begin a captive-breeding program in zoos and other facilities. The breeding program produced more than 300 offspring, which provided ample stock for releasing red wolves into protected areas in the wild.

Although the captive breeding program was a success, conditions in Great Smoky Mountains National Park proved unfavorable to the wolves. The main problems, according to biologists, were a shortage of prey, such as rodents, rabbits, and deer, and a high risk of conflict with humans. In 1999, USFWS biologists were searching for alternative areas in the Southeast to renew the reintroduction program. Their goal was to establish at least three wild populations of red wolves totaling at least 220 animals.

**Lynxes released.** In yet another reintroduction program, Colorado Division of Wildlife officials released 13 Canada lynxes in the San Juan Mountains of southwestern Colorado in February and March 1999. Biologists planned to release several more lynxes during the year and a total of more than 100, captured in Canada and Alaska, over a period of three years. Despite the fact that a number of the animals had died of starvation by mid-1999, the plans continued to move forward.

Lynxes were last seen in Colorado in the early 1970's. Conservationists attributed the cat's decline in the state to such factors as trapping for its pelt and habitat destruction.

**Viagra and endangered species.** Many conservationists expressed hope in 1999 that Viagra, a drug that became available in the United States in 1998 to treat *impotence* (the inability to maintain an erection), would eventually lessen the world's demand for animal body parts sold to increase potency. Unlike preparations made from such body parts as rhinoceros horns and tiger genitals, which are used to treat impotence in some East Asian cultures, Viagra has been proven to be effective in aiding male sexual function. However, in order for the drug to have a positive effect on the preservation of endangered species, conservationists stressed, it needed to become more available in Asia.

## Animals to be removed from Endangered Species List

In 1999, the United States Fish and Wildlife Service planned to remove several animal species from the Endangered Species List or reclassify them from "endangered" to the less serious status of "threatened." Department biologists said the species had reached sufficient numbers to no longer need the protection of the endangered status, which bans hunting and other human activities that could drive a species to extinction.

Gray wolf pups drink from a water pan at a wildlife refuge in Minnesota, *left.* Conservation programs increased the number of wolves in Minnesota, Michigan, and Wisconsin from a few hundred in the early 1970's to about 2,600 in 1999.

An American peregrine falcon rests on a perch, *right.* In the early 1970's, there were only about 320 nesting pairs of these birds in North America. Breeding programs and the banning of pesticides that weakened the falcon's egg shells, killing the young before they hatched, raised the peregrine's numbers up to almost 1,600 pairs by 1999.

### Species slated for list removal or reclassification

| Species | Range |
| --- | --- |
| Aleutian Canada goose | Alaska, Washington, Oregon, California, Canada |
| American peregrine falcon | North America |
| Ash Meadows Amargosa pupfish | Nevada |
| Bald eagle | 48 continuous United States |
| Brown pelican | Texas, Louisiana, Mississippi, Alabama, Florida |
| Columbian white-tailed deer | Washington, Oregon |
| Dismal Swamp southeastern shrew | Virginia, North Carolina |
| Gray wolf | Minnesota, Michigan, Wisconsin, Wyoming, Montana, Idaho |
| Guam broadbill (bird) | Guam |
| Hawaiian hawk | Hawaii |
| Island night lizard | California |
| Mariana mallard | Northern Mariana Islands |
| Oahu tree snail | Hawaii |
| Pahrump poolfish | Nevada |
| Tidewater goby | California |
| Tinian monarch (bird) | Northern Mariana Islands |
| Virginia northern flying squirrel | Virginia, West Virginia |

Rhinos and tigers are foremost among the endangered species whose body parts are valued in Asia as sources of sexual prowess. Conservationists have tried to protect rhinos from *poaching* (illegal killing) by surgically removing their horns. However, some scientists object to this practice because hornless rhinos appear to behave abnormally.

Conservationists believe that rhinos and tigers would be able to recover if poaching could be better controlled. Although Viagra has potential as an unlikely antipoaching agent, observers expected that many Asians would resist using it due to their belief in the effectiveness of traditional remedies. However, conservationists drew hope from the fact that Viagra was much more affordable than powdered rhino horn and other animal-derived potency products.

**Prehistoric Australian extinctions.** An analysis of fossilized egg shells in Australia revealed in January 1999 that *Genyornis newtoni*, a giant flightless bird, suddenly disappeared 50,000 years ago, about the time humans are thought to have arrived on the remote continent.

This finding, by researchers led by geologist Gifford H. Miller of the University of Colorado in Boulder, indicated that humans were the likely cause of the bird's extinction. The researchers said their study also supported the idea that prehistoric humans in various parts of the world may have caused the extinction of many other species of *megafauna* (large-bodied animals).

If true, this conclusion would show that the ability of humans to have a serious impact on the natural environment is not confined to advanced societies and, in fact, predates the technological progress usually blamed for environmental destruction.

Besides flightless birds, the extinct Australian megafauna included horned tortoises the size of cars, giant lizards, flesh-eating kangaroos, and mammoths. Although scientists knew that about 85 percent of Australian animals weighing more than 45 kilograms (100 pounds) disappeared, they did not know precisely when that happened. The lack of a date complicated the search for why the megafauna extinctions occurred.

**Plan to restore the Florida Everglades**
In October 1998, the United States Army Corps of Engineers unveiled a plan to restore the ecological health of the Everglades, a large area of wetlands in southern Florida. Over the previous several decades, the corps had built an extensive system of canals, *below and right,* to prevent flooding and provide water to urban centers in southern Florida. However, this water-control system severely reduced the Everglades' natural flow of water, drastically altering its plant and animal communities. The 20-year, $8-billion plan called for removing many canals and levees and capturing much of the water flow (presently channeled to the sea) in underground reservoirs for later release back into the Everglades. In 1999, the corps was working on details of the plan, which it hoped to begin implementing by 2002.

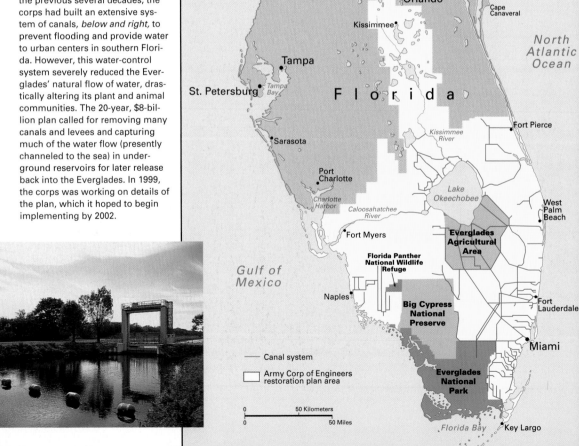

- — Canal system
- ☐ Army Corp of Engineers restoration plan area

To help pin down the date of the extinctions, the scientists focused their research on *G. newtoni,* which stood 1.5 meters (5 feet) tall and left large fossilized egg shells buried throughout central Australia. The researchers used four dating methods, including radiocarbon dating, to arrive at an age for the eggshells. Radiocarbon dating takes advantage of the fact that carbon-14, a radioactive form of carbon present in all organisms, *decays* (breaks down) at an exact, uniform rate over time. By measuring the amount of carbon-14 remaining in a specimen, researchers can determine the age of the specimen.

The dating analysis found that all the eggshells discovered at three separate sites were about 50,000 years old, indicating—because no younger egg remnants have been found—that *G. newtoni* had been extinct for the same length of time. Because scientists estimate that humans first arrived in Australia from Indonesia about 50,000 years ago, the carbon-14 evidence strongly suggested that humans played a role in the extinctions.

The researchers speculated that fires started by humans for such purposes as hunting and signaling may have led to the Australian extinctions. The scientists said human-caused fires would probably have occurred more often than the natural fires to which the native vegetation was adapted. Many of the shrubs and trees eaten by *G. newtoni* may not have been able to survive the frequent fires, leading to the extinction of the giant birds and other browsing animals. With the disappearance of these animals, many creatures that preyed on them would also have perished.

The scientists said additional research would be needed to confirm this theory. They noted, however, that this scenario underscored the harm that even simple human activities can do to wildlife, a finding with implications for primitive societies today. They also said their research promised to shed light on the disappearance of mammoths and other large animals in North America 11,000 years ago—a time when humans were beginning to settle that continent.

**A new look at dams.** In a July 1998 ceremony, U.S. Interior Secretary Bruce Babbitt struck a sledgehammer against the McPherrin Dam on Colorado's Butte Creek, heralding the dam's destruction. The ceremony highlighted an effort by the federal government, state governments, sport-fishing interests, and environmentalists to remove dams that pose a threat to desirable fish such as salmon and trout. After noting the environmental benefits of dam destruction, Babbitt remarked, "The clang of sledgehammer on concrete rings in a new era of watershed restoration."

Babbitt headed a long line of conservationists taking a second look at dams in the United States, where 75,000 large dams blocked 965,000 kilometers (600,000 miles) of what had once been free-flowing rivers. Another 2 million smaller dams blocked lesser waterways.

Although dams have historically been viewed as symbols of progress, they also cause many environmental problems. For example, dams often block salmon migrations to *spawning* (egg-laying) areas. Even when adult salmon successfully bypass dams, many of their *smolts* (offspring) stray into irrigation canals or face the hazards of turbines during their travels downstream to the ocean.

Besides posing a threat to migratory fish, dams also harm many species of freshwater mussels, which are unable to escape to other locations when dams are built. Moreover, dams interfere with the nutrient content of waterways, seasonal patterns of water flow, and water temperatures—all of which result in environmental changes to which many aquatic organisms cannot adapt.

In addition to McPherrin Dam, a number of other dams—including ones on Bear Creek in Oregon, the Neuse River in North Carolina, the Kennebec River in Maine, the Baraboo River in Wisconsin, and the Naugatuck River in Connecticut—were removed in 1998 and 1999. The U.S. Army Corps of Engineers was also studying the possible environmental benefits of eliminating or modifying four large dams on the Snake River in Washington State.

Conservationists noted that the operating licenses for 250 hydroelectric dams built in the 1930's and 1940's were scheduled to expire by 2008. They said this might prompt new questions about the continued existence of these river-altering structures—questions that reach beyond kilowatt hours to the survival of salmon, shad, sturgeon, crayfish, and mollusks.          [Eric G. Bolen]

Notable people of science who died between June 1, 1998, and June 1, 1999, are listed below. Those listed were Americans unless otherwise indicated.

Vladimir Demikhov

**Demikhov, Vladimir** (1916?-Nov. 22, 1998), Russian surgeon who is credited with performing the first animal heart and lung transplants. Demikhov conducted the first heart transplant, on a dog, in 1946. The animal survived, with two hearts, for five months. In 1947, Demikhov performed the world's first lung transplant, again operating on a dog. He conducted the first coronary bypass, on a dog, in 1952. His coronary bypass procedure has since become widespread in human medicine. Christiaan Barnard, the South African physician who performed the first human heart transplant in 1967, credited Demikhov's research and experience as providing the foundation for human transplant surgery. In the 1950's, Soviet officials referred to Demikhov's experiments as "tricks." The value of his work, however, was officially recognized in 1998 when Demikhov was awarded the Order for Services for the Fatherland, one of Russia's highest honors.

Gertrude Belle Elion

**Elion, Gertrude Belle** (1918-Feb. 21, 1999), drug researcher who shared the Nobel Prize in physiology or medicine in 1988 with her collaborator George H. Hitchings. Elion, who did not have a doctoral degree, began working with Hitchings in 1944. Together they developed drugs for the treatment of gout, herpes, leukemia, malaria, and immune disorders. They also perfected drugs designed to block transplanted-organ rejection through suppression of the immune system. Upon her retirement in 1983, Elion became involved in the development of AZT, the first drug created to treat HIV, the AIDS virus. Then-President George Bush described Elion's work as having "transformed the world" when he presented her with a National Medal of Science in 1991.

**Frankel, Sir Otto** (1900-Nov. 21, 1998), Australian scientist who focused worldwide attention on the need for genetic diversity. Genetic diversity refers to the variety of genes present in plant or animal species. Upon his retirement from the Commonwealth Scientific and Industrial Research Organization in 1966, Frankel led an effort, jointly sponsored by the United Nations Food and

Tetsuya Fujita

Agriculture Organization and the International Biological Program, to assess the speed at which genetic diversity was being lost and how that loss in crop plants was affecting the world's food supply. The scientific community credits Frankel with keeping the issue of genetic diversity alive in the 1960's and 1970's through his books and the international scientific conferences he organized.

**Fujita, Tetsuya** (1920-Nov. 19, 1998), Japanese-born meteorologist who invented a scale for rating tornadoes by wind speed and damage. The Fujita scale, devised in 1951, classifies tornadoes on a hierarchy from F0—"light" with winds 64 to 116 kilometers (40 to 72 miles) per hour—to F6—"inconceivable" with winds 513 to 610 kilometers (319 to 379 miles) per hour. Fujita, who directed the University of Chicago Wind Research Laboratory, discovered a wind type that he named the "microburst." Microbursts are spreading downdrafts of air capable of knocking airplanes from the sky.

**Herzberg, Gerhard** (1904-March 3, 1999), German-born Canadian physicist who received the 1971 Nobel Prize in chemistry for his contributions to the knowledge of the electronic structure and geometry of molecules, particularly *free radicals* (fragments of molecules that combine easily with other molecules). Scientists regarded Herzberg as the father of molecular spectroscopy, a technique used to chart the characteristic frequencies of light that molecules emit or absorb. Herzberg's spectroscopic method provided a valuable tool for the study of the chemical composition of objects in the solar system and the rest of the universe.

**Kendall, Henry W.** (1926–Feb. 2, 1999), physicist who shared the 1990 Nobel Prize for physics with Jerome Friedman and Richard Taylor for confirming the existence of the quark, a fundamental particle of matter. Working at the Stanford University accelerator in Stanford, California, Kendall and his colleagues fired electron beams into protons and neutrons and discovered that the beams were being deflected. The deflections proved the existence of quarks—particles of matter more fundamental than protons and neutrons. The existence of these subatomic parti-

Henry W. Kendall

Frederick Reines

Glenn Seaborg

cles had been theorized in 1964 by two American physicists, Murray Gell-Mann and George Zweig. In 1969, Henry Kendall cofounded the Union of Concerned Scientists, which opposes nuclear weapons and power plants.

**Reines, Frederick** (1918-Aug. 26, 1998), physicist who shared the 1995 Nobel Prize for physics for his detection of the neutrino. The existence of the neutrino, a neutral subatomic particle, was proposed in 1931 by Wolfgang Pauli to explain an apparent violation of laws of energy and momentum during nuclear decay. Most physicists believed the neutrino existed but was undetectable. Reines, working with Clyde Cowan at the Los Alamos National Laboratory in New Mexico, fed vast numbers of neutrinos, generated during nuclear reactions, into enormous tanks of highly purified water. When a neutrino struck an atomic nucleus in a water molecule, it produced a flash of light, which was detectable by instruments. Reines in 1956 published proof of the existence of neutrinos, which many researchers thought had no mass. In June 1998, Japanese physicists who had been using methods similar to those developed by Reines and Cowan reported that neutrinos apparently do have a tiny amount of mass.

**Schawlow, Arthur L.** (1921-April 28, 1999), physicist who received the 1981 Nobel Prize for physics for coinventing the laser, a device that produces a narrow beam of light. In 1958, Schawlow and coinventor Charles Townes published a scientific paper, "Infrared and Optical Masers," proposing that the principles of the *maser* (the microwave precursor of the laser) be extended to a device that would amplify light. Light Amplification by Stimulated Emission of Radiation—the laser—became a reality in 1960. It was eventually used for everything from eye surgery and weapons guidance systems to bar-code scanners and compact disc players, producing a multibillion-dollar industry.

**Seaborg, Glenn** (1912-Feb. 26, 1999), chemist and nuclear physicist who headed the team that discovered plutonium, a radioactive element that was used in the atomic bomb dropped in 1945 on Nagasaki, Japan. In 1951, Seaborg shared the Nobel Prize in chemistry with Edwin M. McMillan, an associate at the University of California at Berkeley,

for the discovery of plutonium and for the creation of the artificial elements americium, curium, berkelium, and californium. Seaborg, for whom Element 106 was renamed seaborgium, also contributed to the development of einsteinium and mendelevium. The elements created under Seaborg's direction were produced by bombarding natural elements with neutrons or other kinds of subatomic particles.

**Shepard, Alan B., Jr.** (1923-July 22, 1998), the first American to fly in space and 1 of 12 astronauts to walk on the moon. Shepard, a Navy test pilot who was chosen in 1959 as one of the original seven Mercury astronauts, became a national hero on May 5, 1961. Aboard a Mercury capsule mounted on a Redstone rocket, he was shot 185 kilometers (115 miles) into the air to skirt the edges of space on a near-perfect flight that lasted for 15 minutes. His words on splashing down—"Everything is A-OK!"—became a national catch phrase. On Jan. 31, 1971, Shepard took off for the moon, where he and fellow astronaut Edgar Mitchell spent 33 hours. Author Tom Wolfe describes Shepard in *The Right Stuff* (1979), a book about the original Mercury astronauts, as setting "a standard of coolness and competence that would be hard to top."

**Zoll, Paul** (1911-Jan. 5, 1999), heart specialist who demonstrated that electrical stimulation can restart the human heart. In 1952, Zoll successfully regulated the rhythm of a patient's heart with electrical charges through the chest. In the mid-1950's, a group under Zoll's direction at Beth Israel Hospital in Boston developed a permanent implantable *pacemaker* (an electronic device implanted near the heart to maintain normal rhythm). The pacemaker was first implanted in a human being in 1960. Zoll performed the first external cardiac *defibrillation* (application of electric shock to restore normal rhythm) in 1956. He developed continuous cardiac monitoring with an *oscilloscope* (an instrument that measures oscillations in electrical current) equipped with alarms. The device allowed nursing staffs to detect and treat irregular heart rhythms quickly. Zoll, who won the Albert Lasker Clinical Medical Research Award in 1973, did not profit from his inventions by patenting them. [Scott Thomas]

The United States Food and Drug Administration (FDA) in 1998 and early 1999 approved some 90 new drugs. Thirty of those drugs are referred to as new molecular entities (NME's) because they contain an active substance that had never before been available in the United States. Sixteen NME's received accelerated reviews by the FDA because they were thought to represent major advances in medical treatment.

**New therapy for arthritis.** The FDA in December 1998 approved the use of celecoxib (sold as Celebrex) for the treatment of osteoarthritis and rheumatoid arthritis, conditions that cause the deterioration of joints. Celebrex interferes with the activity of cyclooxygenase-2 (COX-2), an enzyme that plays a role in joint pain and inflammation.

Unlike drugs such as aspirin, ibuprofen, and naproxen, which reduce inflammation in general and are commonly used to treat arthritis, Celebrex works only by interfering with the COX-2 enzyme. Researchers found it to have no effect on an enzyme that protects the stomach lining. As a result, researchers believed that Celebrex would not irritate the stomach, as aspirin and other anti-inflammatory drugs often do.

By 1999, almost 40 million people in the United States—15 percent of the total population—had been diagnosed with arthritis. Many experts in mid-1999 expressed optimism that Celebrex would encourage arthritis patients to continue taking their arthritis medication. Patients have often discontinued taking drugs for arthritis because the drugs caused stomach upset.

However, in April 1999, reports submitted to the FDA did show a possible link between Celebrex and more than 20 cases of *gastrointestinal hemorrhages* (severe bleeding in the stomach or intestines)—including 10 deaths—since the drug was introduced in January. The FDA reported that although additional research into the overall safety of Celebrex would be conducted, the drug was not believed to pose a significant risk.

**More arthritis relief.** The FDA in September 1998 approved the use of leflunomide (sold as Arava) as an alternative to methotrexate, a drug commonly used in the treatment of rheumatoid arthritis in adult patients. Rheumatoid arthritis is an *autoimmune* disease—one

in which the immune system attacks the body's own cells. The effectiveness of Arava is related to the drug's ability to inhibit a key enzyme involved in the autoimmune process. The enzyme is believed to be linked to the overproduction of certain white blood cells, called T cells, involved in the destruction of joint tissue in rheumatoid arthritis.

Researchers cautioned that Arava may produce birth defects. Because it can linger in the blood for up to six months, it may pose a problem for women with rheumatoid arthritis who want to become pregnant.

**Reducing breast cancer risks.** In October 1998, the FDA approved the use of the drug tamoxifen citrate in healthy women to reduce their chances of developing breast cancer. The drug (sold as Nolvadex) has been used in the United States since the 1970's for the treatment of breast cancer.

Research conducted by scientists at the National Cancer Institute in Bethesda, Maryland, revealed that women who had been given tamoxifen every day for up to five years reduced their chances of developing breast cancer by 44 percent. All of the women in the study had been diagnosed as being at a high risk for developing the disease.

Tamoxifen became the first drug to be used in the United States for the prevention of breast cancer. Because of potentially serious side effects, however, including an increased risk of cancer of the uterus and the development of blood clots, the FDA said that tamoxifen should only be used in women with a very high risk of developing breast cancer later in life.

**New breast cancer treatment.** Trastuzumab, sold under the name Herceptin, was approved by the FDA in September 1998 for the treatment of breast cancer. Herceptin, a genetically engineered *antibody* (disease-fighting molecule of the immune system), became the first drug specifically designed to attack cancer at its genetic roots. Herceptin targets breast cancer cells producing abnormally large amounts of a protein called human epidermal growth factor receptor 2, or HER2. Lesser amounts of this protein signal cells to grow and divide normally.

Twenty-five to 30 percent of women with breast cancer that has spread to

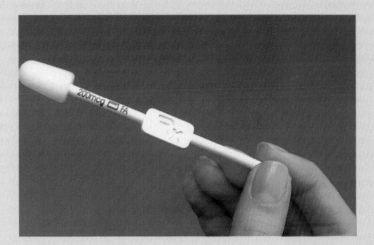

In November 1998, the United States Food and Drug Administration approved a lozenge containing the drug fentanyl to combat the pain experienced by some cancer patients. The stick-mounted lozenge, sold under the brand name Actiq, was designed to dissolve in the mouth for quick entry of the drug into the bloodstream. Because the lozenge resembles a lollipop, health-care professionals warned users of the drug to keep it out of the reach of children.

other tissues and organs also have high amounts of the HER2 protein in their blood. Scientists believed that the overabundance of the HER2 protein makes the cancer spread faster.

Herceptin interacts with the HER2 protein of breast cancer cells, thereby depriving these cells of the stimulation required for growth and development. Following exposure to Herceptin in clinical studies, breast tumors in some women shrank or disappeared.

Although Herceptin is considered generally safe, the FDA cautioned that some patients who take the drug may develop a weakened heart muscle. This condition, known as cardiomyopathy, can lead to congestive heart failure.

**Managing cancer pain.** Fentanyl, a potent drug related to morphine, has been used for several years to treat cancer pain. In November 1998, the FDA approved the use of this powerful painkiller in the form of a raspberry-flavored lozenge on a plastic stick, sold under the brand name Actiq. Patients obtain pain relief by sucking on the lozenge. As the lozenge dissolves, the fentanyl is ab-

sorbed by the lining of the mouth and quickly enters the bloodstream.

The FDA acknowledged that because Actiq strongly resembles a candy lollipop, it may be attractive to a child. However, a large dose of the drug can be fatal to children. As a result, the fentanyl lollipops will come with child-resistant wrapping in an effort to prevent its misuse by children.

Approximately 50 percent of cancer patients experience a type of pain that is so severe that it will not respond to standard pain-relieving medications. Healthcare experts hoped that Actiq would provide relief for many such patients.

**A new drug for obese patients,** orlistat (sold as Xenical), was made available in April 1999 after the FDA approved its use. Xenical works by blocking an enzyme that absorbs dietary fat in the gastrointestinal tract. The drug was recommended for patients who are at least 20 percent overweight and have conditions that are normally aggravated by obesity, including high blood pressure, high cholesterol, or diabetes.

[Thomas N. Riley]

Biologists in 1999 searched for methods to control the spread of a fast-growing species of algae that had been expanding its range throughout the northern Mediterranean Sea during the 1990's. The algae, which biologists believed may have entered the sea in water from an aquarium in the area during the 1980's, had taken over parts of the sea floor that were previously home to diverse communities of sponges, corals, and other organisms.

The sea otter population of western Alaska has declined by 90 percent since 1990, a group of ecologists led by James A. Estes of the United States Geological Survey reported in October 1998. The researchers said that as a consequence of this decline, the area's *kelp forest*—a community of shallow-water organisms dominated by large, brown seaweed—had been devastated. These changes highlighted the interdependence of species in *ecosystems* (groups of organisms and the environment they live in).

According to the report, the crash in sea otter numbers began with overfishing by commercial trawlers, which caused the stocks of perch, herring, and other nutritious fish in the northern Pacific Ocean to decline. These fish are normally the food of marine mammals such as Steller's sea lions and harbor seals. The populations of these mammals declined along with the fish. Next to be affected were killer whales, which usually feed on sea lions and seals but were now forced to prey on sea otters.

The researchers found that the increased hunting of sea otters by killer whales was the direct cause of the otters' decline. They said the kelp forests began to be affected when sea urchins, organisms that the sea otters normally eat, began to multiply and to feed on the kelp in increased numbers.

Many ecologists said that the sea otter study was an unusually vivid example of how human interference in nature can cause unanticipated consequences. Despite the detailed description of the sea otter decline provided by Estes's research team, however, some scientists argued that the chain of events was even more complicated than the report had stated. They raised the possibility that Pacific fish stocks may have plummeted not just because of overfishing but also because of higher ocean temperatures caused by a shift in deep-sea currents since the mid-1970's. According to this scenario, fish populations in the region declined when conditions became too warm for them, and this ultimately led to the drop in sea otter numbers.

**Pleistocene Park.** An international research team released 32 wild horses into a nature preserve in northeastern Siberia in September 1998. The release of the horses, overseen by ecologist Sergei Zimov of Russia's Northeast Sci-

Yakutian horses investigate their new surroundings after being released in September 1998 in a nature preserve in northeastern Siberia. Ecologists envisioned the preserve as a place where the grassland that was present in the region during the Pleistocene Epoch (2 million to 11,500 years ago) might be re-created, replacing the mosses, trees, and other plants of today. They hoped the horses would help alter the plant community by churning up the ground with their hoofs and fertilizing the soil with their dung, much as Pleistocene mammoths did.

entific Station, was the first step in an ambitious effort to re-create the *mammoth steppe,* a type of grassland that existed in northeastern Siberia during the Pleistocene Epoch. That epoch, which lasted from 2 million to 11,500 years ago, was marked by a series of ice ages.

The mammoth steppe was dominated by large mammals, such as woolly mammoths—animals related to modern elephants—woolly rhinoceroses, and bison. These animals helped preserve the grassland by churning up the ground with their hoofs. This made it more difficult for other plants, such as mosses and trees, to take root and spread. In addition, the dung from these animals provided fertilizer for the grasses, which in turn were grazed on by the animals.

Many scientists believe that prehistoric humans hunted the large Pleistocene mammals to extinction. Without the beneficial role provided by these animals, the grassland gave way to a *tundra-taiga* ecosystem—one dominated by mosses, lichens, and evergreen trees. Today, grasses grow only in small patches in northeastern Siberia.

Zimov and his colleagues, from the United States and Canada, planned to bring the steppe back in a nature preserve called Pleistocene Park, encompassing 160 square kilometers (60 square miles) of tundra-taiga. The ecologists hoped that large modern grazing mammals, including horses and bison, would play the same ecological role as the extinct Pleistocene animals and would restore the mammoth steppe.

The scientists faced numerous challenges to the success of their plan, however. First, only a few types of animals can withstand the extremely severe winters of Siberia. Although the horses, a Russian breed known as Yakutians, can survive these conditions, the ecologists believed that American and European bison would be unable to do so. The scientists were thus planning to use woodland bison, a hardy species discovered in Canada in 1959, for their next release. In mid-1999, the researchers were seeking funding to transport a herd of these bison from Canada to Siberia.

A second problem that Zimov faced was the modern climate. Many scientists believe that at the end of the Pleisto-cene Epoch, climatic conditions became warmer—and perhaps wetter—causing a turnover in vegetation that resulted in the mosses, shrubs, trees, and other plants present today.

Zimov, however, disagreed with that view, insisting that it was the extinction of the large Pleistocene mammals that doomed the grassland. He maintained that the soil-churning and fertilizing ac-tivities of the Yakutian horses and wood-land bison would be able to bring back the mammoth steppe.

**Plant-root fungi.** The number of species of fungi living on grass roots in an area influences the number of species of grasses in that area, a team of ecologists led by Marcel van der Heijden of the University of Basel in Switzerland reported in November 1998. The scien-tists said this finding indicated that the preservation of underground fungal communities was a critical factor in the proper management of grasslands and other ecosystems.

This research followed a 1997 Cana-dian study that found that underground fungi called mycorrhizal fungi play an important role in distributing nutrients among different species of trees. Mycor-rhizal fungi form complex networks connecting the roots of different plants. The fungi obtain carbohydrates and other nutrients from the roots that the plants produce through *photosynthesis*, the process by which plants make car-bon-based matter from carbon dioxide and water, using the energy of sunlight. The plants, in turn, obtain nitrogen, phosphorus, and other minerals from the fungi. The Swiss research showed that it is not just the presence of the fungi, but also the diversity of fungal species, that is ecologically important.

Van der Heijden and his colleagues performed two sets of experiments—one in a greenhouse and another in the field. In both sets of experiments, they found that as they added more fungal species to the soil, a greater number of grass shoots and roots developed, and a wider variety of grass species could be sustained. The researchers hoped these findings would aid efforts to conserve land ecosystems.   [Robert H. Tamarin]

## Energy

A greatly improved battery for electric cars and trucks was being used in a Gen-eral Motors (GM) Corporation electric car introduced in December 1998. Pro-ponents of alternative forms of energy hailed the battery as a step toward plac-ing electric vehicles into widespread use for personal transportation. Owners of the GM electric car, the EV-1, expressed great satisfaction with the vehicle's over-all performance.

The EV-1 uses a nickel metal hydride (NiMH) battery, which stores twice the energy of a conventional lead-acid bat-tery of the same weight and volume. The NiMH battery generates an electric current by storing and later releasing hydrogen *ions* (electrically charged atoms). Each battery cell produces 1.2 volts of electricity, with 11 cells needed in each battery to provide the 13.2 volts used by the automobile. The complete battery is sealed in a stainless steel case, requires no maintenance, and can be recycled.

The EV-1 electric car is designed to accelerate from 0 to 100 kilometers (0 to 60 miles) per hour in under 9 sec-onds and can travel a distance of 120 to 225 kilometers (75 to 140 miles) on a full charge.

The manufacturer, the Ovonic Bat-tery Company of Troy, Michigan, pro-jected that battery production would in-crease through 1999 from the initial production schedule of just one per day. Ovonic also planned to introduce a more efficient battery by late 1999.

**Wind energy** continued to gain popularity in late 1998 and early 1999. In September 1998, the Enron Wind Corporation of Tehachapi, California, completed construction of the world's largest single wind-power generation fa-cility, the 107-megawatt Lake Benton I Wind Power Generation Facility near Lake Benton, Minnesota. The 1,900-hectare (4,700-acre) facility, located on a site called Buffalo Ridge, is owned and operated by Enron. The installation was designed to deliver electricity to the Northern States Power Company (NSP) of Minneapolis, the largest power utility in the upper Midwest.

Buffalo Ridge, which is 100 kilo-meters long, runs through southwestern

AeroVironment, a company in Monrovia, California, announced plans in October 1998 to begin testing a device that would reduce the time needed to recharge the batteries in electric vehicles. The PosiCharge system enables motorists to charge an electric-car battery in about 20 minutes, rather than the usual six to eight hours. The system regulates the electrical current so the battery stays at the optimum temperature for recharging, allowing it to charge faster.

Minnesota and is the most prominent natural feature for many miles. Such formations typically provide excellent sites for wind turbines because they are usually exposed to steady winds. Enron officials measured average wind speeds of 26 kilometers (16 miles) per hour on the ridge at a height of 30 meters (100 feet) above the ground.

The Lake Benton wind facility uses 143 turbines, each resembling a giant, three-bladed windmill. Each turbine rotor is 48 meters (157 feet) in diameter. The turbines were the largest ever used in the United States by mid-1999 for commercial energy production and were designed to withstand the rigors of the harsh Minnesota winters. For example, the fiberglass blades contain a black pigment that absorbs a great deal of sunlight, which speeds the melting of any ice that accumulates on the blades.

The annual energy production over the 30-year life span of the project was estimated to be 327,000 megawatt-hours, enough to serve the yearly electrical needs of 43,000 households, or about 120,000 people.

**Solar-powered Ferris wheel.** In October 1998, Pacific Park in Santa Monica, California, began operating the world's first solar-powered Ferris wheel. The nine-story-high wheel runs on electrical power generated by 660 solar panels installed by Edison Technology Solutions of Irwindale, California.

The solar cells generate approximately 200 kilowatt hours of energy a day—71,000 kilowatt hours of solar power annually—which would cost the park about $7,000 if purchased from the local utility. The solar-powered Ferris wheel reverts to using utility-supplied electricity during the evening.

At the dedication ceremony, U.S. Department of Energy Secretary Bill Richardson described the Ferris wheel as being more than just a tourist attraction. Rather, he said, it reflected the emerging market for clean, energy-efficient technologies and would teach visitors about the importance of protecting the environment and saving energy.

**Mining sand for oil.** In August 1998, Suncor Energy, Inc., of Fort McMurray, Canada, opened a processing plant to

In July 1998, the Japanese Marine Science and Technology Center (JAMSTEC) began a two-year test of the Mighty Whale, a structure designed to use ocean waves to generate power. As the device floats in water 8 meters (26 feet) deep, with anchors holding it in position, waves force air into and out of three chambers 12 meters (39 feet) high. The movement of the air drives turbine rotors at the top of each chamber. JAMSTEC engineers said that if successful, the Mighty Whale, shown here in an artist's drawing, could be used to supply energy to isolated areas.

extract petroleum from oil-bearing sand. By late 1998, the Steepbank Mine was producing 105,000 barrels of oil per day. Company officials expected the output to double by 2002. The mine, located on the east bank of the Athabasca River, covers 10 square kilometers (4 square miles) and was leased from the provincial government.

The Steepbank Mine was designed to process deposits of oil sand—a sticky mixture of sand, bitumen, clay, and water—that is found in abundance in the province of Alberta. Once separated from the other material, the bitumen, a black, tarlike substance found in the pores of certain sands, is a rich source of light crude oils used to make various fuels.

Although energy experts once considered it impractical to extract bitumen from oil sand, by the late 1990's petroleum technology had evolved to the point where the production of oil from bitumen was a major part of Canada's energy industry. Such oil accounted for 17 percent of Canada's petroleum needs and 13 percent of all oil produced there.

At the Steepbank Mine, the oil-containing sand is removed from the ground with gigantic power shovels and loaded into huge dump trucks. The sand is later crushed and mixed with water to form a material called a slurry, which is pumped through a pipeline to an extraction plant.

At the plant, the slurry is placed in large drums, where the bitumen is separated from the sand and water. Naphtha, a petroleum solvent, is used to thin the bitumen, which is then spun in a centrifuge to remove any remaining water and particles.

The bitumen is later heated in furnaces, and a heavy *hydrocarbon* (compound containing hydrogen and carbon) residue, called coke, is separated for use as a fuel in processing later batches of bitumen. The remaining, lighter, hydrocarbons are then further separated into naphtha, kerosene, and gas oil.

Energy experts said that as conventional oil and gas supplies continue to be depleted in the 2000's, the reserve of oil sands in Canada may represent an increasingly important energy source for the future.          [Pasquale M. Sforza]

Work crews in July 1998 completed construction on the Millennium Dome, the largest structure of its type in the world. Located 8 kilometers (5 miles) southeast of central London in Greenwich, England, the dome resembles a giant mushroom cap. The dome will first be used to house a year-long millennium exhibition and celebration beginning on Jan. 1, 2000. Plans call for the structure to later be used for various other exhibitions.

The Millennium Dome is the centerpiece of a dozen exhibits on an 8-hectare (20-acre) site. It contains 80,000 square meters (860,000 square feet) of floor space and is designed to accommodate up to 35,000 people. The dome includes both natural ventilation and air conditioning. Fabric panels of Teflon-coated fiberglass cover a superstructure of 2,600 tensioned cables hung from steel masts. The roof cost $60 million to construct. The total cost for the dome and various supporting structures was $429 million.

The first exposition scheduled for the Millennium Dome became the subject of debate in Great Britain in 1998 after some government officials criticized the $1.3-billion cost. British Prime Minister Tony Blair, however, said that the dome and the celebration should be symbols of a modern Britain.

**New baseball stadium.** Crews in Seattle, Washington, in 1999 were finishing construction on a new baseball stadium for the Seattle Mariners. The Mariners were scheduled to play their first game at the new ballpark, called Safeco Field, in July 1999. The team had played in the Kingdome since 1977, but city officials authorized construction of a new facility because of persistent problems with the Kingdome roof in recent years.

Upon its completion, Safeco Field, estimated to cost $498 million, would become the most expensive ballpark ever constructed in the United States. The facility, which will seat 47,000 spectators, was designed with a natural-grass playing field and a three-panel retractable roof.

The roof measures 33,500 square meters (360,000 square feet). When the roof is opened or closed, the three panels, mounted on rails, move simultaneously at varying speeds. It takes about 20 minutes for the roof to fully retract or extend. Structural engineers said that retractable roofs add to design time, compared to the construction of conventional stadiums. Engineers spent some 95,000 hours on the design of Safeco Field—two-thirds of it involving the design of the stadium roof.

The field consists of layers of gravel and sand, under which lies a complex network of more than 32 kilometers (20 miles) of pipes that will circulate hot water beneath the grass to coax it out of dormancy each spring. An underground drainage and ventilation system will also enable maintenance crews to force or draw air through the gravel, sand, dirt, and grass to oxygenate the grass or dry the playing field.

**Hydroelectric power.** Work on the largest privatized hydroelectric power project in Brazil neared completion in mid-1999. The project, called Ita, was expected to be operational in mid-2000.

Conceived in the early 1980's but hampered by financial setbacks, Ita will consist of a hydroelectric dam 125 meters (410 feet) high and 800 meters (2,600 feet) long on a bend in the Uruguay River. Because the structure must withstand the force that may result from sudden, heavy rainfall in the area, concrete slabs about one-half meter (1.5 feet) thick were poured to form the top layer of the dam's face. The structure was scheduled to start supplying power from the first of five 290-megawatt generating units by July 2000.

The dam was being constructed to provide power to Mercosul, a South American trade organization whose members are Argentina, Brazil, Paraguay, and Uruguay. The organization is designed to foster development in the area, but required the ability to generate power to lure companies. The power concession is jointly owned by Gerasul, a private power company, and Ita Energetica SA (Itasa), a holding company whose shareholders include Brazil's largest steel producer, major engineering and construction companies, and a leading cement maker.

**Tongue River Dam.** Engineers in Montana in May 1999 were completing repairs on the 60-year-old Tongue River Dam in southeastern Montana. The repair project, which cost $50 million, was undertaken to remove the structure from a national listing of unsafe dams.

The earthen dam is 380 meters (1,250 feet) long, 28 meters (91 feet) high, and has a concrete channel 182 meters (600 feet) long to handle surplus water. Considered a state-of-the-art structure when it was commissioned in 1939, the dam suffered damage during severe flooding in 1978. After lengthy negotiations among area farmers, coal mine operators, and the Northern Cheyenne Indians who live nearby, work crews began repairs in 1996.

During the reconstruction project, engineers brought the dam up to existing safety standards. They also raised its height 1.2 meters (4 feet) to increase reservoir capacity and improved public recreational facilities.

The rehabilitation project enabled the Northern Cheyenne tribe, whose reservation lies 48 kilometers (30 miles) downstream of the dam on the river's west bank, to divert more water from the reservoir's storage system.

**Everglades restoration.** The U.S. Army Corps of Engineers in April 1999 revised its schedule for the restoration of the Everglades, a large area of wetlands in southern Florida. The Corps of Engineers and the South Florida Water Management District decided to accelerate an $8-billion program to rehabilitate an ecosystem on the verge of destruction. The program called for the removal of some 385 kilometers (240 miles) of canals and levees. In sections of some levees, engineers planned to cut notches to allow excess water to spill into the surrounding wetlands during Florida's rainy season.

Under the Army's revised schedule, 44 of 68 projects in the plan—resulting in the restoration of about 970,000 hectares (2.4 million acres)—would be finished by 2010 in an effort to restore natural water-flow patterns.

Water once flowed in a shallow sheet across the Everglades. But beginning in the 1940's, the Corps of Engineers and the state of Florida disrupted that natural flow of water with a series of canals and levees designed for flood control and farmland drainage. Engineers constructed more than 1,600 kilometers (1,000 miles) of levees and canals. They also built floodgates and pumping stations to provide flood protection and water to the growing number of South Florida residents.

However, such development also created a series of problems for the Everglades, including making water in the wetland too deep in some areas and too shallow in others. The structures eventually affected the Everglades' fragile ecosystem. Engineers theorized that restoring the natural flow of water would alleviate many of the environmental problems plaguing the area.

**Microlock.** Researchers at Sandia National Laboratories in Albuquerque, New Mexico, announced in October 1998 that they had developed a minuscule locking device that could be used to safeguard computer systems against unauthorized users.

The tiny apparatus, about the same size as a shirt button, is called a recordable locking device. Engineers at Sandia designed the device small enough so that it could be implanted on a chip and placed inside a computer. When the device is in place, it locks the connection between the hard drive and the rest of the computer.

The lock consists of a series of six microscopic wheels, each about 0.4 millimeter (0.02 inch) in diameter. Each lock would be encoded with a number between 1 and 1 million that when known by a user would enable that person to open the lock.

The wheels of the lock are turned with computer commands entered on a keyboard. A series of tiny notched gears move to an unlocked position only when the correct code is entered. If an incorrect code is entered into the system, the device "freezes" the computer system, requiring the owner to reset the combination. With the microlock in operation, an unauthorized user would be unable to access any information from the hard drive remotely or even when sitting at the computer's keyboard.

The Sandia research team said that the lock could be used by any business or person requiring a secure computer system. They theorized that some version of the device would be available to consumers by 2001.

The device may also revolutionize the world of microscopic technology. Engineers working on the project discovered that the miniature wheels were capable of turning at incredibly fast rates of speed—up to 350,000 revolutions per minute. They calculated that a gear de-

## Seeing in the dark

In late 1998, the General Motors Corporation (GM) announced plans to market an infrared imaging system similar to that used by the military. The GM system, called Night Vision, was developed for the 2000 Cadillac DeVille and enables motorists to see better at night.

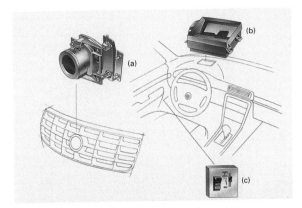

The system consists of an infrared camera (a) mounted behind the front grill of an automobile. The camera detects infrared radiation from objects in the road and translates it into a monochrome picture, which is displayed on a small liquid-crystal screen (b) on the dashboard. A switch (c) enables a driver to turn the system on and off or adjust the image's intensity.

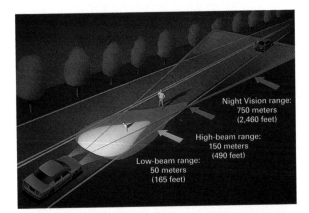

The system enables motorists to view objects in the road ahead of them five times farther away than is possible with most high-beam headlights and 15 times farther away than with standard low-beam headlights.

Night Vision range:
750 meters
(2,460 feet)

High-beam range:
150 meters
(490 feet)

Low-beam range:
50 meters
(165 feet)

The Night Vision system displays the infrared image on the vehicle's windshield, though the image seems to float in space a bit more than 2 meters (7 feet) ahead of the windshield. Objects in the road appear white against the surrounding darkness.

The Storebaelt Bridge, Europe's longest suspension bridge, opened in Denmark in June 1998. The bridge, which is 6.8 kilometers (4.2 miles) long, connects Denmark's largest islands, Sjaelland and Fyn.

signed strictly for speed would, in theory, be able to spin as fast as 10 million revolutions per minute. Such high speeds would ordinarily be impossible due to heat and friction. The engineers reported that tiny gear wheels could be used as high-speed switches in some computers.

**Airport construction.** Construction on Hong Kong's Chek Lap Kok International Airport was completed in 1998. The airport, which sits on a 1,248-hectare (3,084-acre) island that had been constructed from two smaller islands, was opened in July. Engineers described the $20-billion facility as safer, larger, and more modern than the airport it replaced, Kai Tak International Airport, which closed after 73 years of use.

Engineers designed the new airport to handle up to 87 million passengers annually. The facility, which took seven years to build, had to be linked to the mainland of Hong Kong via a series of new expressways, rail lines, an undersea tunnel, and one of the world's longest suspension bridges.

**Denver highway project.** In July 1998, engineers in Denver, Colorado, opened two segments of a privately financed toll road. City officials in Denver said that once the E-470 highway project was fully operational, it would provide rapid access between the Denver International Airport and existing highways in the area.

Engineers said that the E-470 highway project was designed to encircle the eastern suburbs of Denver. The roadway would link the airport with the north and south ends of Interstate 25, which runs through Denver. The highway was scheduled for completion in 2001.

**Longest bridge.** The longest suspension bridge in Europe, the 6.8-kilometer (4.2-mile) Storebaelt Bridge, opened for traffic in Denmark in June 1998. Support pylons rise 254 meters (833 feet) above the water. The Storebaelt Bridge connects the islands of Sjaelland and Fyn and for the first time enables motorists to travel from Copenhagen, located on Sjaelland, to the European mainland entirely by road.

[Andrew G. Wright]

Air pollution from China can travel thousands of kilometers across the Pacific Ocean, covering portions of North America and adding to local air pollution problems. That finding was reported in December 1998 by researchers from the University of Washington at Bothell. They presented the findings at the annual conference of the American Geophysical Union in San Francisco.

The scientists, led by atmospheric chemist Dan Jaffe, examined air pollution originating in China as part of a two-year study. In 1997, Jaffe and his colleagues measured carbon monoxide and other pollutants in Pacific Ocean air sampled on a mountain peak in the western part of Washington state.

They traced the origin of the air flow to Asia and concluded that the pollutants came from Chinese factories, motor vehicles, homes, and dust storms. Such pollutants are caught in strong air currents, particularly in the spring, and carried across the Pacific in 4 to 10 days. The researchers said pollutants from China may be adding significantly to North America's air pollution problems.

**Pesticides and breast cancer.** Exposure to certain chlorinated compounds that are stored in body fat may be responsible for a doubling in the incidence of breast cancer among Danish women since the late 1960's. Researchers at the Copenhagen Center for Prospective Population Studies reported that finding in December 1998.

The investigators measured the concentrations of 46 chlorinated compounds—18 pesticides and 28 polychlorinated biphenyls (PCB's)—in blood samples taken from 720 women. The blood samples had been collected 17 years before the start of the study. Of the women in the original group, 240 were later diagnosed with breast cancer.

The scientists discovered that blood concentrations of two pesticides, lindane and dieldrin, were consistently higher in the women who developed cancer. Women whose blood had the highest levels of dieldrin, which is banned in the United States, were more than twice as likely to develop cancer as women whose blood contained little or none of the pesticide.

The Japanese freighter *New Carissa* burns off the coast of Oregon in February 1999 after the United States Navy detonated onboard explosives. The ship ran aground during a storm, causing damage to the hull that threatened to spill 1.5 million liters (400,000 gallons) of oil into the water. The destruction of the vessel burned most of the oil so that only a small amount of it reached the shoreline. The ship later broke apart, and the Navy sank the portion of it that remained afloat.

Ecuadorian Indians in tribal dress, accompanied by a lawyer, leave a federal court in New York City in February 1999. A group of Ecuadorian Indian tribes had filed a lawsuit claiming that the Texaco Corporation had dumped 115 billion liters (30 billion gallons) of toxic wastewater from oil drilling in the Amazon rain forest since the 1960's. Attorneys representing the oil company claimed that the lawsuit belonged in Ecuadorian, not American, courts.

**Declining amphibians.** Agricultural chemicals, especially insecticides, may be partially to blame for a dramatic decline in frogs and other amphibians worldwide, Canadian researchers reported in September 1998. The scientists—biologist Michael Berrill of Trent University in Peterborough, Ontario, and Bruce Pauli, a biologist with Environment Canada in Hull, Quebec—reached that conclusion after studying the effects of such chemicals on three amphibian species.

Berrill and Pauli exposed eggs and tadpoles of two frog species and one toad species to varying concentrations of endosulfan, an insecticide used on a variety of fruit trees, vegetables, and other plants. They found that although exposure to the pesticide did not impair the eggs' hatching, the tadpoles whose eggs received the highest dose of the chemical exhibited a reduced response to danger. That change would have made the hatchlings more susceptible to predators.

Exposing the tadpoles to endosulfan after they hatched resulted in high death rates among all three species. Up to 30 percent of those tadpoles exposed to the pesticide immediately after hatching died. Among tadpoles exposed to the chemical two weeks after hatching, the rate was much higher—as much as 100 percent.

**Toxic chemical test.** A committee established by the United States Environmental Protection Agency (EPA) announced plans in October 1998 to evaluate thousands of agricultural, household, and industrial chemicals for their potential to disrupt *hormones* (natural regulatory chemicals) in humans and wildlife. The Endocrine Disruptor Screening and Testing Advisory Committee (EDSTAC) called for tests to determine which compounds act as *hormone disruptors*—substances that mimic or block the effects of hormones in the body.

The formation of the committee came after decades of research linking numerous chemicals with reproductive abnormalities and other harmful effects in animals. Scientists have speculated that all these health problems have been

caused by disruptions of the animals' *endocrine* (hormone) systems.

The U.S. Congress in 1996 required the EPA to examine some 87,000 commercial chemicals and determine which of the chemicals are hormone disruptors. EDSTAC recommended eliminating 25,000 of those compounds from the first round of testing for hormonal effects because they had been considered harmless to the body. The committee also recommended that the remaining 62,000 chemicals be tested in the order of their levels in the environment, from highest to lowest. Work on the project was scheduled to begin in mid- to late 1999 and was expected to take many years to complete.

**Air pollutants and fetal death.** Certain forms of air pollution in urban areas may cause an increased death rate in fetuses, according to the results of a Brazilian study published in June 1998. Researchers led by Luiz A. A. Pereira of the University of Sao Paulo studied miscarriages in the third trimester of pregnancy that had been reported in Sao Paulo in 1991 and 1992. The number of miscarriages averaged eight per day. They then compared the fetal deaths to the city's daily recorded concentrations of air pollutants, including nitrogen dioxide, sulfur dioxide, carbon monoxide, *particulates* (tiny particles), and ozone.

The researchers found no connection between miscarriages and high levels of the ozone or particulates. However, they found that reported miscarriages rose three days after increases in the levels of carbon monoxide, sulfur dioxide, and particularly nitrogen dioxide. All three pollutants are produced by the burning of fossil fuels, such as oil and coal, and of products derived from them, such as gasoline.

**Fetal mutations and cigarettes.** Fetuses exposed to even low concentrations of cigarette smoke may develop genetic *mutations* (changes) responsible for some forms of cancer in children, according to a study published in October 1998. The research was conducted by physicians at the University of Vermont College of Medicine in Burlington, led by pediatrician Barry A. Finette. Finette's team studied umbilical cord blood taken from 24 newborn babies. Half of the newborns' mothers had

been exposed to second-hand smoke during their pregnancies. The researchers found an increase in the number of mutations in a gene called HPRT in the blood cells of the babies whose mothers were exposed to smoke.

The researchers theorized that chemicals in tobacco smoke may cause the DNA in a fetus's cells to undergo rearrangements or deletions of its molecular subunits when the cells divide. (DNA—deoxyribonucleic acid—is the molecule genes are made of.) These changes can cause genes, including the HPRT gene, to malfunction.

The HPRT gene produces an *enzyme* (biochemical catalyst) that repairs damage to DNA, so mutations to this gene can lead to steadily accumulating errors in DNA as cells divide. Such damage may result in disease. Research has shown that abnormalities in the enzyme coded for by the HPRT gene are associated with childhood leukemia.

**More on second-hand smoke.** Inhaling second-hand tobacco smoke over long periods places adults at a slightly higher risk for developing lung cancer. However, people who are exposed to second-hand smoke as children but who live in a smoke-free environment when they reach adulthood have no increased risk of that disease. Those were the conclusions of a 10-year study reported in October 1998 by an international team of researchers.

The study, commissioned by the International Agency for Research on Cancer in Lyon, France, was conducted at 12 centers in seven European countries. The study involved 1,542 healthy adults and 650 adults who had been diagnosed with lung cancer. The participants, who ranged in age up to 74, reported having smoked no more than 400 cigarettes in their lifetime. The researchers questioned the participants about their exposure to second-hand smoke as both children and adults.

The investigators found that nonsmoking adults exposed to second-hand smoke from spouses or co-workers had less than a 20 percent increased risk of developing lung cancer. If their exposure to smoke had occurred during childhood, or had stopped at least 15 years before the start of the study, they had no increased risk of developing the malignancy.                    [Dan Chiras]

Most paleontologists support the theory that birds evolved from small meat-eating dinosaurs called theropods, which walked only on their hind legs. In June 1998, a research team from the United States and China reported that fossil specimens with feathers were theropods and, therefore, the strongest evidence yet of a bird-dinosaur relationship.

The new specimens, identified as *Protarchaeopteryx robusta* and *Caudipteryx zoui,* were found in 1996 in fossil beds in China that date from about 120 million to 140 million years ago. Because the specimens' arms were relatively short compared with the size of their bodies, the researchers noted that the creatures would not have been able to fly. They concluded, therefore, that feathers had first evolved for functions other than flying, such as providing warmth.

Because *Archaeopteryx,* the oldest known bird, lived about 140 million to 150 million years ago—before *Protarchaeopteryx* and *Caudipteryx*—the scientists noted that the Chinese species could not have been the direct ancestors of birds. They stated, however, that the evidence suggested that *Archaeopteryx* and the newly identified species shared a common theropod ancestor.

**A trace of ancient animals.** Small branching grooves found on the surface of 1.1-billion-year-old sandstone in India are the trails of wormlike animals, according to an October 1998 report. Paleontologists Adolf Seilacher and Friedrich Pfluger of Yale University in New Haven, Connecticut, and geologist Pradip Bose of Jadavpur University in Calcutta, India, described the impressions as the earliest evidence of multicellular animals. (Any indirect fossil evidence of living creatures, such as animal footprints or the trails of worms, is called a trace fossil.)

The origin of multicellular animals has long been a controversy among paleontologists. Fossil embryos found in 1997 seem to point toward a relatively abrupt appearance of several animal groups about 570 million years ago. Molecular biologists, however, have disputed much of the fossil evidence. Assuming that genetic *mutations* (changes) occur at a constant rate, they have com-

**Fossilized impressions of dinosaur skin**
A rare glimpse of dinosaur skin came from two studies in 1998 and 1999. Researchers at Mesa (Arizona) Southwest Museum discovered impressions from the skin of a *hadrosaur* (duck-billed dinosaur), *below left,* in an ancient lake bed in New Mexico. The ridged bumps are about the size of a dime. A multinational team of researchers working in Argentina found the impressions of the skin of sauropod embryos, *below right,* preserved in fossilized egg fragments. Sauropods were a group of large plant-eating dinosaurs.

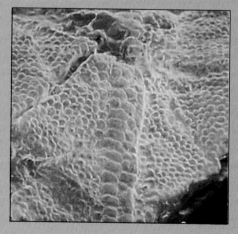

A 140-million-year-old fossilized flower found in northeast China provided researchers with clues to the origin of angiosperms, the largest group of modern plants, all of which produce flowers and encase their seeds in some kind of fruit. A research team led by plant specialist Ge Sun of the Academia Sinica in Nanjing, China, announced the discovery in November 1998. The scientists described the specimen, the oldest known angiosperm, as a relative of the modern magnolia. They noted that details of the specimen (one of several that were found) suggested that angiosperms may have descended from an extinct plant called a seed fern.

pared the genes of many modern animals and estimated that they are all descended from a common ancestor that lived as early as 1.2 billion years ago. Thus, the new report on trace fossils was of great interest.

The fossils described by Seilacher's team are about 5 millimeters (0.2 inch) wide and resemble the marks left by organisms, such as modern earthworms, that move by expanding and contracting their body cavities. The researchers concluded that the trails were preserved because the animals had burrowed through a *bacterial mat*—a leathery film composed of masses of bacteria—in the top layer of sand on an ancient sea floor. The trails were preserved in the mats and then covered by new layers of sand. Over time, all the layers of sand hardened into sandstone.

Some paleontologists disagreed with the findings, suggesting that the grooves were made by nonbiological processes such as gas bubbles moving through the sand or small objects rolling across it. On the other hand, some scientists pointed out that if the grooves were not

so old, few paleontologists would question their identity as trace fossils. Other researchers, in turn, questioned the age of the sandstone.

**Earliest evidence of color.** Many living animals use color to camouflage themselves from their enemies. This type of adaptation, according to a December 1998 report, may be very old, originating during the Cambrian Period (543 million to 500 million years ago) when the predator-prey system first arose in the animal kingdom.

Paleontologist Andrew Parker of the Australian Museum in Sydney described the coloring structures of three fossil animals found in the 115-million-year-old Burgess Shale, a rock formation in British Columbia, Canada. The fossils were of two bristle-covered worms, known as *Wiwaxia corrugata* and *Canadia spinosa*, and a joint-legged animal known as *Marella splendens.*

Parker observed a series of very fine grooves on the surfaces of the animals' protective plates and spines. These grooves, Parker said, acted as *diffraction gratings*, structures that split white light into a rainbow of colors. These natural gratings produced a silvery surface that shimmered with color, much like the hologram on a credit card. Parker speculated that the play of colors served as protection by warning or distracting would-be predators.

**Jurassic flower.** The discovery of fossils of a 140-million-year-old Jurassic Period flower was reported in November 1998 by a team of researchers led by *paleobotanist* (ancient-plant specialist) Ge Sun of the Academia Sinica in Nanjing, China. The fossils, found in northeast China, are the oldest known examples of plants called *angiosperms*, which have dominated the plant kingdom since the middle of the Cretaceous Period (138 million to 65 million years ago).

Angiosperms comprise all the flowering plants, from grasses to tulips to oak trees, which house their seeds in some kind of a fruit. (In botany, anything that encases a seed, such as an acorn, is considered a fruit.) The origin of angiosperms has long mystified paleobotanists because the plants appear abruptly in the Cretaceous fossil record and have

no clear ancestors. Indeed, the British naturalist Charles Darwin, the foremost evolutionary biologist of the 1800's, referred to this problem as an "abominable mystery."

Based on the shape of the newly found fossilized fruits and seeds, Sun's team suggested that the plants were probably related to modern magnolias. They also speculated that the fossil plant may have descended from *seed ferns,* an extinct group of plants with distinctive fernlike foliage that reproduced by seeds. True ferns reproduce with *spores,* single cells that can break free and create new plants.

If the seed-fern theory is true, it runs counter to prevailing wisdom, which argues that angiosperms descended from another group among the *gymnosperms,* "naked seed," or nonfruiting, plants. Modern gymnosperms include conebearing plants such as pine trees. In any case, the new report gave researchers hope that more and possibly older angiosperm fossils would be found in China and that scientists could eventually solve the "abominable mystery."

Paleontologist Paul Sereno of the University of Chicago describes the features of *Sucho-mimus tenerensis,* a previously unknown dinosaur that was found in the Tenere Desert of Niger. Sereno and his colleagues announced the discovery of the dinosaur in November 1998. The researchers chose the name *Suchomimus,* or *crocodile mimic,* because the dinosaur's long, narrow head and sharply pointed teeth resemble the features of a modern crocodile. These features—as well as its long, hooked thumbs—led the scientists to conclude that the dinosaur was adapted for capturing and eating fish.

**Data from dung.** Fossilized *feces* (solid body wastes) might seem an unlikely source of information in paleontology. But, in fact, feces are a type of trace fossil that record information about the ingestion and digestion of food by ancient organisms. Two separate studies reported in 1998 showed that fossil feces, or coprolites, are an important storehouse of information about the behavior of prehistoric animals.

Paleontologist Karen Chin of the U.S. Geological Survey and three associates reported in June 1998 on the discovery of a coprolite measuring 44 centimeters (17 inches) long and 16 centimeters (6 inches) wide. It was discovered in dinosaur fossil beds in southwestern Saskatchewan, Canada, that are between 66 million and 70 million years old.

The researchers determined that the coprolite was from a large predatory animal because it contained numerous bone fragments from small plant-eating dinosaurs. And because the bones of a *Tyrannosaurus rex* had been found in the same fossil beds, the researchers concluded that this huge dinosaur species had been the source of the feces.

The condition of the bones within the coprolite was particularly important and surprising to the researchers. On the basis of evidence from crocodiles, the largest modern predatory reptiles, paleontologists had assumed that the tyrannosaur gulped down chunks of meat with little chewing and that bones of its prey remained relatively whole. They also expected to find bones that were rounded or pitted because strong acids in the stomachs of crocodiles produce those effects on the bones of prey.

The bones in this coprolite, however, were broken into pieces, indicating considerable chewing by the tyrannosaur. The bone fragments showed little signs of strong stomach acids. Hence, the coprolite provided new insights into the diet, eating habits, and digestive system of the long-extinct dinosaur.

A second coprolite study, published in July 1998, was of a much more recent specimen that had not *mineralized* (turned to rock). A research team, led by molecular biologist Hendrik Poinar and geneticist Svante Paabo of the University of Munich in Germany, reported that it had extracted DNA (deoxyribonucleic acid, the molecule that carries the genetic code) from an ancient animal dropping. The 20,000-year-old coprolite from a giant ground sloth was preserved in Gypsum Cave near Las Vegas, Nevada.

Some of the DNA was evidently derived from intestinal cells that had been shed with the feces of the sloth. The sequence of this DNA matched well with DNA that had been extracted previously from bones and dried tissues of another ground sloth, proving the identity of the animal that had produced the feces.

The scientists were also able to identify DNA from plants that had been a part of the sloth's diet. These included grasses, yucca, grapes, and mint. Most of these dietary items were also identified from plant fragments in the coprolite.

These findings were exciting to paleobiologists who study the behavior of ancient animals, as well as to scientists investigating ancient climatic changes. If DNA could be extracted from specimens of many different ages, the information might provide a record of changing vegetation and diets that correspond with major shifts in climate. In turn, this record might yield clues to explain extinction events, such as the extinction of the giant ground sloths in North America about 11,000 years ago.

**Parrots in the age of dinosaurs.** The discovery of the distinctive lower jaw of a parrot from 68-million-year-old sediments in Wyoming was announced in November 1998 by paleontologist Thomas Stidman of the University of California at Berkeley. Bird fossils are not common, and most of them are of water birds, which were sometimes entombed in the sediments of lakes or seas. The oldest known modern birds from the middle of the Cretaceous Period were members of the gull, loon, and duck families.

Although *ornithologists* (bird specialists) had long considered parrots to be primitive birds, they did not have the fossil evidence to support the assumption. The new parrot fossil proved the great antiquity of the parrot family. These land birds predated—and survived—the mass extinction that eliminated dinosaurs about 65 million years ago.                    [Carlton E. Brett]

In the Special Reports section, see FOSSILS, FEATHERS, AND THEORIES OF FLIGHT.

Scientists in Hawaii announced in July 1998 that they had used body cells from adult mice to create dozens of *clones* (genetic duplicates) of the mice. The research, carried out by a team of biologists led by Ryuzo Yanagimachi of the University of Hawaii in Honolulu, was the first published study to confirm the results of the 1996 Scottish experiment that produced the sheep Dolly, the first clone created from an adult mammal.

The method used by Yanagimachi's team was, in many respects, similar to the procedure developed at the Roslin Institute in Scotland, where Dolly was created. First, the scientists extracted the *nucleus* (the part of the cell that contains the genetic instructions) from a body cell taken from an adult mouse. They also removed the nucleus from an egg cell taken from another mouse. Then they placed the body cell's nucleus into the egg cell.

After the resulting cell started to develop into an *embryo* (immature organism) in a laboratory culture solution, the scientists implanted the embryo into the womb of yet another mouse. The embryo then developed into an animal genetically identical to the mouse from which the body cell was obtained.

One major difference between this technique and the one used to make Dolly was that Yanagimachi's team transferred the nucleus from the mouse body cell to the egg cell with a fine needle. The Scottish group had used an electrical impulse to fuse a body-cell nucleus with an egg cell.

Another difference between the techniques was the way in which the scientists reprogrammed the body cell's nucleus to direct the development of a complete organism. (Because each body cell of an adult animal is specialized, it does not naturally have this ability.) Whereas the Scottish scientists did this by starving the body cell for a few days, the Hawaii researchers reprogrammed the body cell's nucleus by allowing substances in the egg cell to interact with the transplanted nucleus for up to six hours before placing the cell in culture.

Yanagimachi tried various kinds of body cells to make the mouse clones. For reasons the researchers did not entirely understand, the method worked best with *cumulus cells,* cells that surround an egg as it matures in an ovary.

Other scientists hailed the work as a breakthrough in cloning for a number of reasons. First, the Scottish scientists had encountered great difficulty in producing Dolly—struggling through nearly 300 failed attempts before successfully creating a single clone. Yanagimachi's team, on the other hand, produced more than 50 cloned mice with relative ease. Furthermore, they were able to create clones of some of the clones.

In addition, because mice are widely used in genetics research, scientists said the availability of cloned mice will open up new avenues of scientific research in many areas of biology and medicine.

**C. elegans genome.** The first virtually complete description of an animal's *genome* (total amount of genetic information) was announced by scientists in December 1998. Two groups of geneticists—one led by Robert Waterston of Washington University in St. Louis, Missouri, and the other by John Sulston of the Sanger Research Centre in England—together determined the sequence of the 97 million nucleotide subunits that make up the genome of the roundworm *Caenorhabditis elegans.* Nucleotides are the individual building blocks of *deoxyribonucleic acid* (DNA, the molecule that genes are made of).

*C. elegans* has long been a favorite research subject for scientists interested in the role of genes in growth and development. By the early 1980's, researchers had tracked the life of each cell in the worm as it develops. With the genetic sequence information in hand, scientists said they would be able to finish putting together all the pieces of the puzzle of how the organism is constructed.

Sequencing the worm's genome was by far the most complex such project yet completed. Other genome sequences that had been completed by late 1998 were all of microorganisms. Scientists said the success of the eight-year project would help pave the way for completion of the human genome project, scheduled to be finished by 2003.

**Syphilis sequence.** The complete genetic code of *Treponema pallidum,* the bacterium that causes the sexually transmitted disease syphilis, was published in July 1998. The work was carried out by a group of geneticists led by Claire M. Fraser of the Institute for Genomic Research in Rockville, Maryland, and

## Cloning research moves forward

Researchers made major advances in 1998 in using body cells from adult animals to create *clones* (genetic duplicates) of the animals. Such clones are important for both medical research and agriculture.

Two calves, among a group of eight clones created by a research team led by biologist Yukio Tsunada of Kinki University in Japan, make their public debut in December 1998. The researchers used a variation of the technique used by Scottish scientists in 1996 to create the sheep called Dolly, the first clone of an adult mammal. Japanese scientists produced several more cattle clones in 1999.

University of Hawaii biologist Ryuzo Yanagimachi points to mouse clones held by postdoctoral student Teruhiko Wakayama. The two biologists reported in July 1998 that their research team had created more than 50 mouse clones. Furthermore, some of the mice they created were clones of clones. The scientists produced the clones from *cumulus cells* (cells that surround the eggs in the ovaries of adult female mice). The ability to clone laboratory mice with relative ease was expected to open up many new areas of scientific research in which mice are used.

# The Genome Race Is On

Imagine starting out in a race that's already more than two-thirds over and declaring that you'll be the winner. That's exactly what geneticist J. Craig Venter, president of the Institute for Genomic Research in Rockville, Maryland, did in May 1998. Venter announced that he would form a private company to challenge an international, government-supported project to decode the *human genome*, the total amount of genetic information in a human cell. Venter said his new company, Celera Genomics Corporation, also in Rockville, would *sequence* (determine the order of) all of the molecular units of the genome by the year 2001. That deadline was four years earlier than the scheduled completion of the Human Genome Project (HGP), a huge undertaking begun in 1990 by the U.S. National Institutes of Health, the U.S. Department of Energy, the Sanger Centre in England, and research institutions in Japan, Germany, and other countries.

In addition to Venter's company, two other private companies—Incyte Pharmaceuticals, Incorporated, of Palo Alto, California, and Human Genome Sciences, Incorporated, of Rockville—announced in 1998 that they too were conducting human-genome sequencing projects.

The stakes in this race were very high. Scientists said the work was likely to lead to many new drugs and other products (worth billions of dollars) that would change the way medicine is practiced in the future. In the words of one of the HGP leaders, geneticist Robert Waterston of Washington University in St. Louis, Missouri, HGP's sequence information "will be the central organizing principle for human genetics in the next century."

The goal of this great race was to determine the order of the approximately 3 billion subunits, called nucleotides, that make up human DNA (deoxyribonucleic acid, the molecule that genes are made of). There are four kinds of nucleotides, which biologists abbreviate as A, G, C, and T. The order of nucleotides in a gene is a code for the information needed to make a particular kind of protein. The thousands of proteins produced by cells carry out many functions in the body and determine or influence everything about us, ranging from hair and eye color to whether we develop certain diseases.

The intense competition in the race to sequence the human genome focuses on genes, even though these protein-coding strings of nucleotides make up just 3 percent of the DNA in our *chromosomes* (the structures that carry the genes). The function of the rest of a cell's DNA is only partly understood. The number of genes we have is also not known with precision—estimates range from 60,000 to 120,000.

One of the reasons that information about the genome is so important is that the U.S. Patent Office allows newly discovered genes to be patented. This means that the researcher who first determines the nucleotide sequence of a gene and the function of the protein coded for by the gene can claim exclusive rights to the use of that information. The scientist can then license those rights to others, collecting a fee or *royalty* (payment based on sales of a product). For example, the researcher might sell the gene information to a pharmaceutical company for use in making a drug to combat a disease caused by the gene.

The great potential for profit was the driving force behind the private-sector efforts to sequence the human genome. Venter said he hoped to patent about 200 genes linked to common diseases. This prospect disturbed many observers, who believed that no one should have the right to "own" human genes.

Although private companies were mainly interested in the protein-coding regions of DNA, the goal of the HGP was to produce a comprehensive description of the human genome that could be used free of charge by anyone. These different objectives were reflected in the different methods used by the competing research teams.

The procedure used by the HGP took into account the fact that many genes cannot be recognized or completely understood simply by determining their nucleotide sequences. Rather, the specific location of a gene in relation to other nucleotide sequences in a chromosome may be important for understanding what the gene does.

Because of this factor, the HGP scientists had planned to produce an extensive map of numerous easy-to-identify genetic landmarks, called markers, before finishing the complete genome sequence. Such a genome map would be equivalent to a city map showing major streets and points of interest but not the actual addresses of houses or buildings on the streets.

To fill in the detail on the genome map, researchers at several HGP-affiliated facilities would then sequence the stretches of DNA between the markers one nucleotide at a time. This was to be accomplished by using DNA sequencing machines, which analyze DNA that researchers have extracted and copied from human cells.

These sequencing machines "tag" the DNA with four differently colored dyes, one color for each of the four types of nucleotides. The order in which these colors are detected by a laser beam on the machine tells researchers the order of

A scientist at the Institute for Genomic Research in Rockville, Maryland, checks data generated by a DNA sequencing machine, a device that determines the order of the *nucleotides* (molecular subunits) in genetic material. The computer monitor displays sections of DNA in which each of the four different types of nucleotides has been designated by a color—blue, green, yellow, or red.

the nucleotides in that particular strand of DNA.

Venter claimed his company would get the sequence data faster and cheaper than the HGP because a detailed landmark map was not necessary. Instead, Celera intended to use a process called whole-genome shotgun sequencing. In this approach, the entire set of genetic material in a cell is broken up into millions of smaller fragments. Then, highly sophisticated DNA sequencing machines developed by Celera's corporate parent, the Perkin-Elmer Corporation of Norwalk, Connecticut, will sequence each piece of DNA. Finally, high-speed computers will assemble the data on all these DNA pieces into a complete picture. Venter said the order and location of the DNA sequences on the chromosomes can be determined by matching up the nucleotide sequences at the ends of one piece of DNA with those at the ends of other pieces.

Although Venter had successfully used the shotgun method in the past to sequence the genomes of a number of microorganisms, which have much less DNA than humans, some HGP scientists doubted that the same method would work as well on the scale of the human genome. However, Venter maintained that the several hundred machines used by Celera would be able to sequence as much DNA in one day as the HGP sequenced in all of 1998.

The other private companies in the race, Incyte Pharmaceuticals and Human Genome Sciences, were using a less ambitious approach than either Celera or the HGP. They were focusing exclusively on finding new proteins and then tracking down the genes that code for them. As of 1999, these companies claimed that they had already sequenced thousands of genes that they intended to patent. However, because they had not released any of their data, other scientists found it impossible to assess their progress.

After the announcement by Venter and the other competing groups, the federal agencies providing funding for the HGP altered their strategy somewhat. They decided to give an increased amount of support to their three most productive research centers—Washington University, the Baylor College of Medicine in Houston, and the Whitehead Institute for Biomedical Research in Cambridge, Massachusetts. In addition, the British government boosted funding for the Sanger Centre.

The HGP also announced that it would give priority to sequencing the gene-rich regions of the genome. HGP officials claimed that these changes would make it possible to have a "working draft" of the genome finished by mid-2000 and the entire genetic sequence completed by 2003. Scientists said it would then take several more decades to determine the functions of all the proteins coded for by the genes.

The information obtained from the human genome is sure to raise many questions concerning not only medical treatments but also ethics, law, and philosophy. Ultimately, those questions may be even more important than the sequencing information itself.                    [David Haymer]

Steven J. Norris and George M. Weinstock of the University of Texas Health Science Center in Houston.

Syphilis, if left untreated, can cause a number of serious problems, including blindness, paralysis, mental illness, and heart failure. Because of the microbe's role in human disease, many medical researchers said the syphilis bacterium was the most important of the several microorganisms that had been sequenced.

The Texas scientists had been working for many years on this project. However, their progress was slow in the beginning, due in part to the relatively large size of the microbe's genome, which contains 1.14 million nucleotides. The scientists also had trouble growing the microbe in the laboratory.

Things speeded up in the mid-1990's, when scientists at the Institute for Genomic Research developed new laboratory methods for working with genomes of microbes. These new procedures, which the researchers then used with *T. pallidum*, made it possible to break up a microbe's genetic material into smaller, more workable pieces. After each of these pieces was thoroughly analyzed, the information could be rapidly reassembled into a complete picture.

The researchers determined that *T. pallidum* has more than 1,000 genes, many of them very different from genes identified in other disease-causing microbes. The scientists hoped that further studies of these novel genes would lead to a better understanding of how the syphilis microbe is able to survive for years in the body after causing an infection. They also hoped to learn how the microbe is able to cause such extensive damage to a person's nervous system.

**Eye-disease gene.** The finding of a gene that causes one form of a serious eye disorder called macular degeneration was announced in July 1998 by a group of scientists led by geneticists Konstantin Petrukhin of Merck Research Laboratories in West Point, Pennsylvania, and Claes Wadelius of the University of Uppsala in Sweden. In macular degeneration, a common cause of vision loss in old age, the *macula* (the part of the retina that is responsible for distinguishing fine detail) deteriorates.

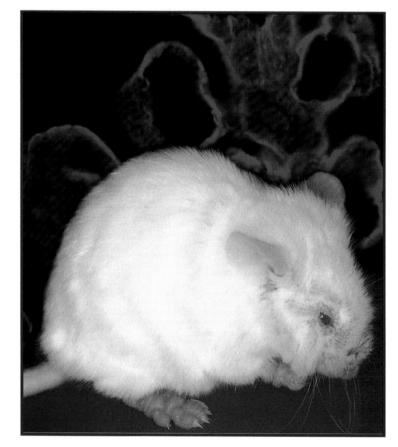

A mouse with an extra-furry coat was one of a group of such mice created by University of Chicago (UC) researchers in 1998 with genetic-engineering techniques. UC biologist Elaine Fuchs reported in November 1998 that she and her colleagues produced the mice by giving them a gene that codes for a protein important to the development of hair follicles. Although the mice had more hair follicles than normal, they also developed many cancerous tumors in their skin. Despite this problem, Fuchs said that additional research on the gene might lead to a way of preventing baldness or to methods for stopping abnormal hair growth.

The scientists found the gene by studying a large family in Sweden in which many of the individuals were afflicted with a form of macular degeneration called Best's macular dystrophy. By analyzing the DNA of these people, the University of Uppsala researchers determined that the macular degeneration gene is located in a particular region of chromosome number 11. (Humans have 23 pairs of chromosomes, the structures that carry the genes.)

The Swedish scientists were stumped, however, in determining which of the many genes in this chromosome region causes the eye condition. To help them find the gene, they teamed up with the Merck group. Together, both groups looked for genetic *mutations* (alterations) that were present only in the family members with macular degeneration.

The geneticists found that one gene in the suspect region of chromosome 11 was defective only in these family members. The gene, which they named bestrophin, was just the second gene known to be associated with macular degeneration. The scientists were not sure, however, what the bestrophin gene normally does or how mutations in the gene cause macular degeneration. Nonetheless, they were confident that further studies of the gene would lead to a better understanding of not only macular degeneration but also a number of other common vision problems.

**Congenital heart defects.** The discovery of a genetic flaw that appears to be responsible for a large number of *congenital* (present at birth) heart defects was announced in February 1999 by a group of researchers led by developmental biologist Deepak Srivastava of the University of Texas Southwestern Medical Center in Dallas. Congenital heart malformations, of which there are more than 30 different types, are the leading cause of death for infants in the first year of life.

The Texas investigators suspected that a missing gene might be responsible for many heart defects because previous researchers had found that 90 percent of infants born with a defect known as DiGeorge syndrome have an abnormality in chromosome 22. In these cases, a small region of the chromosome was known to be absent. However, this region contains a number of genes, and

it was not clear which one of these genes might actually be behind the problem.

To find the gene, the scientists first showed that a mouse gene needed for heart development is similar to a human gene in this region. Both the mouse gene and the human gene, called UFD1, direct the production of substances that connect the heart chambers to nearby blood vessels—a connection that is missing in people with DiGeorge.

To see if DiGeorge patients lack the UFD1 gene, the scientists analyzed DNA from 182 people with DiGeorge syndrome. They found that all of the patients were indeed missing the gene. The researchers also studied genetic material from 100 people without the syndrome and found that none of these people was lacking the gene.

These results strongly indicated that a missing UFD1 gene is the cause of DiGeorge syndrome. However, some scientists noted that other genes might also be involved in causing the problem.

The scientists who found the genetic defect did not completely understand how the lack of the UFD1 gene might cause DiGeorge syndrome. Even so, they expressed confidence that further studies of this and other genes would lead to a better understanding of how congenital birth defects arise. This understanding, in turn, might lead to methods for preventing these problems.

**Bacterial master gene.** California researchers reported in May 1999 that they had discovered a gene in a strain of *Salmonella,* a bacterium that causes food poisoning in humans, that regulates the functions of other genes in the microbe. By using genetic engineering techniques to delete or disable this gene, called dam, the scientists, led by geneticist Michael Mahan of the University of California at Santa Barbara, prevented the bacterium from causing a typhoid-like disease in mice. The researchers also reported that mice vaccinated with the crippled strain were protected from infection by normal salmonella strains.

The scientists said that further research was needed to determine if disabling the dam gene would protect humans from salmonella infections. In addition, they noted that many other types of disease-causing bacteria have dam genes that might also be disabled.

[David Haymer]

New evidence in support of a theory that the Earth was once covered almost completely by ice was published in August 1998 by a team of geologists led by Paul Hoffman of Harvard University in Cambridge, Massachusetts. This hypothesis, nicknamed the "Snowball Earth" theory, had been proposed before, but it received little support because scientists argued that if the entire Earth ever did freeze over, it could never have thawed out again. The new study supported the theory and offered an explanation of how it could, in fact, have occurred.

Geologists have long been puzzled by *tillite-limestone sequences*, alternating layers of two different types of sedimentary rock, which are found only in five widely separated parts of the world. (Sedimentary rock is formed when *sediment*—mineral matter or the remains of tiny organisms—settles out of water and is compressed.) One of the rock types is *tillite*, a fossilized jumble of rock and clay that closely resembles *till*, the material left behind by the retreat of a glacier. The other rock type, limestone, is formed by sediments deposited in warm, shallow waters.

Layers of these rocks were laid down during the Neoproterozoic Era, between 750 million and 550 million years ago. The existence of alternating bands of tillite and limestone indicates a repeating cycle of extreme climatic shifts, each lasting for millions of years. To complicate matters, the *paleomagnetism* (magnetic orientation of minerals in rock) of some tillite layers indicates that they were formed by glaciers near the equator.

Various theories to explain such an unlikely scenario have been proposed, including the possibility that the continents drifted rapidly back and forth between the poles and the equator several times. But Hoffman's team developed a relatively simple explanation by measuring the ratio of carbon-12 and carbon-13, two *isotopes* (variant forms) of carbon within a tillite-limestone sequence in Namibia, a country in southwestern Africa.

On Earth, carbon-12 is normally much more common than carbon-13—the ratio is about 99 to 1. Geologists know that a higher-than-normal amount of carbon-13 in ancient limestone is a sign of increased biological activity in the sea when

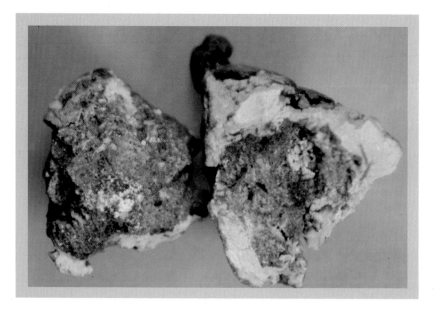

Tiny remnants of what may have been the space rock that scientists think struck the Earth about 65 million years ago, wiping out the dinosaurs and many other life forms, were recovered from Pacific Ocean sediments and analyzed. The fragments contained iron and nickel, indicating that the impacting object was an asteroid rather than a comet, according to a November 1998 report by a University of California geochemist. Comets are typically composed of ice and rock.

the rock formed. That is because marine plants use up more carbon-12 than carbon-13 when they build their tissues. When plants are using large amounts of carbon-12, there is less of it to go around, so limestone (calcium carbonate) that forms in the ocean at the time contains a higher proportion of carbon-13. Hoffman's group found that the Namibian tillite-limestone sequence showed sudden and large changes in the ratios of carbon-12 to carbon-13—the largest shifts ever seen in geologic history. This became the key to explaining how Earth could have thawed after becoming frozen over.

The investigators theorized that, during most of the Neoproterozoic Era, Earth's atmosphere contained high levels of carbon dioxide. Because carbon dioxide helps keep Earth warm by preventing solar heat from escaping, the average temperature on the planet must have been much higher than it is today. Then, just prior to the ice ages, the ancient continent of Rhodinia, which contained all of today's major land masses, broke apart. Fragments of the gigantic land mass drifted apart and began to erode. This erosion converted huge amounts of atmospheric carbon dioxide into carbonates and also released minerals into the seas. In response to the rich nutrient supply provided by these minerals, *algae* (simple plantlike organisms) in the oceans boomed. The algae then began absorbing atmospheric carbon dioxide and incorporating it into their tissues.

When the organisms died, they sank as sediment to the ocean bottom. The organic carbon in their tissues was thus trapped and prevented from returning to the atmosphere. Global levels of atmospheric carbon dioxide declined, and the Earth became less efficient at retaining heat. The planet cooled and ice sheets began to form. The ice reflected sunlight back into space, causing the planet to cool even more. Finally, the oceans froze, wiping out nearly all life on Earth.

Over a period of about 4 million to 10 million years, Hoffman and colleagues speculated, volcanic gases from within the planet slowly released carbon dioxide into the atmosphere. With almost no algae to absorb it, the carbon dioxide accumulated until it eventually reached 350 times its present level. This enabled the atmosphere to trap huge amounts of heat, causing the ice to melt. As the ice

receded, it reflected less sunlight, and global temperatures rose faster. The oceans became warm again, resulting in the rapid deposition of large amounts of limestone, which caused atmospheric carbon dioxide levels to return to normal. Eventually, Earth's average temperature dropped to today's levels and marine plants flourished again. Then, the whole cycle was repeated—at least four times. The researchers speculate that "Snowball Earth" conditions never recurred after the Neoproterozoic Era primarily because the supply of nutrients to the ocean decreased. This most likely occurred as the fragments of Rhodinia drifted farther apart, and some became submerged beneath the sea.

**Geology's influence on marine life.** Geologists have long known that the chemistry of carbonates in the sea has undergone a series of major shifts, a cycle controlled by changes in the rate of sea-floor spreading. In February 1999, paleobiologist Steven M. Stanley and geochemist Lawrence A. Hardie of Johns Hopkins University in Baltimore reported that changes in the carbonate content of certain marine organisms are also linked to sea-floor spreading rates.

Calcium carbonate is the most common material precipitated as sediment out of seawater. It can occur as either of two minerals, calcite or aragonite, which are formed by different arrangements of the same atoms. For the past 40 million years, aragonite has been the more common form of calcium carbonate. However, from 200 million to 40 million years ago, calcite was more common. From 350 million to 200 million years ago, aragonite predominated, and prior to that, calcite predominated.

Shifts in carbonate mineralogy are related to changes in the chemistry of seawater, which occur due to changes in the rate at which plates of the Earth's crust move away from one another. Most of this spreading occurs beneath the ocean. New sections of sea floor are formed when molten rock from deep within the Earth's crust wells up to fill the gaps created when plates drift apart. At the same time, the overlying seawater reacts with the molten rock. As seawater percolates into the molten rock, minerals dissolved in the water are exchanged with those in the rocks. One of these exchanges takes magnesium from the water and replaces it with calcium. This alters the ratio of calcium to magnesium *ions* (electrically charged atoms) in the water. It is this ratio that dictates whether the carbonate will be calcite or aragonite. If the rate of sea-floor spreading increases, the proportion of calcium in the water increases, which favors the production of calcite.

Most modern reefs are built by corals that form their skeletons of aragonite, and most shallow-water plants secrete aragonite. When these organisms die, their skeletons become sea-floor sediment. Studies of fossil coral reefs and sedimentary rock showed that, during periods when calcite was predominant, reefs were built by clams that had calcite skeletons, and shallow-sea sediments were formed from calcitic skeletal remains of tiny marine organisms.

The researchers concluded that the shifts between calcite and aragonite in marine organisms have coincided with the same shifts in seawater chemistry triggered by the rate of sea-floor spreading. In this way, sea-floor spreading rates have undoubtedly played a crucial role in the evolution of life in the sea.

**Mass extinction and recovery.** In another study concerning the link between marine geochemistry and marine life, researchers led by paleoceanographer Steven D'Hondt of the University of Rhode Island estimated how long it took for marine life to recover after a catastrophic asteroid impact. Considerable evidence indicates that about 65 million years ago, at the end of the Cretaceous Period, an asteroid hit the Earth, hurling tons of dust and other debris into the air. Huge debris clouds hung in the skies for years, blocking out sunlight and killing off many plant and animal species.

By analyzing layers of sedimentary rock dating back to the time of the impact, the investigators noted that the level of carbon-13 in carbonate deposits declined sharply immediately after the impact, indicating a major loss of life in the ocean. Although some forms of marine life survived the impact, the fossil evidence showed that large-scale marine geochemical and biological cycles did not recover for 3 million years.     [William W. Hay]

In the Special Section, see WHAT HAS CAUSED MASS EXTINCTIONS?

### Seeing beneath the ocean floor

In mid-1998, U.S. geophysicists released a series of images revealing geological activity along a portion of the midocean ridge, where the sea floor is pulling apart to form new crust.

A *sonar* (sound-wave) map of a portion of the ridge (arrow) in the Pacific Ocean reveals that there is more volcanic activity on the westward-moving Pacific plate than on the eastward-moving Nazca plate, *top panel*. Satellite measurements of tiny variations in gravity in the rocks below the sea floor showing differences in density, indicate that there is also a greater flow of *magma* (molten rock) beneath the westward-moving plate, *middle*. That finding was supported by *seismic* (earth tremor) data provided by instruments placed on the sea floor. An image made from the seismic data, *bottom,* shows bands of magma moving toward the west, perhaps explaining the greater volcanic activity on that side of the ridge.

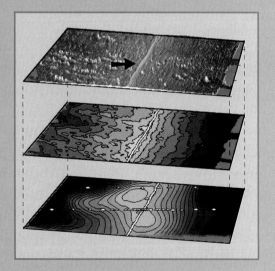

The top medical news of 1998 and 1999 included significant progress in efforts to use stem cells to grow living human tissues in the laboratory. Researchers also accomplished the first successful transplant of a human hand from a donor to another person.

**Advances in stem-cell research.** Teams of investigators at the University of Wisconsin at Madison and Johns Hopkins University in Baltimore announced in November 1998 that they had independently succeeded in isolating human primordial stem cells and growing them in the laboratory. Primordial stem cells are immature cells that have not yet become *specialized* (genetically programmed to become a particular type of cell). Therefore, they have the ability to become any of the different types of cells that make up the body, such as skin, muscle, or nerve cells.

Both teams obtained the stem cells from human *embryos* (organisms at an early stage of development). The embryos were donated by fertility clinics with the consent of the patients from whose egg and sperm cells the embryos had been created.

By exposing stem cells to certain biological and chemical triggers in laboratory cultures, the researchers said, it should be possible to direct the cells to develop into particular types of human tissue. These tissues could be used to replace damaged or diseased tissue in the various organs of the body.

Scientists at Osiris Therapeutics, a small biotechnology company in Baltimore, took stem-cell research a step further in April 1999, when they reported that they had isolated another type of stem cell and used the cells to grow bone, cartilage, fat, tendon, and muscle tissue. This team used human mesenchymal stem cells, a type of stem cell found in the bone marrow of all human beings. Because the use of mesenchymal stem cells does not rely on donated embryos as a source, the development of this branch of stem-cell research would avoid the controversy that has surrounded the field. In addition, experts speculated that the risk of rejection in transplants would be eliminated because this type of stem cell could be taken directly from a patient and used to create new, living tissues that the body would recognize as its own.

**The origin of AIDS.** A particular subspecies of African chimpanzee carries a virus closely related to the human immunodeficiency virus (HIV), the virus that causes AIDS, and so was probably the original source of HIV. That finding was reported in January 1999 by virologist Beatrice H. Hahn and colleagues at the University of Alabama at Birmingham. As part of the ongoing worldwide effort to find a cure for AIDS, the researchers were tracing the roots of HIV-1, the form of the virus responsible for the vast majority of the world's AIDS cases. (Another virus strain, HIV-2, is found mostly in Africa and has not been a major factor elsewhere in the world.)

For many years, scientists suspected that HIV-1 originated from SIV (simian immunodeficiency virus), a similar virus found in monkeys and apes. But Hahn's team found that individual chimp subspecies carry different strains of SIV, which they dubbed SIVcpz (for "chimpanzee"). In addition, they discovered that a particular chimpanzee subspecies, *Pan troglodytes troglodytes,* carries a previously unknown strain of SIVcpz.

This newly identified strain so strongly resembles HIV-1 that only a minor genetic *mutation* (change) would have been needed to make the chimp virus capable of infecting humans. Samples of SIVcpz were isolated from three specimens of *P. t. troglodytes*—and each sample strongly resembled HIV-1.

*P. t. troglodytes* is found in western equatorial Africa, where the first human AIDS cases surfaced. The researchers theorized that humans were probably first infected while butchering chimpanzee meat, which is consumed by people in many parts of Africa.

AIDS researchers hope that by learning why SIVcpz infection does not cause illness in chimps, they might find a way to block HIV-1 infections from triggering AIDS in humans. Similar studies could also lead to the development of tests to identify other viruses that could potentially cause disease in humans.

**A show of hands.** Successful operations to transplant a human hand were reported in September 1998 and January 1999. Clint Hallam, a 48-year-old Australian, became the first person to receive a successful hand transplant after undergoing a 13-hour procedure at Edouard Herriot Hospital in Lyon,

France. Four months later, Matthew David Scott, 37, of Absecon, New Jersey, underwent a similar procedure at Jewish Hospital in Louisville, Kentucky.

Although surgery to reattach a person's own severed limb had become almost routine by 1998, previous attempts to transplant a limb from a donor had all failed due to rejection of the donated limb. (Rejection is a biochemical reaction in which the recipient's immune system attacks the transplant as though it were an invading organism.) The only previous attempted hand transplant took place in Ecuador in 1964, but the patient's body rejected the donor hand after only two weeks. In later years, however, sophisticated new drugs that suppress the immune system gave doctors new hope for success.

Several months after the operation on Hallam, who had lost his right hand in a sawing accident, his doctors reported that he had developed a grip strong enough to hold a glass in the transplanted hand and drink from it. In addition, they said, Hallam had experienced significant nerve regeneration from the point of the transplant down to the beginning of the palm, and the transplant showed no signs of rejection.

Scott, who had lost his left hand in a 1985 fireworks accident, underwent a 15-hour procedure to receive his new hand. A 17-member surgical team attached the hand at a point 5 centimeters (2 inches) above the wrist. Scott's doctors said that the ability of the hand to function would not be known for some time and that more than a year of physical therapy would be needed. Early reports after the surgery indicated that Scott had acquired movement in the tips of his new fingers.

To prevent rejection, Hallam and Scott, like other transplant patients, will have to take immunity-suppressing drugs for the rest of their lives. Because these drugs weaken the immune system, transplant patients become more susceptible to infections, cancer, and other potentially life-threatening conditions. This fact contributed to an ethical debate over the value of transplanting so-called nonvital body parts. Unlike a heart or liver transplant, a limb transplant is not a lifesaving procedure. Therefore, some people argued, the benefits of receiving such a transplant do not justify the potential health risks posed by having to take immunity-suppressing drugs.

**Lyme disease drug.** In December 1998, the United States Food and Drug Administration (FDA) approved the sale of LYMErix, the first vaccine to prevent Lyme disease. A bacterial infection spread by the bite of a deer tick, Lyme disease afflicts some 16,000 people in the United States each year. (The disease is named for the city in Connecticut where it was first identified in 1975.) A person infected by a tick bite typically develops a red, ring-shaped rash, accompanied by fever and chills. If left untreated by antibiotics, the bacterium, *Borrelia burgdorferi,* moves through the bloodstream, causing such potentially serious and chronic effects as arthritis, hearing loss, heart problems, and inflammation of the optic nerve.

The new vaccine was derived from OspA ("outer surface protein A"), a protein that is present on the exterior surface of the Lyme bacterium. The vaccine causes the immune system to produce disease-fighting molecules called antibodies that recognize and attack OspA on Lyme bacteria from an infected tick. When a bite occurs, the antibodies enter the tick's body, latch onto the OspA on the bacteria it is carrying, and destroy the microbes.

LYMErix is given in a series of three injections over a one-year period—the first two a month apart and the third 11 months later. In clinical trials reported to the FDA, the vaccine was 78 percent effective in adults who received all three injections.

The vaccine was not recommended for people under age 16 or over age 70, because it was not tested in those age groups, or for people with arthritis or certain heart conditions. It is most appropriate for individuals who work outdoors for extended periods in areas where Lyme disease is most prevalent, including the Northeastern and upper Midwestern United States. Because LYMErix does not produce complete protection, the FDA advised people who received the vaccine to use a tick repellent containing 30 percent diethyl toluamide (deet) on exposed skin while outdoors.

**Major cancer-treatment changes.** In February 1999, officials at the National

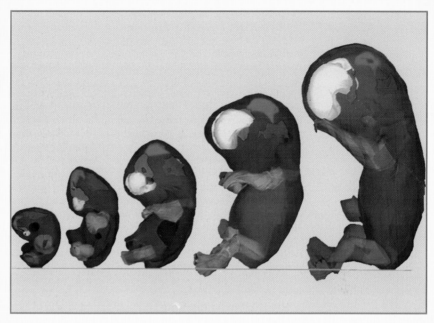

A system that combines calculations of volume with traditional ultrasound imaging, which uses high-frequency sound waves, produced views of five human embryos 7 to 10 weeks after conception. The data were fed into a computer, which used the information to create three-dimensional digital simulations of the tiny embryos and their internal structures. The simulations could then be manipulated in the computer to provide views from a variety of angles. The images, published in February 1999, were created by obstetrician Harm-Gerd Blaas of Trondheim University Hospital in Norway.

Cancer Institute (NCI) announced that new research findings had prompted the agency to make major changes in its recommendations for the treatment of advanced cervical cancer. Five new studies had revealed that treatment plans combining radiation and chemotherapy could reduce death rates from cervical cancer by 30 to 50 percent, compared with using radiation alone. The new recommendations were considered the first important change in guidelines for treating cervical cancer since radiation therapy was introduced in the 1950's.

The studies involved more than 1,800 women at a number of medical centers in the United States. In three of the studies, researchers evaluated treatment with radiation alone; with radiation in combination with a cancer-fighting drug called cisplatin; and with radiation and two drugs, cisplatin and an anti-cancer compound called fluorouracil (5-FU). The results of all three studies showed that a combined radiation-drug treatment for cervical cancer is much more effective than radiation alone.

In the other two studies, all the pa-tients received both radiation and che-motherapy. Half the women in these studies received drug therapy consisting of either just a drug called hydroxyurea or a combination of hydroxyurea and cisplatin. Here too, the use of both radiation and drug resulted in the highest survival rates, with cisplatin proving to be the more effective drug.

In announcing the results of all five studies, the NCI said that the best chemotherapy regimen for advanced cervical cancer had not been determined. It added, though, that "significant results were seen using cisplatin alone or cisplatin in combination with 5-FU and other agents."

Approximately 12,800 women in the United States develop cervical cancer each year. Since the 1950's, surgery and radiation therapy have been the standard treatment strategy for cervical cancer that had spread *locally* (within the cervix) or *regionally* (within the pelvis). The new guidelines apply to about 25 percent of women with cervical cancer—those with "locally advanced" or "invasive" disease, which means that either the tumor is large or that cancerous cells have spread to other parts of the pelvis or to the nearby lymph nodes.

NCI Director Richard D. Klausner said that the new research was "likely to change the standard of care for invasive cervical cancer." Ironically, as recently as 1996, the National Institutes of Health had issued a "Consensus Statement on Cervical Cancer" which concluded that no evidence existed at that time to support the use of chemotherapy in standard practice.

**Brain cell regeneration.** The first evidence that the brains of adult humans develop new cells was reported in November 1998 by researchers from the Salk Institute for Biological Studies in La Jolla, California and the Sahlgrenska University Hospital in Sweden. The belief that dead brain cells are not replaced was commonly accepted by researchers, but only because of a lack of scientific evidence to the contrary.

The investigators discovered that *neurons* (nerve cells) in a part of the brain called the hippocampus are constantly dividing and producing new cells. The hippocampus is the region where learning occurs and memories are processed. It is also an area that is affected by Alzheimer's disease and other degenerative brain disorders.

The scientists injected five cancer patients with a chemical substance called bromodeoxyuridine (BrdU). In the body, BrdU attaches to the DNA (deoxyribonucleic acid, the molecule that genes are made of) of dividing cells and can thus serve as a marker to identify newly formed cells.

When these patients later died (at ages ranging from 57 to 72), slices of their hippocampus were examined using advanced imaging techniques. The researchers detected the presence of BrdU in the cells of each person's hippocampus, indicating that neurons had been formed after the chemical was injected into the bloodstream. This showed that neurons in the adult brain, and even in the brains of elderly individuals, continue to divide and generate new cells. In the five test subjects, cell division occurred at a rate that produced several hundred new cells per day.

The study's findings may lead to treatments for patients with disorders such as Alzheimer's disease and Huntington's disease, both of which are characterized by cell injury or death of cells in the brain and nervous system. One possibility is that neurons could be grown in the laboratory for transplantation in patients with this type of disease.

**The cause of Parkinson disease.** In January 1999, two years after the identification of a gene believed to be the cause of many cases of Parkinson disease, a study concluded that most cases of the nerve disorder are actually triggered by exposure to chemicals in the environment. Researchers at the Parkinson's Institute in Sunnyvale, California, reached this conclusion after a study of almost 20,000 male twins.

The subjects were drawn from a list of twins who were veterans of World War II (1939-1945). The results showed that Parkinson disease developed most often in only one member of a pair of twins, regardless of whether the twins were *identical* (the result of the split of a single fertilized egg) or *fraternal* (the result of two separate fertilized eggs). Because identical twins have the same genetic code, both twins would be expected to develop Parkinson disease if the condition had a genetic basis. But because the identical twins of Parkinson patients were found to be at no more risk of developing the disease than the general population, the researchers concluded that no genetic component was responsible for developing the condition.

Although the specific substances responsible for causing Parkinson disease remained unidentified, physicians speculated that risk factors might include exposure to pesticides or living close to industrial plants. The study did not completely rule out genetic factors in some cases of Parkinson disease. In fact, a small percentage of patients in the study were found to have a familial type of the disease, which develops before age 50. When this form of the disease was found, it occurred in both twins.

Between 500,000 and 1 million Americans in 1999 had Parkinson disease. Symptoms of the condition include trembling hands, rigid muscles, slow movement, and balance difficulties.

[Richard Trubo]

See also SCIENCE AND SOCIETY. In the Special Reports section, see TISSUE ENGINEERING—FROM SCIENCE FICTION TO MEDICAL FACT.

The 1998 Nobel Prizes in science were awarded for the development of methods for the theoretical study of molecular processes, the discovery of unexpected behavior in *electrons* (negatively charged particles in the atom), and discoveries about the role of nitric oxide in the body.

**The Nobel Prize for chemistry** was shared by Walter Kohn of the University of California at Santa Barbara and John Pople of Northwestern University in Evanston, Illinois. The two physicists independently developed computational methods that led to the development of *quantum chemistry*, the application of quantum mechanics—a branch of physics that deals with the behavior of atoms and subatomic particles—to chemical reactions.

Chemists have attempted since the early 1900's to understand and predict chemical reactions by describing how bonds between atoms function. The mathematical framework for understanding these bonds involved equations that were often too complex to be solved. Walter Kohn developed a theoretical system that simplified the computations. His "density functional theory" maps the spatial distribution of electrons around atoms. Chemists use powerful computers and Kohn's method to predict the shape and reactivity of atoms. The calculations allow chemists to design and manipulate "virtual" molecules on computer screens.

John Pople developed computational tools to aid in the theoretical study of the properties and chemical interactions of molecules. Those tools include a library of energy components for electrons in various molecular settings. He incorporated his tools into a software computer program, Gaussian, that allows users to predict the outcome of highly complex chemical reactions. The program can also calculate the stability and geometry of any particular molecule. Pople's software, which is used worldwide by research chemists, unlocked quantum chemistry for broad segments of the scientific community.

**The prize for physics** was shared by Horst Stormer of Columbia University in New York City, Daniel Tsui of Princeton University in Princeton, New Jersey, and Robert Laughlin of Stanford University in Stanford, California. In 1982, Stormer and Tsui found that electrons, under high magnetic fields and low temperatures, collapse into "quasiparticles" with electrical charges that are fractions of the original charges. This phenomenon—named the "fractional quantum Hall effect"—appeared to shatter the scientific rule that all electrons have the same electric charge. Laughlin was awarded the physics prize for explaining the phenomenon. He demonstrated that electrons in a powerful magnetic field can condense into a substance called a quantum fluid—Stormer and Tsui's quasiparticles—that conducts electricity without resistance. According to Laughlin, the fractional quantum Hall effect shows the unpredictability of the laws of quantum mechanics.

**The prize for physiology or medicine** went to Robert Furchgott of the State University of New York in New York City, Louis Ignarro of the University of California at Los Angeles, and Ferid Murad of the University of Texas Medical School in Houston. The three pharmacologists completed important work on the function of nitric oxide in the human body. In the 1980's, Furchgott recognized that an unknown factor in the body could counteract the effects of various drugs on human blood vessels. He eventually discovered that cells lining the inside of blood vessels produce a substance that relaxes smooth muscle and causes the vessels to widen. In 1986, Furchgott and Louis Ignarro, working independently, discovered that the substance was nitric oxide.

This simple gas, commonly associated with pollution produced by burning fossil fuels, functions as a *signal molecule* (biological messenger), carrying important information through the body. Furchgott and Ignarro's announcement sparked intense research on nitric oxide, which revealed that the gas initiates functions in the immune, nervous, and cardiovascular systems. Ferid Murad, also working independently, discovered that nitroglycerin used to treat heart disease releases nitric oxide, which causes blood vessels to relax, lowering blood pressure. Nitric oxide research also disclosed that the gas triggers erection of the penis. This discovery led to the development of Viagra, the first drug for *impotence* (the inability to maintain an erection).                [Scott Thomas]

## Nobel Prize winners in physiology or medicine

Pharmacologists Ferid Murad of the University of Texas Medical School in Houston, Robert Furchgott of the State University of New York in New York City, and Louis Ignarro of the University of California at Los Angeles were awarded the 1998 Nobel Prize for physiology or medicine. The three researchers were recognized for their investigations into how nitric oxide functions as a messenger in the human body, initiating changes in the immune, nervous, and cardiovascular systems.

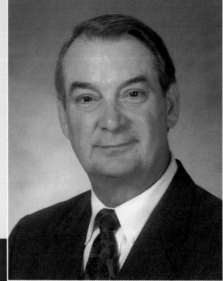

Ferid Murad

Robert Furchgott

Louis Ignarro

A large amount of fiber in a person's diet may not help prevent colorectal cancer, medical scientists reported in January 1999. Researchers led by *oncologist* (cancer specialist) Charles Fuchs of Brigham and Women's Hospital in Boston, found no evidence that women who eat large amounts of high-fiber foods—such as bran, fruits, vegetables, and whole-wheat bread—lower their risk of colon cancer. That conclusion contradicted many years of dietary advice that fiber helps prevent such cancer.

Fiber is a complex carbohydrate that cannot be broken down and absorbed by the body. It passes through the intestinal tract and is excreted in the *feces* (solid waste matter). Experts had long believed that dietary fiber prevents colorectal cancer, in part because that disease is rare in Africa, where people typically eat a high-fiber diet. They offered several theories to explain how dietary fiber might prevent colorectal cancer, including the possibility that it dilutes or absorbs fecal *carcinogens* (cancer-causing agents), or reduces the time that potential carcinogens remain in the colon.

The Boston researchers tracked the eating habits of 88,757 women who had participated in the Harvard Nurses Health Study from 1980 to 1996. The women had no history of cancer or inflammatory bowel disease and were between the ages of 35 and 59 when the study began. The participants completed questionnaires about their diet in 1980, 1984, and 1986.

During the 16-year study, 787 of the women developed cancer of the colon or rectum. After adjusting for age and other dietary factors, such as energy intake, Fuchs and his colleagues concluded that the women who consumed the most fiber—about 25 grams daily—were no less likely to develop colon or rectal cancer than women who ate the least amount of fiber—about 10 grams daily. The researchers said that their conclusion would undoubtedly also apply to men, even though men had not been included in the study.

Previous studies had been inconclusive about the possible link of dietary fiber and colorectal cancer. These studies had included only limited data on other dietary factors, making it difficult to separate the effects of fiber from other components of plant foods. More-

over, those studies examined people's previous dietary intake after colorectal cancer had been diagnosed.

Even with the results of Fuchs's study, nutritionists said that fiber was still an important part of the diet. Many studies have found that a diet rich in fruits, vegetables, and whole grains, which are good sources of fiber, have other health benefits, including reducing the risks of heart disease and high blood pressure.

**Cholesterol and dietary fiber.** An analysis of 67 controlled clinical trials that had investigated the effects of dietary fiber on cholesterol in the blood concluded that fiber can reduce cholesterol levels in some people. That finding was published in January 1999 by researchers led by nutritionist Lisa Brown at the Harvard School of Public Health in Boston.

In each of the studies analyzed, participants had received fiber supplements or food with fiber-enriched ingredients such as oat bran. All of the types of fiber used in the various studies were water-soluble. The analysis conducted by Brown and her colleagues showed that these kinds of fiber resulted in a small but significant reduction in blood cholesterol, specifically low-density lipoprotein (LDL) cholesterol. Elevated LDL cholesterol, often called "bad cholesterol," is considered a contributor to heart disease.

When whole foods, such as fruits and vegetables, are the source of fiber, the researchers said, the effect on cholesterol may be greater. Moreover, they noted, a diet rich in fruits and vegetables provides the body with a number of other substances that are known to lower the risk of disease.

**Tomato-based foods and cancer.** A study published in February 1999 found that people who eat tomatoes and tomato-based products lower their risk of developing numerous cancers.

Scientists led by Edward Giovannucci, a nutritionist at the Harvard School of Public Health, examined the results of 72 prior studies that had investigated the link between tomato consumption and cancer. Some of the studies also looked for blood concentrations of lycopene, the compound that gives tomatoes their red color. Lycopene is known to protect cells from chemicals called free radicals, which can cause cancer.

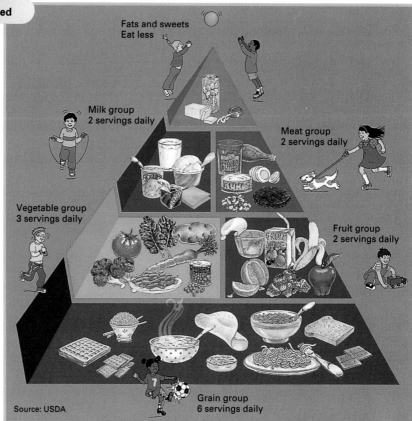

Fats and sweets
Eat less

Milk group
2 servings daily

Meat group
2 servings daily

Vegetable group
3 servings daily

Fruit group
2 servings daily

Grain group
6 servings daily

Source: USDA

The United States Department of Agriculture (USDA) in March 1999 released a version of the Food Guide Pyramid designed to more clearly explain the average daily nutritional needs of children. Aimed at children from 2 to 6 years old, the pyramid—similar to one that the USDA had previously introduced for all age groups—featured simplified language and illustrations. It recommended smaller servings of meat, dairy products, and candies, while emphasizing larger servings of grains, fruits, and vegetables.

Giovannucci reported that 57 of the 72 studies linked tomato intake with a reduced cancer risk. In 35 of those studies, the connection was considered strong enough to be statistically significant. The link was strongest for cancers of the prostate gland, lung, and stomach. The findings also suggested a link between tomatoes and lower levels of several other cancers, including pancreatic, colorectal, esophageal, oral, breast, and cervical cancers.

The individual studies included in the Harvard summary study included tomatoes in various forms, including raw and in ketchup, spaghetti sauce, tomato paste, soup, and salsa. Benefits were found from all those forms, and cooking and processing tomatoes did not diminish the positive effects. In fact, the scientists found that cooked tomatoes were more potent, possibly because cooking them breaks down the skin and releases more lycopene.

Most of the studies examined by Giovannucci and his colleagues targeted the dietary differences between healthy individuals and people who had been diagnosed with cancer. In many instances, the primary difference between the two groups appeared to be either the consumption of tomatoes or tomato products or the presence in the blood of high levels of lycopene. Although none of the studies explained how tomatoes lower the risk of cancer, the ability of lycopene to neutralize free radicals presumably plays a part in their protective effect.

**Organic meat.** Officials at the United States Department of Agriculture (USDA) announced in January 1999 that the agency would permit meat and poultry products to be labeled organic. The new labels were expected to be in use by mid-1999. Organic certification would mean that the animals had not been exposed to antibiotics or confined in small feeding areas.

Until January, the USDA had permitted only naturally grown fruits and vegetables to be labeled organic. USDA officials in spring 1999 were still developing specific national standards to cover the organic meat industry.

[Phylis Moser-Veillon]

Scientists at the National Oceanic and Atmospheric Administration (NOAA) in June 1998 began studying a condition in the central Pacific Ocean that turned the tropical waters abnormally cool. The phenomenon, called La Niña, is the opposite of El Niño, which is a warming of the central Pacific Ocean. A major El Niño in 1997 and early 1998 affected global weather patterns.

A La Niña (Spanish for *the little girl*) often develops after a strong El Niño (Spanish for *the child,* referring to the Christ child, because it usually occurs around Christmas). A La Niña occurs about every three to seven years—not always after an El Niño—and lasts one to two years.

Scientists used satellite measurements of sea level, or the height of the ocean, to observe La Niña. Because water expands when it gets warmer and contracts when it grows cooler, the sea level tells scientists how much heat is contained in a region of the ocean. The world's oceans are reservoirs of heat, and can influence climatic conditions by cooling or heating the atmosphere.

A La Niña is generated by stronger-than-normal winds along the equator. These winds skim away sun-drenched surface waters, exposing the cooler waters beneath. The normal surface temperature of the ocean near the equator is about 30 °C (86 °F). Oceanographers estimated that the 1998-1999 La Niña dropped surface temperatures by 2 to 3 °C (3.6 to 5.4 °F).

Like El Niño, La Niña can affect global weather patterns. Scientists said that the La Niña of 1998 and 1999 was associated with storms and flooding in South Africa, the Philippines, and Indonesia and droughts in Kenya and Tanzania. Oceanographers theorized that in the United States, the phenomenon may also have caused heavy rain and snow in the Pacific Northwest and upper Midwest. Based on satellite measurements taken in January 1999, scientists believed that La Niña would end by the following summer.

**Coral bleaching** spread extensively throughout the world's reef communities in the late 1990's and often infected coral that had been free of the problem. That finding was reported in September 1998 by the International Society for Reef Studies, a scientific organization dedicated to protecting coral reefs.

Bleaching occurs when algae living in coral *polyps*—the small, hollow-bodied animals that make up the living outer layer of a reef—die or leave the coral for some reason. Because these algae are often brightly colored, their absence may turn the coral a stark white. Although coral reefs can often recover from moderate bleaching, severe bleaching can eventually kill a reef. The white skeleton that is left behind is then overgrown by underwater plants.

Scientists theorized that several factors, including pollution and a gradual warming of tropical waters in recent decades, may be causing the bleaching. They said the problem might have been made worse in 1999 by higher-than-normal water temperatures resulting from the El Niño of 1997 and 1998.

The International Society for Reef Studies reported that coral bleaching had been described by scientists in 32 countries in 1997 and 1998. The Pacific and Indian oceans, the Mediterranean and Caribbean seas, the Red Sea, and the Persian Gulf were the areas hardest hit by the bleaching. In some sections of the Indian Ocean, coral mortality from bleaching reached 90 percent in late 1998.

Some oceanographers predicted that up to 70 percent of all coral reefs worldwide may be destroyed by 2050 if the rate of decline that existed in 1999 continues.

**Coral reefs and $CO_2$.** An accumulation of carbon dioxide ($CO_2$) in the world's oceans may also pose a threat to coral reefs. An international team of researchers led by Joan Kleypas, a biologist at the National Center for Atmospheric Research in Boulder, Colorado, reported that finding in April 1999.

The researchers reviewed early experiments that had been conducted by scientists at French aquariums and at the Biosphere 2 facility in Arizona. In those experiments, the scientists discovered that adding extra carbon dioxide to water tanks at the facilities disrupted chemical reactions that corals use to produce their limestone skeletons, from which reefs are built.

The oceans absorb $CO_2$ from the air as part of a natural cycle. After reviewing the findings of the earlier experiments, the Boulder researchers theo-

# Spying on the Seas

The rapidly advancing science of oceanography took another step forward in mid-1999 with the scheduled launch of an advanced National Aeronautics and Space Administration (NASA) satellite designed to study ocean winds from orbit. The spacecraft, the Quick Scatterometer (QuikSCAT), was just one of many satellites that were enabling scientists to monitor the oceans from space.

Satellite technology revolutionized the study of the Earth during the 1990's. Scientists used satellite images for everything from tracking schools of tuna to watching volcanic eruptions. By 1999, satellites were making almost continuous measurements of ocean temperature, heat content, color, and sea-surface height.

QuikSCAT, orbiting at an altitude of about 800 kilometers (500 miles), follows a path that takes it almost directly over the North and South poles. On this course, the satellite views some 90 percent of the world's oceans in a single day. The satellite collects data over oceans, land, and ice in a continuous band 1,800 kilometers (1,118 miles) wide. Its observations enable scientists to gather information about near-surface wind speeds and wind direction faster and more completely than had been possible.

QuikSCAT, like many satellite instruments, views Earth by emitting a beam of *microwaves*—high-frequency radio waves—and then detecting the portion of the beam that is reflected back to it. The microwaves bounce like rubber balls off tiny wrinkles called capillary waves on the surface of the ocean. Because capillary waves are made only by winds, the reflection pattern captured by the satellite tells scientists how strong and in what direction the winds were blowing when the measurements were made. With better wind data, scientists can improve weather forecasts on land and give advanced warnings for natural disasters such as hurricanes and typhoons.

QuikSCAT was the successor of another sea-monitoring satellite, the NASA Scatterometer (NSCAT). NSCAT began transmitting data in 1996 but was rendered useless in 1997 when its solar panel broke. QuikSCAT filled the void created by the unexpected failure of NSCAT.

Like QuikSCAT, NSCAT used microwaves to gather data from the ocean. The main advantage in using microwaves for monitoring the planet from space is that they are able to penetrate the atmosphere and clouds. Because they have short wavelengths, they can also detect fine oceanographic features such as capillary waves or changes in the tides. Visible light, which like microwaves is a form of electromagnetic radiation but has much longer wavelengths, does not have these properties.

Microwave technology was first developed for space probes being sent to explore other planets in the solar system. As the technology advanced, scientists created instruments with microwave sensors that could be aimed down at the Earth.

Scientists use the TOPEX/Poseidon satellite, shown in an artist's drawing, to determine the height of the sea surface by measuring how long it takes microwaves sent by the satellite to bounce off the water below and return. Where the ocean is higher, it is warmer. Data gathered by the satellite allow oceanographers to track average seasonal sea levels and determine how much heat is stored in the oceans.

Another microwave satellite is TOPEX/Poseidon. The TOPEX/Poseidon satellite is operated jointly by the space agencies of France and the United States. TOPEX/Poseidon determines ocean height by measuring the time it takes for the microwaves sent by the satellite to travel down to the ocean and back again. The satellite houses an instrument called an altimeter, which uses the time interval to calculate the distance between the satellite and the ocean surface. Where the distance is shorter, the water has formed a bulge, meaning that it is warmer than the surrounding waters.

In 1998 and early 1999, oceanographers used TOPEX/Poseidon to study a well-known but mysterious shift in the temperature of the central Pacific Ocean. A strong El Niño, a warm ocean current that can trigger far-reaching changes in weather patterns, had given way to a cool ocean current called La Niña. El Niño and La Niña were the first major oceanic events ever captured by satellite imagery.

Scientists in 1999 also continued to gather ocean data from an older satellite technology called the Advanced Very High Resolution Radiometer (AVHRR). The AVHRR senses infrared radiation—"heat rays"—as it is radiated from the ocean's surface. The warmer the water, the more infrared is radiated. Thus, the AVHRR indirectly measures ocean temperatures. Since the 1970's, scientists have used AVHRR data to monitor global warming and examine natural changes in climate that, researchers theorize, may have played a role in the onset of major climatic events, such as the ice ages that have occurred repeatedly during the past few million years. The AVHRR is operated by the National Oceanic and Atmospheric Administration (NOAA) and has been collecting temperature measurements since 1978. It currently is carried on NOAA's Polar Orbiting Environmental Satellites.

Not all satellite instruments use microwaves or infrared light to observe oceanographic phenomena. A few, like the Sea-Viewing Wide Field-of-View Sensor (SeaWIFS), launched in 1997, are sensitive to visible light. SeaWIFS measures ocean color, which is determined by *phytoplankton* (microscopic plantlike organisms). Phytoplankton turn water green because they contain chlorophyll. Hence, the greener the water, the greater the population of phytoplankton.

Oceanographers in spring 1999 used SeaWIFS to document an intense phytoplankton bloom in the North Atlantic Ocean. Each spring, as sunlight intensifies, the phytoplankton populations in the Atlantic explode. It was only the second time this bloom had been observed from space.

Phytoplankton are also an important component of the global carbon cycle, as they remove

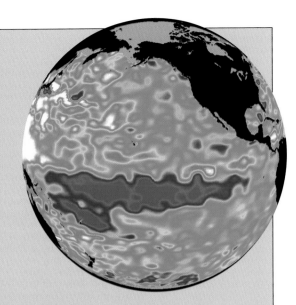

An image of the Earth created with data obtained by the TOPEX/Poseidon satellite in November 1998 shows water temperatures in the Pacific Ocean. The blue and purple band in the central Pacific is a region of cooler-than-normal water known as La Niña.

carbon dioxide from the air. Carbon dioxide is a greenhouse gas—one that traps heat in the atmosphere. SeaWIFS's observations of ocean color help scientists estimate the rate at which oceans are removing carbon dioxide from the atmosphere. This has important consequences for understanding how human activities, such as the burning of fossil fuels (which produces carbon dioxide), may be changing the global climate.

Oceanographers in 1999 were also working on other methods for monitoring the seas from space. For example, they were deploying floats, instruments that ride on ocean currents and radio their positions to satellites. By tracking the floats, oceanographers learn how the currents move. However, these floats are only capable of one-way communication and cannot take commands from shore.

By adding two-way communication systems to such floats, oceanographers would be able to go one step further than simply gathering information from the devices—they would be able to use communication satellites to control the floats' paths. In this way, oceanographers would greatly enhance their ability to explore the open ocean. Such systems were already under development in 1999. Some scientists predicted that such systems would be ready for testing in the early 2000's.

[Christina S. Johnson]

## How fishing can destroy the sea floor

Fishing boats that drag their nets along the ocean floor to snare shrimp and other bottom-dwelling sea creatures may be destroying an area twice as large as the continental United States each year, according to a December 1998 study by the Marine Conservation Biology Institute in Redmond, Washington.

A portion of undisturbed sea floor in the Georges Bank area off the coast of Massachusetts contains a thriving sea-floor community.

A second area, just 500 meters (1,640 feet) away from the one shown above, was left bare after a trawling net passed through it. Marine biologists predicted that it would take decades or even centuries for life to return to the area.

rized that $CO_2$, which has been increasing in the atmosphere as a result of human activities such as the burning of fossil fuels, has been building up in the oceans. Although corals use carbon in the form of calcium carbonate (limestone) to build their skeletons, too much carbon dioxide in the water has a negative effect on them. Carbon dioxide in the water creates carbonic acid, which can damage coral. Carbon dioxide also combines with calcium carbonate, thereby reducing the amount of the calcium compound available to corals.

The scientists predicted that the extra $CO_2$ would not kill coral but would slow its growth and make it more susceptible to storm damage and erosion.

**More on oceanic $CO_2$.** The oceans not only absorb carbon dioxide, they also release much of it back into the atmosphere as the gas wells up to the surface from great depths. In April 1999, NOAA scientists reported that El Niños greatly reduce the release of $CO_2$ in the tropical Pacific, which is a major area for $CO_2$ upwelling.

The researchers, led by oceanographer Richard Feely, studied an El Niño that occurred from 1991 to 1994. They found that 30 to 80 percent less carbon dioxide was released into the atmosphere in the Pacific during each of those four years, compared with non-El Niño years.

According to the NOAA study, the Pacific along the equator released about 816 million metric tons (900 million tons) of carbon in 1996, a non-El Niño year. During an El Niño episode, however, the ocean released 272 million metric tons (300 million tons) in 1992, 544 million metric tons (600 million tons) in 1993, and 635 million metric tons (700 million tons) in 1994.

An El Niño prevents the normal upwelling of $CO_2$ to the surface, the NOAA scientists said. In addition, prior research on El Niños showed that the phenomenon also causes an increase in the growth of underwater plant life, which absorbs carbon dioxide.

Many oceanographers said these studies would help researchers better understand how changes in the release of oceanic carbon dioxide affect the Earth's climate. $CO_2$ is a so-called greenhouse gas, which helps retain heat in the atmosphere.

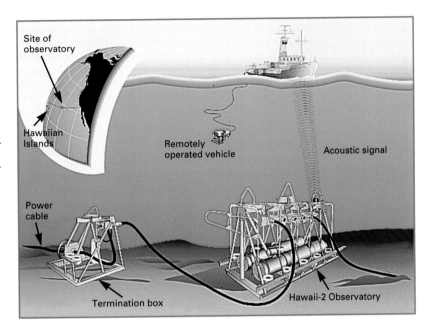

Scientists at the Woods Hole Oceanographic Institution in Massachusetts began in September 1998 to operate the first deep-ocean observatory. The Hawaii-2 Observatory sits in 5,000 meters (16,400 feet) of water between Hawaii and California. Data on sea-floor phenomena, such as earthquakes, are conveyed by an underwater cable. The cable also supplies electrical power through a termination box, which functions much like a phone jack. Scientists on a surface ship use an *acoustic* (sound) signal to locate the observatory. A remotely operated vehicle is used to connect instruments to the observatory.

Site of observatory

Hawaiian Islands

Remotely operated vehicle

Acoustic signal

Power cable

Termination box

Hawaii-2 Observatory

**Targeting harmful species.** United States President Bill Clinton in February 1999 signed an executive order establishing an Invasive Species Council. The council was charged with coordinating a federal plan to slow the spread of non-native plants, insects, and other animals in the United States, including in bays and waterways along the coasts.

The council, which was to be led by representatives from the Commerce, Agriculture, and Interior departments, was ordered to present a comprehensive plan in 2000 to combat the spread of invasive species. Foreign plants and animals can harm native species, which have few defenses against the interlopers. For example, the Japanese shore crab, which was first sighted in New Jersey in 1988, had spread and inhabited shorelines along the East Coast from Massachusetts to North Carolina by 1998. The shore crab reproduces in dense colonies and eats native species, including blue mussels and soft-shell clams. Scientists believed that the shore crabs' expansion can decimate such species.

Oceanographers and marine biologists in 1999 believed that the increase in marine bio-invasions in the 1990's was linked to an increase in the amount of shipping traffic into U.S. ports. Ships pump ballast water into their hulls to make themselves heavier after unloading their cargo or to provide stability in rough seas. That water is swarming with organisms.

When the ships discharge the water after loading cargo or as they enter calmer seas near port, eggs, larvae, and adult animals are likely to pour out into the new, foreign marine environment. The world's cargo ships may carry tens of thousands of marine species in their ballast waters at any given time, research has shown.

Studies have also found that busy ports and harbors worldwide are among the areas most vulnerable to invasions by exotic species. Although most foreign organisms die when introduced into a new habitat, the few species that manage to survive may thrive at the expense of the native species.

[Christina S. Johnson]

A giant particle detector located deep in a mine in western Japan revealed the first strong evidence that neutrinos, the most elusive of the fundamental particles of matter, may possess a tiny amount of mass. The preliminary reports on the evidence were announced in June 1998 by a team of more than 120 Japanese and American scientists, led by physicist Yoji Tatsuka of the University of Tokyo's Institute for Cosmic Ray Research.

**How to detect a neutrino.** A neutrino belongs to the same family of particles as the electron, which is a negatively charged particle. Neutrinos, however, carry no electric charge. Because particle-detecting equipment is sensitive only to charged particles, scientists are not able to observe neutrinos directly.

But researchers can detect neutrinos indirectly. If a neutrino collides with a proton or neutron, the particles in the nucleus of an atom, the neutrino is transformed into a charged particle. These reactions, however, are rare—a neutrino can pass all the way through the Earth with only a small chance of such a collision.

There are three kinds of neutrinos, which are identified by the kind of charged particle into which each can be transformed. Electron neutrinos, or e-neutrinos, become electrons. The other two types, called mu-neutrinos and tau-neutrinos, transform into particles called muons and taus, which are sometimes referred to as heavy electrons.

The neutrinos studied in Japan originated high in the Earth's atmosphere. The neutrinos were produced from reactions induced by high-energy cosmic-ray particles colliding with air molecules. (Cosmic-ray particles are the only matter from outside the solar system that reaches the Earth.) Physicists had learned in other studies that these upper-atmosphere reactions produce roughly two mu-neutrinos for every e-neutrino, and only an occasional tau-neutrino.

These three kinds of neutrinos are constantly bombarding the Earth. (An estimated 10 trillion neutrinos pass through a human body every second.) But to study these neutrinos closely, physicists had to create sophisticated instruments, such as Japan's Super-Kamiokande, or Super-K, detector.

The primary feature of the Super-K is a tank containing 45 million liters (12.5 million gallons) of ultrapure water. The tank's inside surfaces are covered with 11,200 photomultiplier tubes, or electronic "eyes," that are wired to a computer.

If a particle passing through the water collides with a proton or neutron, the reaction generates a tiny flash of light. The photomultiplier tubes detect these flashes, and the electronic equipment translates the readings into information, such as the path of the original particle through the water and what kinds of particles were the result of the reaction.

Another important feature of Japan's Super-K detector is that it is located about 1,000 meters (3,280 feet) underground. Most particles bombarding the Earth are stopped at or near its surface. By building the detector so far underground, the researchers could screen out almost all other particles and therefore observe almost exclusively the reactions of neutrinos.

**In search of "missing" neutrinos.** At the time of the June 1998 report, the scientists had recorded more than 4,500 transformations of neutrinos into charged particles inside the Super-K. The scientists observed, however, fewer muons than the expected two-thirds of the transformations. Therefore, some of the expected mu-neutrinos seemed to have disappeared.

A neutrino, however, cannot simply disappear. It must be transformed into another particle. The most likely explanation for the "loss" of mu-neutrinos is that they were being changed into tau-neutrinos by *neutrino oscillation,* a process in which a neutrino switches from one form to another. Tau-neutrinos, in turn, are difficult to detect because they very rarely have enough energy when colliding with a proton or neutron to transform into the heavy tau particle. Therefore, the "lost" mu-neutrinos most likely were transformed by oscillation into tau-neutrinos and then went undetected.

There was additional evidence to back up the neutrino oscillation explanation. Among the neutrinos that had traveled a relatively short distance, directly from the sky and through 1,000 meters of the ground, there were close

## Weighty neutrinos

A team of physicists from Japan and the United States reported in June 1998 on evidence that an elusive subatomic particle, the neutrino, may possess a tiny amount of mass.

The neutrinos they studied had been produced high in the Earth's atmosphere by reactions between cosmic rays and air molecules, *below*. These neutrinos bombard the Earth from all directions and are capable of passing all the way through the Earth, only rarely interacting with other particles.

Researchers check 1 of the 11,200 *photomultiplier tubes,* or electronic "eyes," inside the cavernous tank of Japan's Super-Kamiokande (Super-K) neutrino detector, *above.* When the detector is operating, the tank is filled with ultrapure water.

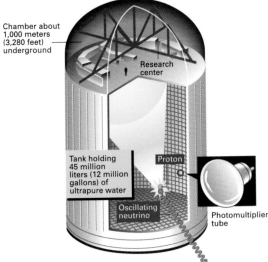

Neutrinos are virtually the only particles able to penetrate the ground and pass through the Super-K's tank, *left.* Occasionally, one of the neutrinos collides with a proton or neutron in the water. This collision transforms the neutrino into another particle. The energy released in this process creates a tiny flash of light that the photomultiplier tubes can translate into electronic data. After analyzing the data, physicists deduced that the neutrinos that had traveled all the way through the Earth had oscillated, or switched back and forth, between two different forms of neutrinos. Oscillation, physicists know, can only occur if at least one of those forms has mass.

Source: University of Hawaii.

to the expected two-thirds count of mu-neutrinos. In contrast, the neutrinos that had traveled all the way through the Earth—and had much more time to oscillate into the hard-to-find tau-neutrinos—appeared to have undergone a greater "loss" of mu-neutrinos.

**The massive neutrino.** The evidence of neutrino oscillation is what led to the conclusion that neutrinos may have mass. Physicists know that in order for the oscillation process to take place, one or both of the neutrinos (the original neutrino and the neutrino it is transformed into) must have mass, and their masses must be different. The bigger the difference in masses, the faster the oscillation.

Particle masses are measured in units of energy—the amount of energy that would be produced if the total mass of a particle were converted to pure energy. These units are called electronvolts (eV). An eV is 1.6 trillionths of an erg, a metric unit of energy. (An erg has often been compared to the amount of energy transferred when a slow-moving mosquito bumps into a person.)

The researchers could not measure the masses of the neutrinos, but they could measure the rate of oscillation and then use a mathematical formula and predictions based on other particle research to make a rough estimate of the neutrino mass. According to their calculations, the neutrino masses should be around a few hundredths of an eV. That means that neutrinos are millions of times lighter than an electron, which has a mass of 511,000 eV.

The finding that neutrinos may have mass could help physicists resolve a long-standing puzzle—the "missing" solar neutrinos. The nuclear reactions that power the sun should produce a steady and predictable stream of e-neutrinos. But detectors on Earth observe only 30 to 60 percent of the number predicted. This could be the result of e-neutrinos oscillating to one or both of the other two types of neutrinos on their way to Earth.

Another area where the Super-K data might shed some light is the problem of so-called dark matter. Observations of the motions of stars in nearby galaxies

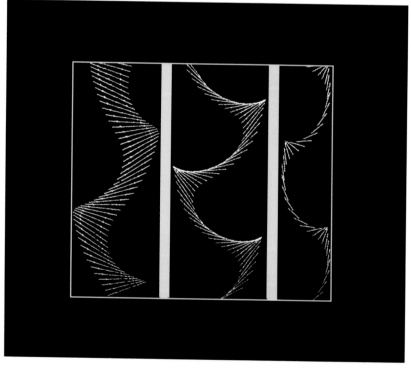

A series of video images record the path of flat objects as they fall through liquid-filled containers. *From left to right:* The first object flutters gently, while the second object tilts at larger angles. The third object pitches almost to a vertical, tumbling position. Researchers at the Weizmann Institute of Science in Rehovot, Israel, reported in July 1998 on how the properties of various objects affect how they fall. Their research on more than 30 metal and plastic strips of different weights and sizes contributed to a growing body of research that could have implications for various fields of study, including airplane design.

suggest that there is far more mass in or around a galaxy than can be seen by astronomers. The stars and clouds of dust and gas that can be seen in a galaxy can account for only about a tenth of the calculated mass. There have been many suggestions as to what the invisible dark matter might be. Neutrinos with mass might be part of the answer. A number of experiments capable of revealing more about neutrino masses were in preparation in 1999.

**As time goes by.** Physicists reported in December 1998 the first observation ever of a basic natural process in sub-atomic particles that is not symmetric in time. The discovery, which ends a quest begun in 1964, was reported by an international team of scientists at the CERN laboratory, a research center near Geneva, Switzerland. Physicist Panagiotis Pavlopoulos of the University of Basel, Switzerland, led the investigation.

Symmetry in time means that a physical process occurs at the same rate as the reverse of that process. Therefore, a violation of time symmetry occurs when the reverse process is faster or slower. At the level of things that we can observe, there are many violations of time symmetry. For example, a cup of coffee cools to room temperature in a certain amount of time. Because a countless number of atoms are involved—not only in the coffee, but also in the cup and the air—the probability of the coffee warming up again is practically nonexistent, no matter how much time passes. At the level of individual atoms or particles, however, a reverse action is highly probable, and the two processes should be symmetric in time.

In the CERN experiments, the researchers were studying a particle called a K-zero meson and its *antiparticle,* the K-zero-bar. An antiparticle has the same mass as its corresponding particle, but otherwise its properties are the exact opposite. For example, if a particle has a negative charge, the corresponding antiparticle has a positive charge.

K-zero mesons, or K-zeros, are very unstable and last only a few billionths of a second before they break up into other particles. But this fraction of a second is a long time by the standards of particle physics—plenty of time for K-zeros to transform into their own antiparticles or for K-zero-bars to transform into K-ze-

ros. Physicists do not know of another particle-antiparticle combination that can do this.

Normally, particle physicists would expect such a transformation and its corresponding reverse reaction to happen in the same amount of time. However, Pavlopoulos's team observed that the change of a K-zero into its antiparticle takes about 0.66 percent longer than the transformation of a K-zero-bar into a K-zero.

Oddly enough, the CERN researchers not only expected this violation of symmetry, they were hoping to observe it. The observation held true to a rule of particle physics called the CPT theorem.

This theorem relates to what physicists call the three symmetries of nature. The symmetry of matter and antimatter (or particle and antiparticle) is called charge symmetry or C-symmetry. Parity symmetry, or P-symmetry, is the symmetry of right and left, or what might be thought of as the mirror images of the various measurable properties of particles and antiparticles. T-symmetry is the symmetry of time.

The theorem states that these three symmetries as a whole must not be violated. In other words, if one symmetry is violated, then the other symmetries must also be violated in order to maintain a kind of balance. An analogy might be a waiter balancing a tray with three glasses. If one glass is moved to a different spot on the tray, then one or both of the others must shift to stay in balance.

In 1964, researchers observed a combined violation of CP-symmetry in K-zeros. They assumed that to preserve CPT-symmetry, the K-zeros must have also violated T-symmetry. Before the recent CERN study, physicists had only been able to predict what the violation of T-symmetry ought to be. Pavlopoulos's measurement of a 0.66-percent slower rate was remarkably close to the predicted rate of 0.62 percent.

To conduct the recent experiment, the CERN researchers collided protons with antiprotons. These collisions produce an equal number of particles and antiparticles, including an equal number of K-zeros and K-zero-bars. By measuring the electrical charges of other particles produced in the collision, the researchers could identify the K-zeros and K-zero-bars and use a formula to calculate the rate at which they were transformed.

The result of the T-symmetry study was significant because it confirmed predictions about the early history of the universe. Normally, when a particle collides with its antiparticle, they destroy each other in a burst of energy. One would then expect that all the matter and antimatter produced in the burst of energy at the beginning of the universe would have been destroyed. Physicists had theorized, therefore, that a violation of symmetry must have occurred in the first fractions of a second to create slightly more matter than antimatter, thus allowing our universe of stars and planets to come into being.

**The "sweet spot" revisited.** As the St. Louis Cardinals' Mark McGwire and the Chicago Cubs' Sammy Sosa staged a duel for the major league home-run record in the summer of 1998, many physicists were as captivated as any other fans. In fact, at least one physicist was able to follow the race with a somewhat expert eye. In September, in the midst of the home-run battle, an article appeared in the *American Journal of Physics* reporting a study of the so-called "sweet spot" of a baseball bat, the zone responsible for launching many a home run.

The author, Rod Cross of the University of Sydney in Australia, was by no means the first physicist to take a professional interest in the American pastime, for the sport naturally lends itself to scientific study. Wind tunnels at leading aeronautical laboratories have been used to track the flight of curve balls and have confirmed that a knuckleball really does take a fluttery path to home plate. Particle physicist Robert Adair of Yale University in New Haven, Connecticut, even served for a time as "official physicist" of major league baseball.

The "sweetness" of the sweet spot is how the bat feels in the player's hands when a ball is hit solidly. When the ball makes contact at the sweet spot, very little force is exerted on the batter's hands. In contrast, if the ball is struck by another part of the bat, the forces acting against the hands are strong and the batter feels a "sting." The sting is more than just an unpleasant sensation. It represents wasted energy—energy that could have gone into propelling the ball toward the fences.

The wasted energy is transmitted to the hands in two ways. First, the bat is flexible. When the ball makes contact, the bat vibrates. Second, the impact of the ball exerts a strong force on the barrel end of the bat, causing the handle of the bat to push in the opposite direction. If the batter were not gripping the handle, the force of the impact would send the bat into a rotating motion like the blades of a helicopter. The farther the center of that rotational force is from the batter's hands, the greater the force pushing against them.

To better understand these forces and the physics of the sweet spot, Cross taped lightweight piezoelectric sensors to his bat, an 84-centimeter (33-inch) wooden Louisville Slugger. A piezoelectric sensor generates an electric signal when a force is exerted on it. Very fine wires carried the signals to the end of the bat handle and on to recording and display instruments. Experienced players swung the bat at a ball hung from a string rather than at a moving ball to give greater control over the impact.

Cross found that the sweet spot is actually a small zone that starts 15 centimeters (6 inches) from the end of the barrel and extends about 3 centimeters (1.2 inches) toward the handle.

This zone has two characteristics that make it sweet. First, the "sweet zone" contains a *node*, a point in a vibrating bat where no movement occurs. If the ball hits the node, very little vibration is generated through the rest of the bat. Second, when the ball hits the sweet spot, the bat's center of rotation, where the least motion takes place, lies directly under the batter's hands. Therefore, there is no opposite force to push against the hands.

Although the design of the baseball bat was arrived at by trial and error through generations of batters and batmakers, Cross found that the design was pretty close to the optimum for "sweetness," guaranteeing that the vibrational node and the optimum rotational point would be in harmony. He also learned that a second vibrational node is located under the batter's hands, minimizing the effect of any vibrations. Cross demonstrated that practical skill had perfected the baseball bat long before scientists figured out why it works as well as it does.                    [Robert H. March]

A newly developed questionnaire that could help primary-care physicians quickly diagnose major depression was reported in December 1998 by a team of scientists led by internal medicine specialist David Brody of the MCP Hahnemann University School of Medicine in Philadelphia. The researchers proposed that the screening questionnaire would help physicians identify the often overlooked symptoms of a disorder that can result in significant disability or death.

Major depression affects nearly one-third of the people in the United States at some time in their lives. According to the *Diagnostic and Statistical Manual of Mental Disorders*, Fourth Edition (DSM-IV), the disorder has nine symptoms: (1) low, or sad, mood; (2) *anhedonia*, a diminished interest or pleasure in the person's usual activities; (3) sleep disturbance; (4) low energy; (5) appetite change; (6) poor concentration; (7) feelings of guilt or low self-esteem; (8) *psychomotor retardation* (slowed movements); and (9) suicidal thoughts.

A physician usually diagnoses major depression if several, but not necessarily all, of these symptoms are present for at least one month. Unfortunately, there are no laboratory tests that can reliably identify people with major depression. The diagnosis is based on "taking a history" and asking a series of detailed questions that test the patient's mental functioning. Primary-care clinicians may not have sufficient time or training to perform such an evaluation.

Brody's research team, therefore, set out to develop an effective diagnostic method that would require little training and time. First, the researchers asked 1,000 adult patients in four different cities the following questions: During the past month, have you often been bothered by (1) little interest or pleasure in doing things? or (2) feeling down, depressed, or hopeless?

Nearly 50 percent of the patients answered yes to one or both of these questions. However, after a thorough examination, only 16.5 percent of the people actually had enough of the DSM-IV symptoms to be diagnosed with major depression.

The researchers then examined the

**Not just another pretty face**
Computer-enhanced photos of a man's face helped a team of psychologists led by David Perrett at the University of St. Andrews in Scotland, draw conclusions about what features people find attractive. Participants in the 1998 study preferred images of male faces with smaller, more rounded features, *left,* which are generally considered feminine, over the broad, square features that are considered masculine, *right.* People also associated "feminine" faces with honesty, cooperation, and an interest in being a good father.

patients' responses to the full set of diagnostic questions to determine which of the symptoms most reliably indicated depression. The researchers found that the symptoms identified in nearly all of the patients with major depression were sleep disturbances, anhedonia, low self-esteem, and appetite change. They referred to these symptoms as the "core" symptoms of depression. Brody noted that the symptoms could be easily remembered by the initials S-A-L-S-A.

Brody and his colleagues, therefore, proposed a two-step diagnostic process. If the answer to either of the primary questions is "yes," then the physician should follow up with questions about the four core symptoms.

**ADHD consensus-panel report.** Children diagnosed with attention-deficit/hyperactivity disorder (ADHD) often receive inconsistent care, according to a November 1998 report by a panel of experts convened by the National Institutes of Health (NIH) in Bethesda, Maryland. The panel of psychiatrists, pediatricians, family practitioners, and psychologists also noted that the system for treating children with ADHD was not well coordinated or cost-effective.

ADHD affects between 3 and 5 percent of school-age children. Children with the disorder have more difficulty paying attention and concentrating than do other children of the same age or developmental level. They also exhibit inappropriate levels of activity, are easily distracted, and tend to act without thinking. Children with ADHD usually have difficulty at school, at home, and in other settings. Their behavior often leads to rejection by their peers and trouble with authorities. The disorder may also predispose children to other problems later in life, such as depression, alcohol abuse, and work problems.

Parents and teachers are usually the first people to recognize that a child exhibits symptoms of ADHD, but a physician (most often a family practitioner) usually makes the diagnosis. The condition is treated with medications known as psychostimulants and with *behavioral psychotherapy* (talk therapy focused on changing behaviors).

The NIH panel evaluated the scientific evidence on diagnostic methods and treatments and presented a "Consensus Document" to guide physicians and re-

searchers. The panel concluded that although it appears that ADHD has been correctly classified as a disorder and that it causes significant problems for children, the diagnosis and treatment of the disorder are controversial. Some physicians and educators, according to the report, believe that ADHD is diagnosed in children more often than it should be. Additionally, the panel found that the family practitioners, for reasons not yet determined, tend to prescribe medication more often than psychiatrists and pediatricians do.

The NIH panel members surveyed studies on the safety and effectiveness of psychostimulants. They found that most studies had followed children's treatment for three months or less. Therefore, though the medications appear to be safe and effective, the panel called for studies on the long-term effects of psychostimulants.

The panel also evaluated the overall effectiveness of treatment programs. Although a physician prescribes the treatment, school personnel often monitor the child's response to the therapy and recommend the need for further services. The panel concluded, however, that there was inadequate communication between physicians and schools. The NIH group called for better training for teachers in identifying ADHD-related symptoms and evaluating the progress of treatments.

**Ticklish touch.** Have you ever wondered why you cannot tickle yourself as easily as someone else can? A group of neuroscientists led by Sarah Blakemore at University College London set out to answer that question. In November 1998, they reported on the use of a brain-scanning technique called functional magnetic resonance imaging (fMRI) to determine which part of the brain distinguishes between self-produced nonticklish touches and the tickling sensations caused by other people. The results of the study also indicated how the brain, in general, sorts out various stimuli.

The researchers noted that a part of the brain called the somatosensory cortex is constantly bombarded by information about touch sensations on the skin. Most of this sensory information, such as the feel of our clothing as we move our limbs, is expected and irrelevant.

We feel these sensations all the time and think nothing of them. But some touch sensations indicate a change in the environment that might be harmful, such as a sharp or hot object touching the skin. Therefore, the brain must be able to distinguish between a multitude of expected, harmless sensations and a few important or "warning" sensations that ought to prompt a reaction.

Blakemore and her colleagues theorized that a "central monitor" exists somewhere in the brain that can sort out the unexpected and expected sensations. The central monitor, the scientists suspected, must be able to "cancel" the expected sensory messages and send along the sensations that should draw our attention.

The researchers hypothesized that the reason another individual's touch may cause a tickling sensation is that it registers in the central monitor as an unanticipated stimulus. Furthermore, they thought they could locate the central monitor by using fMRI technology while people either tickled themselves or were being tickled.

To test this theory, the scientists used a remote-control arm with a soft brush attached. The subjects of the study were instructed to move the arm in a back-and-forth motion to tickle the palms of their hands. Then, at an unexpected moment, an examiner used the same machine to tickle the subjects. The fMRI machine captured images showing which areas of the brain were activated during the experiment. One area of the brain was activated during the self-produced tickling but not when the examiner did the tickling. The scientists deduced that this area was the central monitor and that the brain activity was the monitor stopping the sensation in its tracks.

The suspected central monitor was in the cerebellum, a large structure in the brain that lies at the base of the skull. Blakemore's team concluded, therefore, that among the cerebellum's major functions is anticipating sensory stimuli and allowing a stimulus that does not match expectations to pass through and activate other regions of the brain.

[Michael Murphy]

## Public Health

In 1998, because of the effects of the human immunodeficiency virus (HIV, the virus that causes AIDS), the average life span in five African nations fell to less than 40 years. The United States Census Bureau reported that finding in March 1999 for Botswana, Malawi, Swaziland, Zambia, and Zimbabwe. In comparison, the average life span in the United States in 1998 was 76 years. Before the AIDS epidemic began in the early 1980's, people in Zimbabwe—one of the most developed nations in sub-Saharan Africa—could expect to live until age 65.

The report also noted that more than 10 percent of adults in 13 African nations are infected with HIV. In Botswana, Swaziland, and Zimbabwe, more than 25 percent of all adults are infected. Twenty-one African nations have the highest AIDS death rates in the world.

**AIDS deaths in the United States** have fallen dramatically, from 43,000 deaths in 1995 to 16,685 deaths in 1997, researchers at the National Center for Health Statistics, an agency of the U.S. government, reported in October 1998.

By 1997, there were fewer than 9 deaths for every 100 infected individuals, compared to nearly 30 deaths per 100 in 1995. Experts attributed the drop in mortality to new drug treatments that have delayed the onset of AIDS-related conditions in HIV-infected people.

However, public-health officials pointed out that efforts to change the behaviors that lead to HIV infection have not been so successful. The incidence of new HIV infections remained stable—about 40,000 per year. Helene Gayle, director of the National Center for HIV, Sexually Transmitted Diseases, and Tuberculosis Infections at the Centers for Disease Control and Prevention in Atlanta, Georgia, noted that HIV-infected people now live longer and feel healthier. Thus, they may be more apt to transmit AIDS to others through sexual intercourse and shared needles used for injecting drugs.

**Teen homicide.** A rash of shootings at schools in the United States in 1998 and 1999 raised fears among the public that the number of adolescent homicides was increasing. In each case, adolescent boys fired on classmates. Killings committed by young people have always been a factor in U.S. homicide statistics. However, most previous cases had been gang related or involved money or rivalry over a girl. The widely publicized 1998 and 1999 incidents involved boys who fired into a crowd of students. In addition, most of the boys communicated their intentions ahead of time and had relatively easy access to high-powered firearms.

Although the nation focused on these school killings, youth homicide in general—which had risen from the mid-1980's to the early 1990's—was actually on the decline. The National Center for Health Statistics reported in August 1998 that from 1993 to 1995, firearm homicides by people 15 to 24 years of age fell 4.4 percent in counties bordering large metropolitan areas and 15.3 percent in medium-sized counties with populations of 250,000 to 1 million. Similarly, non-firearm homicides declined by as much as 8 percent from 1990 to 1995. In 1995, 84 percent of adolescent homicides involved firearms, compared with 66 percent in 1985. Preliminary data for 1996 and 1997 showed that the homicide rates continued to fall. The greatest decrease was among African Americans. Black males have traditionally had the highest rate of homicide.

**Listeria outbreak.** By mid-1999, eight recalls of food products possibly contaminated with the rare bacterium *Listeria monocytogenes* had been announced in the United States. These measures followed the December 1998 recall by Chicago's Sara Lee Corporation of 16 million kilograms (35 million pounds) of hot dogs and luncheon meats. The meats had been processed by Sara Lee subsidiary Bil Mar Foods at its Zeeland, Michigan, plant. Nearly 100 people in 22 states became ill with flulike symptoms, and 21 people died.

Scientists investigating the outbreak were not able to determine how the listeria contaminated the processed meats. The organism can be found throughout the environment: in animals, soil, dust, water, and many different foods. It can survive refrigeration and freezing,

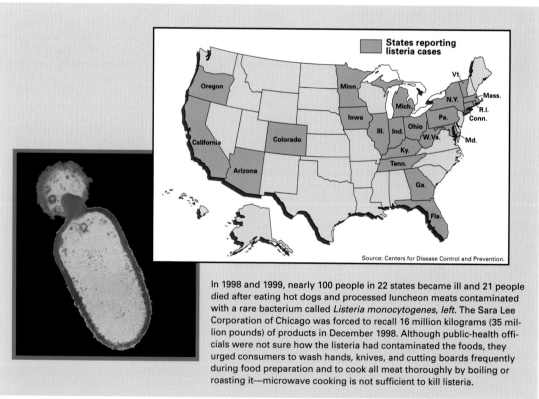

States reporting
listeria cases

Source: Centers for Disease Control and Prevention.

In 1998 and 1999, nearly 100 people in 22 states became ill and 21 people died after eating hot dogs and processed luncheon meats contaminated with a rare bacterium called *Listeria monocytogenes, left.* The Sara Lee Corporation of Chicago was forced to recall 16 million kilograms (35 million pounds) of products in December 1998. Although public-health officials were not sure how the listeria had contaminated the foods, they urged consumers to wash hands, knives, and cutting boards frequently during food preparation and to cook all meat thoroughly by boiling or roasting it—microwave cooking is not sufficient to kill listeria.

though not cooking with intense heat.

Public-health officials recommended that consumers cook all meats thoroughly and wash fruits and vegetables that will be eaten raw. They urged food preparers to wash their hands, knives, and cutting boards frequently, particularly after handling raw meat. Pregnant women, their unborn children, and people with weakened immune systems were at greatest risk. They were advised to take more extensive measures, including avoiding all soft cheeses—such as feta, brie, and Camembert—and cooking hot dogs and deli meats by roasting or boiling. The researchers cautioned that microwave cooking does not destroy listeria.

**How much exercise is enough?** Two clinical studies reported in January 1999 supported previous claims that modest physical activity can reduce the risk for a number of diseases generally associated with a sedentary lifestyle. The studies showed that the risk of such conditions as coronary heart disease, diabetes, and colon cancer can be lowered by 30 minutes of moderate exercise on most days

of the week. Both studies noted, however, that only about 20 percent of Americans got even that much exercise.

One study was conducted by researchers led by Andrea L. Dunn of The Cooper Institute for Aerobics Research in Dallas. For two years, 235 healthy—but sedentary and slightly overweight—men and women between the ages of 35 and 60 years engaged in either moderate physical activity for 30 minutes on most days of the week or in more vigorous exercise at least three times a week. Both groups of participants significantly improved their fitness and reduced their risk of disease.

Similar findings were reported by a team of researchers at the Johns Hopkins University School of Medicine in Baltimore. That group compared the effects of a structured aerobic exercise program with more moderate activity in 40 sedentary, obese women between the ages of 21 and 60. The researchers noted that moderate physical activity may be more easily incorporated into Americans' lifestyles than structured, vigorous exercise.               [Deborah Kowal]

## Science and Society

Cell biology took center stage among bioethics issues in 1998 and 1999. In November 1998, biologists James A. Thompson at the University of Wisconsin in Madison and John D. Gearhart at Johns Hopkins University in Baltimore separately announced that they had succeeded in establishing long-lived cultures of human stem cells. Such cells can develop into any type of tissue, including bone, heart, liver, or blood cells. Scientists hoped to use stem cells to grow replacement cells for treating a number of conditions, such as spinal cord injury, juvenile diabetes, Parkinson disease, and Alzheimer's disease.

However, under a law passed by the United States Congress in 1995, federal funds may not be used to support research in which human *embryos* (organisms in the first stages of development) are destroyed. One method of procuring human stem cells involves the destruction of embryos provided by fertility clinics. At a Senate hearing held in December 1998, scientists argued that the law should not apply to research on stem cells themselves because a stem

cell cannot develop into a human being.

The National Institutes of Health (NIH), a U.S. government agency, had not allowed its funds to be used for human stem cell research. (Thompson and Gearhart's work was privately funded.) In early 1999, however, NIH Director Harold Varmus stated that the 1995 law did not apply to research on already extracted cells. Varmus said the NIH will set guidelines for the research it would support and screen proposals.

In April 1999, researchers at Osiris Therapeutics in Baltimore reported that they had isolated stem cells from adult bone marrow. Some researchers speculated that such a technique could eliminate the need for human embryos as a source for the cells.

**Mentally ill research subjects.** In November 1998, the National Bioethics Advisory Commission (NBAC), established in 1996, called on the federal government to adopt new regulations to protect mentally ill patients participating in medical research. Patient advocates throughout the 1990's had questioned whether people with mental disorders are capable of judging the risks of taking part in clinical trials, particularly those that induce or worsen patients' *psychotic* (delusional) symptoms and those that withhold medication from psychotic or schizophrenic patients. Such studies can benefit society by leading to new treatments, but they often expose participants to risks without offering them any benefits. Researchers argue that without volunteers, such studies cannot be carried out.

In its report, the NBAC called for the creation of a government body to screen proposals for high-risk studies. It also suggested that the Institute of Medicine (IOM), an arm of the National Academy of Sciences, review the ethical issues involved. In January 1999, Steven Hyman, director of the National Institute of Mental Health—the largest source of federal funds for clinical research in psychiatry—suspended 29 studies at U.S. institutions until the institute could review the studies' objectives and protections for their subjects. In February, Hyman announced the formation of a panel to screen high-risk human studies before they are funded.

**Medical research priorities.** In July 1998, the IOM issued a report on how the NIH allocates nearly $15 billion in annual federal government funds for biomedical research. The U.S. Congress requested such a report in 1997, when the NIH was besieged by requests for additional funds from lobbyists and special-interest groups for various diseases and conditions.

Throughout the 1990's, for example, AIDS activists gained an increasing share of NIH research money, stimulating a debate on the roles of public pressure and scientific judgment in setting priorities for medical research. Groups advocating research on breast cancer, heart disease, and Alzheimer's disease also called for increases, arguing that the NIH was spending far more per patient on AIDS than on those diseases.

The IOM recommended that the NIH create a Council of Public Representatives that would include members of "disease-specific interest groups, ethnic groups, public-health advocates, and health care providers."

**The remains of Kennewick Man**— one of the oldest and most complete skeletons of a prehistoric human ever found in North America—finally became available for study in February 1999. The U.S. Department of the Interior announced that a team of scientists had been named to analyze the remains.

When the 9,300-year-old bones were discovered in Washington state in 1996, they were seized by the U.S. Army Corps of Engineers under the Native American Graves Protection and Repatriation Act. Although scientists wanted to study the skeleton because of its age and reports that it had *Caucasoid* (Europeanlike) features, a coalition of Indian tribes claimed the right to rebury it. The skeleton remained in storage at a Richland, Washington, laboratory while a federal judge sorted out the conflicting claims.

Eventually, the Corps turned matters over to the Interior Department, and the judge allowed scientists to begin an examination. The purpose of the analysis, which was to take place at the Burke Museum in Seattle, was to determine whether Kennewick Man meets the legal definition of a Native American. If so, scientists would look for a possible "cultural affiliation" with a modern Indian tribe.

**Census 2000.** In a January 1999 decision, the U.S. Supreme Court ruled that the federal government could not use statistical *sampling* to supplement its traditional head count in the 2000 census. (Sampling uses surveys of small groups of people to estimate the total population.) The court left the decision of how to conduct the census up to Congress and the president.

The U.S. Bureau of the Census had argued that it could make a more accurate and economical count of the U.S. population by using statistical methods to make inferences about the number of households that do not return forms by mail and are also missed by house-to-house census workers. Many statisticians supported that view. Critics, however, argued that sampling would be vulnerable to error and, possibly, political tampering. They favored a traditional, person-by-person count and noted that the Constitution calls for an "actual enumeration."

The stakes were high. The once-a-decade counts serve as the basis for the distribution of $180 billion in federal aid to states and localities. The data are also used to allocate seats in the House

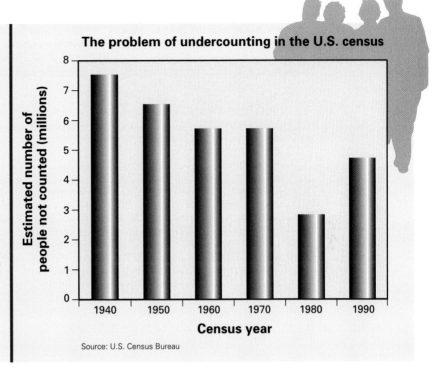

## The problem of undercounting in the U.S. census

The number of people not counted by the United States census, taken every 10 years, has generally been declining. The estimated number of people missed by the 1990 census was considerably less than the number missed in 1940. Nonetheless, more than 4 million people—mainly minorities, immigrants, and low-income individuals—were uncounted. Census officials considered this a serious problem and were looking for ways to solve it.

Estimated number of people not counted (millions)

Census year

Source: U.S. Census Bureau

of Representatives and to divide the states into congressional districts. The 1990 census reportedly missed several million Americans, many of them African Americans and Hispanics. Most analysts said sampling would increase the political representation of minorities, immigrants, and low-income people.

The Supreme Court decision affects only the use of sampling for congressional reapportionment. It left open the use of sampling data for other census applications, such as policymaking and social science research.

**Cryptography battle continues.** In February 1999, legislators in the U.S. House of Representatives reintroduced the "Security and Freedom Through Encryption Act" (SAFE). The bill would remove the Clinton Administration ban on the export of the most powerful forms of computer software used for *cryptography* (encoding or scrambling messages to protect privacy). The SAFE Act was introduced in the House in 1997 and again in 1998 but was defeated both times.

Most companies doing business on the Internet favored free use of cryptography to ensure the confidentiality of financial transactions such as credit card purchases or bank transfers. Public-interest groups regarded government restrictions as infringing on personal freedom. The government, however, was concerned about the use of cryptography by terrorists and other criminals and insisted on having a "key" that would allow law enforcement agencies to break into messages when necessary.

Enforcing the government's policy had become difficult by the late 1990's, when powerful encryption software was available on the Internet from sites outside the United States. In May 1999, a U.S. court of appeals ruled that the government's limits on encryption code export violate the constitutional right of freedom of expression. However, the Clinton Administration remained adamant in its opposition to changes in encryption policy.    [Albert H. Teich]

See also ANTHROPOLOGY (Close-Up). In the Special Reports section, see TISSUE ENGINEERING—FROM SCIENCE FICTION TO MEDICAL FACT.

After 15 years of planning and work on the ground, 16 nations began the assembly of the International Space Station in low Earth orbit in late 1998. The enterprise marked a historic milestone as the largest multinational science project in history.

In the United States, the National Aeronautics and Space Administration (NASA) had formally begun the project in 1984. NASA had enlisted Japan, Canada, and most of the member nations of the European Space Agency as partners. In 1993, Russia was added as a partner as well. When it is completed around 2004, the International Space Station will weigh more than 455 metric tons (500 tons) and sprawl over an area the size of two football fields. As many as seven astronauts will live and conduct scientific research at the station for several months at a time.

Phase 1 of the space station program, which began in 1993 and ended in mid-1998, featured flights by U.S. astronauts and Russian cosmonauts on each other's spacecraft. On June 2, 1998, the space shuttle Discovery flew the last of nine U.S. shuttle missions to Mir, the smaller space station that Russia began building in 1986. Discovery's crew of six included Russian cosmonaut Valery Ryumin. On board Mir was NASA astronaut Andrew S. W. Thomas. He returned to Earth on Discovery, having spent 128 days aboard the Russian space station.

**Space station assembly begins.** In the fall of 1998, the station program entered Phase 2—the assembly in orbit of the new international facility. The first component of the International Space Station, a Russian-built control module called Zarya (Russian for "sunrise"), was launched on a Proton rocket from the Baikonur Cosmodrome in Kazakhstan on Nov. 20, 1998. The 20-metric-ton (22-ton) module, which had its own propulsion, guidance, power, and thermal control systems, was to be the initial core of the space station.

Two weeks later, the U.S. space shuttle Endeavour carried the second station component into orbit. Endeavour lifted off from the Kennedy Space Center in Florida on Dec. 4, 1998, with a crew of six, commanded by U.S. astronaut Robert D. Cabana and including Russian cosmonaut Sergei Krikalev. In the shuttle orbiter's cargo bay was a "node," a 12-metric-ton (13-ton) module whose chief function was to connect larger station modules. NASA had named the node Unity.

After Endeavour caught up with Zarya—which had been orbiting Earth under the control of Russian ground crews—the in-orbit construction work began. As the two spacecraft circled 390 kilometers (240 miles) above the Earth, astronaut Nancy J. Currie used a Canadian-built manipulator arm 15 meters (50 feet) long aboard the shuttle to lift Unity out of the cargo bay and mate it to Zarya. Over the next week, the crew connected the two modules. Astronauts Jerry L. Ross and James H. Newman donned space suits and conducted three space walks spread over six days outside the embryonic space station.

Although station assembly in orbit began smoothly, the entire project was on less than an even keel. Even before Zarya was launched, the planned launching of the station's third component had been delayed.

Called the Service Module, the third component was to provide the station's first living quarters. Like Zarya, the module was being built in Russia. But whereas the United States had paid Russian contractors to build Zarya, the Russian government was paying for the Service Module. Still struggling after the collapse of the Soviet Union in 1991, Russia faced daunting economic and political problems. As a result, government money to industrial space contractors slowed to a trickle. NASA began building its own backup module in case the Russian vehicle was never completed. By the spring of 1999, NASA reported only that the Service Module would be launched by the end of the year.

The second mission to the space station, flown by Discovery on May 27, 1999, was also the first U.S. shuttle flight of the year. The crew consisted of five American astronauts, Julie Payette of the Canadian Space Agency, and Russian cosmonaut Valery Tokarev. Over a six-day period, the Discovery crew unloaded components for the station that had been too heavy to include in the original launch of Zarya and Unity.

**Glenn flies again.** During 1998, NASA flew only five shuttle missions— the lowest number since the shuttle resumed flying in 1988, 2½ years after the

A new satellite-launching system called Sea Launch is prepared for testing in March 1999 in the Pacific Ocean south of Hawaii. Developed by a U.S.-Russian-Ukrainian-Norwegian partnership, Sea Launch consists of two vessels—a launch platform and a command ship. Sea Launch can travel to the ideal launch site for a particular mission—for example, a spot on the equator where the Earth's rotation will add maximum momentum to a rocket. This system requires less rocket fuel than is needed for conventional launches, allowing rockets to carry heavier payloads.

shuttle Challenger exploded. The agency had been forced to decrease the number of missions because neither the Service Module nor a new orbiting telescope were ready to be launched on schedule. However, one 1998 mission drew more attention than any other in recent years. On Oct. 29, 1998, John H. Glenn, Jr., one of the original seven U.S. astronauts selected in 1959 and the first American to orbit Earth, made his second trip into space.

Glenn had left NASA decades earlier and launched a career in politics, serving as a U.S. senator from Ohio. As he neared retirement, Glenn lobbied NASA to let him fly again so that scientists could study the effects of space on his body to learn more about the physiology of aging. Although some critics claimed that Glenn's flight was just a publicity stunt, most Americans seemed to feel that it was justified—or at least that Glenn deserved a second flight.

When Discovery lifted off in October with Glenn aboard, the 77-year-old former astronaut became by far the oldest person ever to fly in space. But his sec-

ond flight was much different than his first. On Feb. 20, 1962, Glenn had flown alone, strapped into a cramped Mercury capsule, and traveled around the planet just three times. In 1998, Glenn floated inside a relatively spacious spacecraft with six crewmates, including Pedro Duque, the first Spaniard in space, and Japanese physician Chiaki Mukai, who was making her second space flight. He orbited Earth 134 times.

**Advanced-technology space probe.** On Oct. 24, 1998, NASA launched a spacecraft called Deep Space 1 (DS-1), bound for an encounter with an asteroid called 1992 KD in July 1999. However, as the first of a series of spacecraft in NASA's New Millennium program, DS-1's main mission was to test a dozen new space technologies. Chief among those was a novel form of propulsion.

DS-1 was propelled by an engine that accelerated the spacecraft very slowly by separating electrons from the element xenon to create *ions* (electrically charged atoms) and ejecting them in a high-speed beam from the back of the probe. Although the thrust produced by this method was tiny, the engine could run for long periods. Just 4.5 minutes into an engine firing in November, however, the ion engine shut down. Controllers determined that specks of debris in the engine had caused electrical short circuits. Such problems had been seen in earlier ion engines, so NASA engineers knew what to do. They transmitted changes to the spacecraft's engine-control software to avoid a recurrence of the problem. After passing 1992 KD, DS-1 may visit two comets.

**Mars missions.** Japan launched its first spacecraft to Mars on July 4, 1998, from its Kagoshima launch site. Known at first as Planet-B, the small spacecraft was later renamed Nozomi, Japanese for "hope." Nozomi was to reach Mars in late 1999 and study the effect of the solar wind on the planet's atmosphere.

On Dec. 20, 1998, Nozomi returned to within 1,000 kilometers (620 miles) of Earth, so that it could use the planet's gravity as a slingshot to gain speed. However, an engine valve had become stuck, causing the spacecraft to waste fuel. Even with the gravitational boost, Nozomi was unable to accelerate enough to reach a proper *trajectory* (curved path) for Mars. Engineers executed corrective maneuvers to put the vehicle on course for a delayed rendezvous with the planet in 2003.

NASA launched two space probes to Mars just a few weeks apart in 1998 and 1999. The first probe, Mars Climate Orbiter, was launched from Cape Canaveral Air Force Station in Florida on Dec. 10, 1998. The spacecraft, set to arrive at Mars in September 1999, carried instruments to study the thin atmosphere of the red planet. Equipment aboard the Orbiter would also serve as a communications link for the second spacecraft.

Mars Polar Lander was also launched from Cape Canaveral, on Jan. 3, 1999. The vehicle carried two "microprobes" that were to be fired into the surface of the planet, as well as a device to record sounds on Mars. Polar Lander was to set down near the planet's south pole. Previous spacecraft—the two Viking missions of 1976 and Mars Pathfinder, which landed on the planet in 1997— had explored other areas of Mars, so scientists expected Polar Lander to find different landscapes and conditions than those seen before. Polar Lander

was to reach Mars in December 1999.

Another NASA spacecraft, in orbit around the red planet since 1997, experienced troubles in April 1999. Mars Global Surveyor, sent to map the entire surface of the planet, shut down its science instruments when the side-to-side movement of its main antenna became blocked. Controllers resumed the spacecraft's mapping functions in May. However, because the antenna still could not move freely, Global Surveyor was not expected to be able to send as much data as scientists had originally hoped.

**Collecting stardust.** On Feb. 7, 1999, NASA launched a spacecraft called Stardust on a seven-year journey to intercept a comet, collect some of the material in its tail, and return those samples to Earth. The only previous missions ever to bring back samples were those to Earth's moon. Stardust was expected to reach the comet Wild-2 (pronounced *VILT-2*) in 2004. Small cubes of aerogel—a silicon-based material best described as "frozen smoke"—aboard the vehicle would then capture matter spewing from the comet. En route to the comet, the spacecraft is to collect samples of *interstellar* (between the stars) dust. Stardust is to return to Earth in 2006, ejecting a small capsule that will make a parachute landing in Utah.

**NEAR miss.** The first spacecraft sent to extensively explore an asteroid failed to brake properly and flew right by its target on Dec. 20, 1998. The Near Earth Asteroid Rendezvous spacecraft (NEAR) began what was to be a 10-minute firing of its engine to slow down so it could enter an orbit around the asteroid Eros. But the engine briefly exceeded the thrust limits set in its on-board computer, causing it to shut itself down. Then the spacecraft started to tumble.

Controllers spent more than a day trying to contact the spacecraft, which was 383 million kilometers (238 million miles) from Earth. When they did, they instituted a back-up plan. The spacecraft photographed the asteroid, 40 kilometers (25 miles) long, as it flew by. Engineers commanded the spacecraft to execute another engine firing on Jan. 3, 1999, which they expected would bring it past the asteroid again in February 2000 for another rendezvous attempt.

**Rocket failures.** The U.S. aerospace industry suffered serious setbacks in

## Assembling the International Space Station

Construction of the International Space Station in orbit began in late 1998, when the first two station modules were launched and connected.

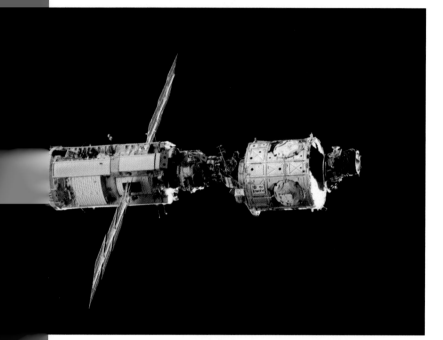

The modules Zarya (left) and Unity orbit the Earth after being joined by the crew of the space shuttle Endeavour in December 1998. Zarya, a Russian-built unit financed by the United States, will serve as a control module. It had been launched the previous month and was guided by Russian ground crews until the U.S.-built node, Unity, was brought up by Endeavour and connected to it. Unity will function as a passageway to other, larger units.

U.S. astronaut James H. Newman installs cables on Unity (foreground) in preparation for connecting the node to Zarya, whose solar arrays appear below. Newman and fellow astronaut Jerry L. Ross executed three spacewalks over six days in December 1998 to connect the modules. The station is expected to be completed by about 2004.

A blue tail of *ions* (electrically charged particles) streams from the spacecraft Deep Space 1 as it heads toward its planned mid-1999 encounter with an asteroid in this artist's rendering. The most important aspect of Deep Space 1's mission is the testing of new technologies, such as the solar-powered ion engine that requires only one-tenth the fuel of conventional engines. The spacecraft, which was launched aboard a rocket in October 1998, may also pay a visit to two comets.

1998 and 1999 when six rocket launches in nine months failed. The first flight of the new Delta 3 launch vehicle, built by the Boeing Company of Seattle, ended 71 seconds after liftoff from Cape Canaveral on Aug. 26, 1998. The rocket, carrying a commercial communications satellite, veered out of control and was destroyed. In a Delta 3 launch on April 22, 1999, the engine did not ignite. On May 4, that same Delta 3 placed a communications satellite in too low an orbit.

Lockheed Martin of Bethesda, Maryland, experienced problems with its launch vehicle, Titan 4, as well. A Titan 4 exploded 40 seconds after launch on Aug. 12, 1998. Other Titan 4's launched on April 9 and April 30, 1999, placed satellites in low, useless orbits. Industry experts were uncertain whether the accidents had anything in common. Still, the string of launch failures raised government concerns about quality control in the space industry.

**Europe's new large launcher,** Ariane 5, made its first fully successful flight on Oct. 21, 1998. The first Ariane 5 had exploded seconds after liftoff from the launch complex at Kourou, in French Guiana, in 1996. The second flight, in 1997, was declared a success, but it placed a test satellite in too low an orbit. The third Ariane 5 performed perfectly as it launched another test satellite. The rocket is big enough to simultaneously launch two of the largest, most powerful communications satellites on the market or three smaller ones. The European Space Agency was considering using Ariane 5 to carry supplies or astronauts to the International Space Station.

**North Korea** claimed to have been attempting to place its first satellite in orbit when it launched a new rocket, the Taepo Dong, on Aug. 31, 1998. However, international observers regarded the flight as a step by the authoritarian regime toward developing long-range ballistic missiles. The move was seen as especially provocative by Japan, because the rocket passed directly over the northern part of the island nation.

[James R. Asker]

See also ASTRONOMY. In the Special Reports section, see THE WIRELESS WORLD.

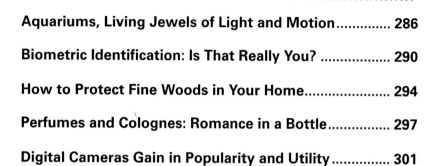

Topics selected for their current interest provide information that the reader as a consumer can use in understanding everyday technology or in making decisions—from buying products to caring for personal health and well-being.

SCIENCE YOU CAN USE

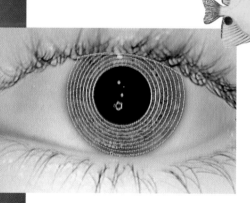

# Aquariums, Living Jewels of Light and Motion

Aquariums are popular decorative features in homes, offices, and even restaurants. They can be beautiful facsimiles of exotic underwater environments, such as the Amazon River or the Great Barrier Reef. And besides being fascinating to look at, aquariums provide a tranquilizing element in a world filled with stress. But before you decide to add an aquarium to your home or office, you should learn as much as possible about their care and maintenance.

Most people who have aquariums—or aquarists, as they are known—keep *tropical* (warm-water) fish, such as angelfish, rather than cold-water fish, such as goldfish. That is because tropicals are generally more colorful and exotic-looking than cold-water fish.

There are two basic types of tropical fish aquariums, freshwater and saltwater. Most experts advise against starting with a saltwater aquarium, because it is more expensive to set up and stock and more difficult to care for than a freshwater aquarium. If you begin with a freshwater aquarium, you can use most of the same equipment later if you decide to switch to a saltwater setup.

The first thing to consider is the size of the aquarium you want. If you are trying to fill a specific location in your home, you will have to buy what fits best in that spot. To get some advance idea of the size aquarium you need, keep in mind that one of the most common 76-liter (20-gallon) rectangular tanks has proportions of about 60 centimeters long

Aquariums make fascinating additions to homes, offices, and other places. They can be exotic, fanciful, and full of color. However, stocking and caring for an aquarium can be tricky.

by 30 centimeters wide by 40 centimeters high (24 x 12 x 16 inches).

If space is not an issue, buy the biggest tank you can afford—at least 76 liters. You may be tempted by inexpensive 38-liter (10-gallon) starter aquariums available in many pet stores. But these usually come with inferior equipment and do not offer enough room for even a small number of fish, and certainly not for the variety of fish you are likely to want.

When selecting a tank, keep in mind that the larger it is, the heavier it will be when it is full and operating. A liter (0.26 gallon) of water weighs 1 kilogram (2.2 pounds). So, a 76-liter aquarium, once filled, weighs more than 76 kilograms (170 pounds), counting its own weight and the weight of anything in it heavier than water. It is important to consider such great weights when deciding where to place your aquarium. You might want to buy a special stand for it. Some stands are made of black wrought iron, and others are specially constructed wood cabinets.

The next consideration in selecting an aquarium is whether to buy a glass tank or one made of plexiglass, a strong plastic material. Because of the great transparency of plexiglass, a tank made of this material can actually look clearer than a glass tank. In addition, a plexiglass tank is usually constructed so that the front face curves gracefully backward at the corners to form the sides of the aquarium. This design thus eliminates the sharp vertical lines of a glass aquarium, in which the front panel meets the sides. With the absence of edges, a plexiglass aquarium looks more like a magical block of water than a container holding water.

Plexiglass also has other advantages. It is much lighter than glass, it maintains water temperature better than glass, and it is virtually unbreakable.

On the other hand, a plexiglass tank is more expensive than one made of glass. A typical 76-liter glass tank costs

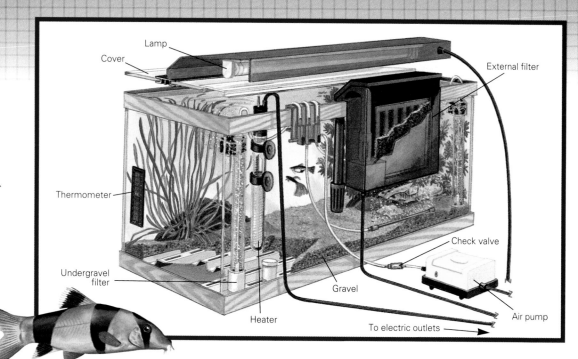

Lamp

Cover

External filter

Thermometer

Check valve

Undergravel filter

Gravel

Heater

To electric outlets

Air pump

$25 to $35, while a similar-size plexiglass tank can cost twice as much. Plexiglass also scratches more easily than glass, so an aquarium owner must take extra care when cleaning the tank or adjusting the contents of the aquarium.

When you have finally chosen a tank and brought it home, be sure to find an appropriate place for it. One consideration is the amount of natural light that the aquarium will get. It surprises many novice aquarists that window light—especially direct sunlight—can be a great menace to an aquarium. Too much light can promote the growth of algae in the tank, turning the water green. It can also raise the water temperature to levels that kill the fish. In addition, streams of hot or cold air can affect water temperature, so protect your aquarium from winter drafts and avoid placing it near open heating and air-conditioning vents.

Having set up your tank on a strong base and in a good location, you are ready to fit it out. The basic equipment for an aquarium includes a cover or hood for the tank, a heater, a filter, a light source, a thermometer, and various decorative items, such as gravel, plants, and rocks.

A heater is an essential element of an aquarium because you will probably be

keeping tropical fish that require constant water temperatures of 24 to 27 °C (76 to 80 °F). Aquarium heaters are rated in watts of power. Figure on 1.3 watts of power for each liter of water in an aquarium (5 watts per gallon). So, for example, a 76-liter tank requires a 100-watt heater.

Aquarium heaters are adjustable and controlled by a thermostat. The thermostat and heating element are enclosed in a long glass tube that either hangs into the water from the side of the tank or is fully submerged in the water.

A filter is another important piece of equipment for your aquarium. The best filters perform both mechanical filtration to remove particles and chemical filtration to remove harmful chemicals resulting from the body wastes excreted by the fish. Filters also accelerate the exchange of oxygen for carbon dioxide (released when fish breathe) at the water's surface.

Many experts believe that the best of the new filters are those that hang on the outside-back of the tank. These filters take water in from a tube that hangs in the tank, pass it through a fiber mesh to filter out particles, then pass all or some of the water over a revolving wheel. The water then returns to the

A well-equipped home aquarium has an air pump and one or more filters to keep the water clean. It also has a heater to maintain the water at an ideal temperature, and a cover to keep heat and fish in and cats and dust out. Plants and gravel create a tropical setting and provide a healthy environment for the fish.

# Components of freshwater aquariums and saltwater aquariums

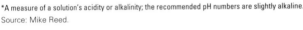

| | Freshwater | Saltwater |
|---|---|---|
| **Best tank size** | 76 liters (20 gallons) | 152 liters (40 gallons) |
| **Water temperature** | 24-27 °C (76-80 °F) | 22-27 °C (72-80 °F) |
| **Water pH*** | 7-7.6 | 8.1-8.3 |
| **Water quality** | Medium-hard tap water containing 100-200 mg/liter $CaCO_3$ (calcium carbonate) | Tap water with nitrates removed; add marine salt to create a salinity of 27.2-33.7 parts per thousand |
| **Heater** | 100 watts | 200 watts |
| **Filter** | Undergravel filter driven by a pump, or a power filter (filtration with foam cartridges) | Undergravel filter driven by a pump, and a power filter |
| **Lighting** | Standard-output fluorescent tubes | High-output bulbs or fluorescent tubes |
| **Decorations** | Gravel, rocks, wood, plants (such as hairgrass, Amazon swordplant, dwarf swordplant, Congo anubias) | Coral sand, coral chips, gravel, live rock, tufa rock, anemones |
| **Suitable fish** | Angelfish, guppies, platies, swordtails, tetras, barbs, catfish, "sharks," rainbowfish | Clowns, hogfish, parrotfish, pufferfish, butterflyfish, gobies, damselfish |
| **Maintenance** | Use a siphon or vacuum to change 10-15% of water twice a week or 25% once a week; take pH readings twice a week before changing water; clean filter once or twice a week; remove algae from sides of tank if it builds up, but not from rocks; check temperature daily to see that thermostat is working; check pump every three months. | Use a siphon or vacuum to change 10-15% of water twice a week or 25% once a week; use a hydrometer to measure salinity once or twice a week; take pH readings twice a week; clean and check filter once or twice a week; remove algae from sides of tank if it builds up, but not from rocks; check temperature daily to see that thermostat is working. |

*A measure of a solution's acidity or alkalinity; the recommended pH numbers are slightly alkaline.

Source: Mike Reed.

aquarium across a broad lip that creates a sort of small waterfall just above the surface of the water in the tank.

The entire filter, and in particular the wheel, becomes inhabited by beneficial bacteria that occur naturally under water. These microbes break down the deadly ammonia and nitrites that form in an aquarium and convert them into relatively harmless compounds called nitrates. And the water's trip in a relatively thin sheet over the spinning wheel and through the waterfall maximizes the exchange of carbon dioxide for oxygen.

The best kind of light source for your aquarium is one that is built into the hood or cover that fits over the top of the tank. A hood is needed to keep fish from jumping out of the tank and to prevent cats and dust from getting in. It also minimizes evaporation and heat loss from the tank.

Although most aquarium hoods are equipped with fluorescent lights, there are now many kinds of high-output bulbs available to aquarists. The new bulbs provide brighter light with spectrums that aid plant or coral growth and intensify the colors of the fish.

Aquarists also have different kinds of thermometers to choose from. One of the most convenient to use has flat magnets mounted on its ends. It is placed against an inside surface of the aquarium and is held in place by another set of magnets on the outside of the tank. Sliding the outside magnets causes the inside magnets to move the thermometer, a feature that is particularly useful when you are cleaning the inside of the tank.

Aquarium gravel comes in many colors. Most aquarists prefer natural colors, but some like to use brightly colored gravel that you would never see in nature. When making your own selection, just remember that you want the fish to be the main attraction, not the gravel.

The rocks and plants for your aquarium should come from a pet shop, not from a streambed or a vacant lot. The wrong types of rocks can cause changes in the chemical composition of the water that are harmful to the fish. The wrong kinds of plants can introduce diseases and foul your aquarium water. Many aquarists avoid real plants altogether and decorate their tanks with plastic plants. These artificial plants are remarkably lifelike and make an excellent substitute for natural vegetation.

Stocking your aquarium should be a gradual process. It is best to add only a few fish at a time to avoid overstocking the tank and jeopardizing the health of the fish. No hard and fast rules exist about how many fish you can put in your tank for a given volume of water, but it is better to be understocked than overstocked. A rough estimate is 2.5 centimeters (1 inch) of fish body length per gallon of water, meaning you can put about ten 5-centimeter (2-inch) fish in a 20-gallon tank. This number, however, can be safely doubled to 2 inches per gallon in a well-filtered and well-aerated tank.

Most aquarists start with a so-called community tank, which is a mixture of compatible fish selected without much regard to what parts of the world they come from. There are a few basic rules for picking such a mixture. You should consider the behaviors of the species you will mix, such as their aggressiveness or sense of territory; the level of the tank at which the fish eat (top, middle, or bottom); the size each type of fish may reach as it matures; and the chemical composition of the water you will be using. A knowledgeable pet-store employee can help you pick the right mix of fish for your aquarium and tell you how to introduce them to your tank.

Once you have your aquarium up and running, there are several potential problems and hazards you should look out for:

**Overfeeding the fish.** The number-one reason new aquarium owners fail is that they give their fish too much to eat. Excess food can foul the water and harm the fish. Feed your fish no more than they will eat in five minutes. And be sure they are actually eating the food rather than simply taking it in their mouths and then spitting it out.

**Incorrect water temperature.** Because fish are so sensitive to changes in water temperature, it is advisable to check the aquarium thermometer every day to make sure your water heater is functioning properly.

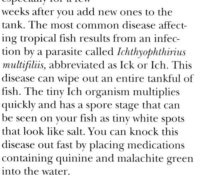

**Fish diseases.** Fish can get sick just as people do, so look at your fish carefully every day, especially for a few weeks after you add new ones to the tank. The most common disease affecting tropical fish results from an infection by a parasite called *Ichthyophthirius multifiliis,* abbreviated as Ick or Ich. This disease can wipe out an entire tankful of fish. The tiny Ich organism multiplies quickly and has a spore stage that can be seen on your fish as tiny white spots that look like salt. You can knock this disease out fast by placing medications containing quinine and malachite green into the water.

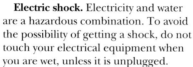

**Electric shock.** Electricity and water are a hazardous combination. To avoid the possibility of getting a shock, do not touch your electrical equipment when you are wet, unless it is unplugged.

**Skin irritations.** If you have your hands and arms in the aquarium a lot, you may find that they start to feel dry and itchy, and you may even develop a rash. If this problem occurs, the best solution is to buy a pair of elbow-length waterproof gloves to protect your skin from the water.

Many aquarists become so interested in one aspect or another of their aquariums that they begin to keep specialized varieties of fish. Some concentrate on one type of fish, such as angelfish or guppies. Others specialize in fish from specific locations, such as South America or Africa. Some aquarists even begin to breed their fish.

If you become interested in this approach to owning an aquarium, you can find specialized equipment and other aids at most pet stores. Among the newest aids are sophisticated chemical water supplements that have only recently become available to amateur aquarists. Sold under trademarked names like Instant Amazon and Rift Lake Vital, the supplements alter your aquarium's water chemistry to closely resemble that of the waters from which specific kinds of fish come.

Setting up a home aquarium can be a complicated and expensive venture. But the rewards can be priceless in terms of adding beauty and tranquility to your personal environment. [Mike Reed]

# Biometric Identification: Is That Really You?

The scenarios are straight out of a James Bond movie filled with high-tech gadgets, yet all the situations are real:

• The National Security Agency, a division of the United States Department of Defense, was testing computers in 1999 that operate only when an authorized user faces the computer screen and is recognized.

• At thousands of buildings around the world, entrance requires placing one's hand on a device that examines its shape.

• Some automated teller machines (ATM's) dispense cash only to customers who have been identified by a system that recognizes patterns in their eyes.

• Some personal computers have fingertip touch pads that serve as personal identifiers.

Who are you? Can you prove it? Answering those questions has long required memorizing passwords and carrying various kinds of identification cards. But as the 1990's drew to a close, new automated technologies based on analyzing people's physical characteristics promised to make establishing one's identity a quick, simple, and foolproof procedure.

The applications for this new technology range from replacing computer passwords to adding security to bank transactions. That's good news for computer users who forget their passwords, store owners who would like a way to identify known shoplifters, people wanting to cash checks without opening a bank account, or health-club members tired of having to present an ID card to gain entrance to the club. Meanwhile, the potential for using these new electronic identification systems to monitor people's every move has raised concerns among defenders of privacy.

The technologies used to identify people from their physical traits are known collectively as biometrics. These technologies are based on the idea that every individual has certain identifying characteristics that are unique and unchangeable. Moreover, unlike passwords and key cards, biometric identifiers cannot be lost, forgotten, or forged.

By 1999, a small industry had sprung up offering biometric identification systems to computer-system operators, financial institutions, building managers, and security agencies. Experts said that as these technologies became common, they would make it much more difficult for impostors to get into restricted areas or gain access to private information.

All biometric identification systems work on roughly the same principle. A person enrolls in a system by submitting a sample of a particular physical feature. The individual might have a fingerprint taken, for example, or be photographed in close-up. The particular feature being used is then translated electronically into digital data (a string of 0's and 1's). Lastly, a special computer program creates a *template*—an abbreviated description—of the feature. The template contains enough information to identify the person in the future.

A biometric identification system can operate in either of two modes. The more challenging mode is that of true identification—that is, determining who an individual is out of a large population. For example, a system might be used to identify someone crossing a border by photographing the person's face, converting the image to digital data, and then comparing the information with stored templates to find a match.

More commonly, however, biometrics operates in a simpler mode, performing the more limited task of verifying that people are who they claim to be. The personal identification numbers (PIN's) that people use to gain access to ATM's are an example of a numerical verification system. With a biometric system, the PIN would be replaced by a scan of a physical trait, which would be compared with the stored template for that trait.

While technology is making possible a variety of biometric systems, the inked fingerprint remains one of the most valuable means of identification. But now, with new technology, people's prints can be analyzed automatically, without the use of ink.

In most automatic fingerprint readers, an optical scanner takes a picture of the fingertip. But a more recent development is the so-called finger chip—a computer chip that translates touch into a print pattern. When someone places a fingertip on one of the tiny silicon squares, the chip produces an electric charge wherever the tiny ridges of the fingertip contact the chip's surface.

Both kinds of fingerprint readers convert the pattern in the skin to a digital format. The digitized pattern is stored in an electronic file and can be swiftly compared with other prints on file. Fingerprint readers began to appear on some computers in 1999, enabling users to be identified without typing in a password.

Another biometric system identifies people not by the tiny lines on their fingers but by the overall shape of their hands. More than 20,000 of these hand-geometry readers were in use in 1999, ac-cording to Recognition Systems, Incorporated, a manufacturer of such devices. A hand-geometry unit examines the profile of a hand, calculating such values as the relative length of the fingers and the thickness of the knuckles. Hand readers regulated access to the Olympic Village during the 1996 Summer Games in Atlanta and later were widely adopted for blocking unauthorized entrance to many kinds of facilities. These facilities range from nuclear power plants to health clubs to day-care centers.

Another body part that is ideal for biometric identification is the eye. For example, the pattern of blood vessels in the *retina*, the light-sensitive layer of cells at the back of the eye, is as unique as a fingerprint. The trouble with retinal identification is that it requires people to position the eye within a few centimeters of a device that shines a beam of light onto the retina to analyze it. Because most

## Iris recognition

In 1999, many automated teller machines (ATM's) were using a type of biometric identification system called iris recognition. In this system, a computer analyzes a person's *iris* (the colored part of the eye), a feature that differs greatly from one individual to another.

Close-ups of three different human eyes, *left*, show how distinct the patterns of the iris can be. An ATM's scanning device divides an image of a bank customer's iris into eight concentric circles, *below*, enabling the system to better map particular features of the iris. Within a matter of seconds, the ATM's computer compares this image to a *template* (stored description) previously made of the customer's iris to determine if the person being scanned should gain access to the account.

An ATM capable of iris recognition focuses three video cameras on the eyes of a customer. Two cameras scan a wide-angle view to locate the customer's head and eyes. The third camera captures a detailed image of one eye's iris.

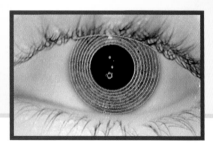

## Types of biometric identification systems

| Type | Availability | Accuracy | Uses |
|---|---|---|---|
| Iris recognition | Available now. | Highly accurate. Can be fooled by scratches on eyeglasses but not by contact lenses. | Customer identification, ATM's, access, security. |
| Electronic fingerprint ID | Available now. | Highly accurate. Computerized systems can even identify damaged fingerprints and account for skin changes caused by aging. | Forensics, government databases, identification, entrance to restricted areas, access to computer control. |
| Hand-geometry readers | Available now. | Limited accuracy because many people have similar hand geometry. | Access to buildings. |
| Face recognition | Available now. | Highly accurate. Some systems can compensate for changes in facial features, such as beards and eyeglasses. | Identification, access to personal computers and ATM's, security. |
| Signature recognition | Available now. | Less accurate than other biometric systems. | Identification. |
| Voice recognition | Available now. | Highly accurate in a controlled environment with no background noise. | Identification, access, phone-based identification systems. |
| Typing recognition | Under development. | Not known. | Identification access for computer passwords. |

Source: Cartech Securities Company.

people are reluctant to submit to this sort of procedure, retinal identification has not found widespread acceptance.

A less invasive process takes advantage of individual variations in the *iris,* the colored part of the eye. An ordinary camera can capture an image of the iris at distances up to 1 meter (3 feet). In one of the first major tests of this system, the Nationwide Building Society (NBS) in England used iris recognition in place of PIN numbers to identify more than 1,000 of its customers. Cameras at each of this savings-and-loan's cash machines looked into customers' eyes and completed each transaction only if the person's iris pattern matched the one on file. NBS reported high customer satisfaction with the new system, with more than 90 percent of the test participants saying they would choose iris identification over conventional PIN numbers or signatures.

When humans recognize one another, they usually do so by looking at each other's entire faces. Another biometric technology uses computers to do the same thing. A camera takes a picture of a person and compares it to a set of stored images. If the picture matches, the system grants access. One such system, developed by Miros, Incorporated, of Wellesley, Massachusetts, uses a type of software known as a neural network, which operates in a way roughly similar to the network of nerve-cell connections in the brain. Miros engineers train the system to recognize faces by showing it thousands of pairs of pictures. Some pairs are of the same person, while others are of two different people.

Faces change, of course—people age, put on or take off glasses, and get different hair styles. But companies that make facial-recognition systems claim that their computer programs are intelligent enough to avoid being fooled by such superficial alterations in appearance.

In a test of face-recognition systems conducted by the National Security Agency in 1999, workers with secret information on their computer screens stayed in view of a video camera. If an unauthorized person tried to access the data, the face-recognition system detected the impostor and refused to display any information. Indeed, the screen would go blank whenever the authorized user's face disappeared.

Another type of face-recognition system uses an infrared camera to record the heat being radiated by the major blood vessels in a person's face. Using this information, the system then creates a *thermogram* (heat pattern) of the face. Every person's thermogram is unique and does not change with surgery, age, or other changes in appearance.

Face-recognition systems were starting to appear in public places in the late 1990's. One of the most extensive applications of face-recognition biometrics was its adoption by a Fort Worth, Texas, check-cashing company, the Mr. Payroll Corporation. Mr. Payroll was installing face-recognition systems developed by Miros in many of its check-cashing *kiosks* (small freestanding structures) in the Southern and Western United States.

Users of the Mr. Payroll service—many of them low-income workers without bank accounts—simply insert their paycheck into an ATM-like machine. A camera in the kiosk captures an image of the person attempting the transaction, and if the face matches that of a person who has enrolled with the system, the machine dispenses cash.

Technology was also being used to update another long-standing method of identification: the handwritten signature. A system developed by PenOp, Incorporated, of New York City, analyzes the way a signature is written to verify a person's identity. The system requires signatures to be made with a stylus on an electronic pad that plugs into a computer. Software in the computer examines the order of the strokes and the acceleration and deceleration of the pen at various points in the writing of the signature. This information is compared with data from an original signature that the person made when enrolling with the system. Even if someone were able to produce a perfect visual replica of another person's signature, says a PenOp spokesperson, failure to duplicate the rhythm of the original would betray the individual as a forger.

A major application for such a system would be in authenticating signatures on electronic documents. Many companies see value in being able to create contracts, invoices, purchase orders, and other documents on computers, and transmit and receive them back electronically. The ability to verify customers' signatures on such documents would make many companies more comfortable with buying and selling products over the World Wide Web. Thus, the stylus and electronic pad could become as essential as a mouse and keyboard for a Web-surfing computer user in the early 2000's.

Just as people develop an identifiable style of signature, they also type on keyboards in characteristic ways. A company called Net Nanny Software was developing a computer program in 1999 that takes advantage of these differences to make typed passwords more secure. The system calculates the time a person takes between keystrokes when typing a password. If the rhythm deviates substantially from that of the password holder, the Net Nanny software denies access.

In perhaps the most natural biometric identification system of all, high-tech equipment can identify people merely by the character of their voice. A person first trains the system to recognize his or her voice by reciting a few phrases into a microphone. The sound signal from the microphone is converted to a digital format, and a computer then analyzes the tonal patterns of the voice and stores them as a template. Experts say this system is highly reliable because even someone with a knack for imitating voices generally achieves only a superficial resemblance to a person's real voice.

While all this whiz-bang technology made its way into society, skeptics cautioned that biometric systems raise thorny legal and privacy issues that have yet to be addressed. Many consumers were already concerned about having their personal credit information widely distributed through computer databases. How much more suspicious would they be of data banks that record data on physical features and voice patterns?

But unless such issues sidetrack this latest march of technology, experts said in 1999, we could soon expect to see biometric devices wherever we turn—at building entrances, cash machines, and computer terminals, and perhaps even in automobiles. Will the devices be primarily a step toward greater efficiency and security or a further intrusion into people's privacy? Only time will tell.                    [Herb Brody]

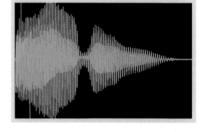

**Voice recognition**
Sound waves from a person's voice are converted into a digital image by a voice-recognition system that uses an electronic speech analyzer. The wave patterns are then stored as a template in a computer. To identify that individual at a later time, the system compares the person's voice to the template.

A secretary desk made of fine wood is a decorative and functional piece of furniture that will last for generations if given proper care. Simple precautions will protect such furniture from damage resulting from moisture, sunlight, scratches, dents, and other causes.

Nothing adds beauty and warmth to a home more than the richness and sheen of furniture and floors made of fine wood. Wood is one of the most versatile, decorative, and durable materials available. But even with its great strength, wood can easily be damaged. Keeping it free from harm and looking new requires giving it special care.

Wood, of course, comes from trees. Technically speaking, it is the hard, fibrous substance found between the bark and the *pith* (the spongy tissue in the center of a tree). Wood is not uniform. Rather, it consists of a number of different types of cells, all held together by an organic substance called lignin.

Woods are classified as either hardwoods or softwoods. Hardwoods come from broad-leafed trees, such as mahogany and maple. Softwoods come from *conifers* (cone-bearing trees), such as pine and cedar. The terms hardwood and softwood aren't always accurate, though, because some softwoods are actually harder than some hardwoods.

Woods selected for furniture and floors are usually hardwoods. Two of the most popular hardwoods for flooring are oak and maple. Furniture is often made of cherry, mahogany, walnut, oak, or maple. These hardwoods are plentiful because they come from trees that grow in many parts of the world. Also, since these woods come from trees that often grow to a very large size, their boards are wide, making them ideal for furniture and flooring.

These woods have other advantages as well. They are simple to work with, because the wood can be easily shaped and drilled. They are also stable. If they are dried slowly after being sawed into boards, they won't crack as they age.

Hardwood furniture and floors can last for decades—even centuries—but they must be protected from damage. One of the major concerns, wood experts say, is moisture—either too much or too little.

Wood itself is moist, and wood furniture normally has a water content of 4 to 12 percent by weight. If wood were completely dry, explains Mickey Moore, director of technical services of the National Oak Flooring Manufacturers Association, it would be very brittle. Too much moisture, on the other hand, can cause mildew and rot.

The air in a home must have a proper balance of moisture to keep wood in good condition. For most of the year, this is not much of a problem. But during the winter months, when indoor air becomes very dry, it may be necessary to add moisture to the air with a humidifier to prevent wood from shrinking.

Wood absorbs or gives off moisture unevenly, says Mark Knaebe, a chemist at the U.S. Department of Agriculture's Forest Products Laboratory in Madison, Wisconsin. He explains that when the surface areas of wood get wet or dry, it takes a long time before any damage occurs in the underlying layers.

When the surface of a piece of wood becomes drier than the interior, cracking occurs. If the moisture content of wood fluctuates, the wood may go through cycles of swelling and contracting. These changes put stress on the wood, again causing cracks. The process is roughly similar to the freezing and thawing of roads that produces potholes.

Sometimes the surface of wood may be wet for prolonged periods, as from a leaking potted plant. Then the surface cells become softened and can be crushed if a heavy object is placed on the wood. Such damage will mar the appearance of the wood.

Sunlight can be another problem. Direct sunlight can break down the lignin holding wood together.

Putting a piece of wood furniture in the right place is crucial to protecting it, Knaebe notes. He says that people should not place wood furniture near a heating pipe or vent because the heat can cause the wood to dry out. Nor should furniture be placed in front of a window receiving direct sunlight unless the window is fitted with heavy curtains.

Caring for fine wood also involves maintaining its finish. There are a number of liquid finishes that can be applied

# How to remove stains or marks from hardwood furniture

| | |
|---|---|
| **Water marks or rings** | Rings or water marks are most often in the wax and not the finish. Cover the stain with a clean blotter, press down with a warm iron, then repeat. Or try rubbing with salad oil, mayonnaise, or white toothpaste. Then wipe dry. |
| **White marks** | Rub with a cloth dipped in a mixture of cigarette ashes and lemon juice or salad oil. Or rub with a cloth dipped in lighter fluid, followed by a mixture of *rottenstone* (a powder made of decomposed limestone) and salad oil. Wipe dry. |
| **Milk or alcohol** | Use your fingers to rub liquid or paste wax into the stain. Or rub in a paste of boiled linseed oil and rottenstone with the grain, substituting pumice for dull finishes. Or rub with ammonia on a dampened cloth. Then wipe dry. |
| **Cigarette burns** | Minor burns can be remedied by rubbing with scratch-concealing polish or with a paste of linseed oil and rottenstone, working with the grain until the burn mark disappears. |
| **Heat marks** | Rub gently along the grain using a dry steel wool soap pad, extra-fine (0000) steel wool, or a cloth dampened with camphorated oil or mineral spirits. Wipe clean. |
| **Nail polish** | Blot the spill immediately, then rub with fine steel wool (0) dipped in wax. Wipe dry. |
| **Paint marks** | Remove fresh latex paint with water, and oil-based paint with mineral spirits. If the paint is dry, soak the spot with boiled linseed oil, wait until the paint softens, then lift carefully with a putty knife or wipe with a cloth dampened in boiled linseed oil. Residue can be removed by rubbing the grain with a paste of boiled linseed oil and rottenstone. Wipe dry and wax or polish. |
| **Scratches** | For scratches in dark wood or wood with a dark stain, fill the scratches with shoe polish that matches the lightest shade of the finish, or rub with walnut or Brazil nut meat in the direction of the scratch. For cherry wood, fill the scratches with cordovan or reddish shoe polish that matches the wood, or apply darkened iodine with a cotton swab or a thin artist's brush. For light wood or wood with a light stain, fill the scratches with a tan or natural shoe polish, or apply a mixture of 50 percent iodine and 50 percent denatured alcohol. |
| **Stuck-on paper** | Dampen the paper thoroughly with salad oil, wait five minutes, then rub along the grain with extra-fine (0000) steel wool dipped in mineral spirits. Wipe dry. |
| **Wax or gum** | Harden the substance by holding an ice cube wrapped in cloth against it. Pry off with a fingernail or credit card. Rub the area with extra-fine (0000) steel wool dipped in mineral spirits. Wipe dry. |

Source: Hardwood Information Center of the Hardwood Manufacturers Association.

to wood furniture and floors to protect them. These finishes leave a durable, transparent film over wood that resists water and enhances the appearance of the wood without changing its color or texture. Most of these finishes are made of natural oils or substances called resins, which can be either natural or synthetic. Natural resins are secreted from trees and plants. Synthetic resins are chemical imitations of natural resins.

A finish should not be confused with a stain. Stains are often used to change a wood's color, but finishes are intended only to protect wood.

One natural resin that has long been used to protect wood is linseed oil, which is derived from the seeds of flax plants. Linseed oil penetrates deeply into wood and forms chemical links that hold the wood molecules together. Those bonds help protect wood against damage caused by spills of milk, juice, or other liquids.

But oils alone are not enough, Moore cautions. He says that furniture should receive a second finish, such as shellac, and floors should get the same plus a wax.

Shellac is a natural finish that seals the pores in the surface of wood. It is a type of varnish—a resin mixed with alcohol. Shellac is an alcohol-based varnish made from lac, a resin secreted by *Laccifer lacca*, an insect found in Southeast Asia. A coating of shellac will protect floors and furniture from abrasions and dents as well as from moisture damage.

Oil-based varnishes also provide excellent protection for woods. For example, polyurethane varnish, a synthetic resin combined with linseed oil, is much more effective than linseed oil alone, says Sal Marino, a wood-care consultant and au-

To remove stubborn stains and smudges, such as flower sap or hair spray, from wood furniture, *far left,* make a buffing compound from a 50-50 mixture of baking soda and toothpaste, *top.* Then apply the compound to the spot with a wet cotton cloth and rub until the spot disappears, *bottom.* If the mixture starts to dry out, add a little water to the cloth and continue rubbing.

thor. When applied to a wood surface, it creates a barrier between air and the wood itself.

Polyurethane is one of the toughest finishes available. In fact, argues Marino, there is no need to apply an oil first when using a polyurethane finish, because the oils in the polyurethane penetrate into the wood. At the same time, the resins build up on the surface, protecting the wood from abrasions, spills, and other kinds of damage.

A polyurethane finish is often the best choice of varnish for a floor, but it is not recommended for fine furniture. High-quality furniture is usually hand-rubbed after the finish is applied to bring out the beauty of the wood, and polyurethane makes this step much more difficult.

Nitrocellulose lacquer is the most commonly used varnish for furniture, and has been since about 1920, Marino says. Nitro-cellulose lacquer is a natural product made of nitric acid compounds and *cellu-lose,* the major constituent of plant cell walls. This finish is more durable, more water-resistant, and more abrasion-resistant than shellac, says Marino. Moreover, like other fine finishes, it is clear and does not change the color of the wood or the stain that has been applied to the wood.

Wood experts disagree on whether it is advisable to treat furniture with the liquid polishes found in the home-care aisles of many supermarkets. Some authorities in-

sist that these products are horrible to use, while others contend that they are all right, Marino says. His personal opinion is that 95 percent of all the liquid polishes on the market are safe to use. He says that if the furniture's finish is in good condition, they can't really harm the wood. Marino cautions, however, that most commercial polishes contain silicones. If silicones are applied to bare wood after the original finish has been stripped off, they can prevent a new finish from adhering properly.

In addition to being subject to moisture and sunlight damage, wood is also vulnerable to scratches, dents, and gouges. Fortunately, most such damage can be repaired. For instance, minor mars or scratches on a finish can be smoothed away with fine steel wool.

If a scratch goes deep into the wood itself, a lacquer burn-in stick can be used to repair it. These sticks, which come in a range of furniture colors, are easy to use. The stick is held over the damaged wood and melted with a hot spatula or a specially designed electric knife. A bit of the liquid lacquer is dripped into the scratch, and the lacquer is then smoothed out. A fine paintbrush can be used to restore the grain of the wood, but this can be a difficult job that should probably be done by a professional.

Wood furniture and floors enhance the livability of any home. Wood has a look and feel that no other material can match. But fine woods require care and attention from their owners to keep looking their best.       [Harvey Black]

# Perfumes and Colognes: Romance in a Bottle

Perfumes and colognes: Some women pour them on, some use them sparingly, and others never touch a drop to their skin. But regardless of how they're used, there's no denying that these fragrances are among the most romantic and mysterious products ever to come out of a chemist's lab. After all, what other chemical creations bear such intriguing names as Adoration, Obsession, or Secret of Venus?

The word *perfume* itself has an exotic origin. It means *through the smoke* in French, and refers to civilization's oldest known source of fragrance—the burning of incense.

Choosing from the bewildering array of perfumes and colognes that line the counters of department stores can be a very difficult task. They all look wonderful in their crystal flasks and artful packaging. Yet, why does one fragrance cost $300 an ounce and another just $10? And why does a scent that smells fantastic on one person smell ho-hum on another? Perfume chemists refuse to reveal their closely guarded formulas for mixing the 200 or more ingredients (out of some 2,000 that are commonly used in the industry) that may go into creating a scent. However, a basic lesson in chemistry can go a long way toward demystifying the world of perfumes and colognes.

There are four main types of fragrances: perfume, eau de parfum (perfume water), eau de toilette (toilet water), and cologne. Each of these fragrances is distinguished by its concentration of scent-imparting components, known as essential oils or essence. A fragrance is a combination of essential oils dissolved in alcohol and usually (except for most perfumes) containing a certain amount of water.

Perfume has by far the highest concentration of essence, anywhere from 15 to 30 percent of the liquid in the bottle. Because perfumes contain such a high percentage of essence, they can be applied in tiny dabs and are very long-lasting on the skin.

Eau de parfum is the next-strongest kind of fragrance, containing from 8 to 15 percent essence. Although it is still highly concentrated and long-lasting, eau de parfum has to be applied to the skin in larger amounts, and more frequently, than perfume to get the same effect.

Eau de toilette contains 4 to 8 percent essence and is usually sold in spray bottles. Spraying the fragrance makes it easy to use over a wide area of the body.

Cologne, also called eau de cologne (cologne water), is the least powerful category of fragrance, containing from 2 to 5 percent essence. A newer type of cologne, known as splash cologne, is even weaker, with just 1 to 3 percent essence. Deriving its name from the city of Cologne, Germany (where it

Perfumes and colognes come in a wide range of fragrances with a variety of price tags. Choosing the right scent for your own style and taste can be a bewildering but pleasant experience.

## The essence of the perfumer's art

Fragrances contain various oils, collectively called essence, that give them their scent. The essence is dissolved in alcohol, which makes up most of the product. Most fragrances also contain varying amounts of distilled water.

■ Alcohol and distilled water

■ Essence

| Type of fragrance: | Perfume | Eau de parfum | Eau de toilette | Cologne | Splash cologne |
|---|---|---|---|---|---|
| Percentage of essence: | 15-30% | 8-15% | 4-8% | 2-5% | 1-3% |

was first made, in 1709), cologne has a cool, refreshing feel because of its high alcohol and water content.

Besides concentration, the most important factor to consider when choosing a fragrance is what type of scent it contains. Most of today's fragrances fall into one of three basic scent groups: florals, fruits, or modern blends.

Floral scents, as their name implies, smell like flowers. They are made from essential oils derived from either whole flowers or petals. Such oils can be costly because a huge number of flowers are needed to make a small amount of oil. For example, it takes 360 kilograms (800 pounds) of jasmine flower blossoms—about 1 million flowers—to produce slightly less than ½ kilogram (1 pound) of jasmine essence. One kilogram (2.2 pounds) of jasmine oil can cost more than $25,000. Because flowers used in making scents must be picked by hand, labor represents about 60 percent of the cost of flowers for perfumes and colognes.

Fruit scents can either have a clean, fresh smell or a sweet aroma. They are often made from oils distilled from the peels of citrus fruit. But, as with the production of natural floral scents, the production of fruit fragrances can be time-consuming and expensive because of the large amounts of fruit required to distill the oils. For example, 1,500

kilograms (3,300 pounds) of a sweet-smelling but inedible Italian citrus fruit called a bergamot are needed to yield just 1 kilogram of essence.

Modern blends encompass a wide variety of spicy, Oriental, and forest scents. Spicy fragrances contain the scents of vanilla, clove, or ginger, sometimes in combination with a floral scent. Oriental fragrances also contain spicy scents, usually combined with exotic floral scents and the scent of amber, a sharp-smelling oil distilled from fossilized plant resin. Forest scents are earthy aromas that feature oils, gums, or resins from trees such as balsam, cedar, rosewood, or sandalwood.

Many of the natural oils used in modern blends, like those of floral and fruit fragrances, are quite expensive. A kilogram of essential oil of eucalyptus, for example, can cost more than $550.

Due to the high cost of natural essence, a growing number of perfume makers use two groups of chemical compounds, called aldehydes and esters, to imitate natural aromas or develop intriguing new scents. Chanel No. 5 and White Linen are two famous perfumes that are scented with floral aldehydes. In the 1930's, perfumes typically consisted of about 85 percent natural ingredients and 15 percent synthetic compounds. In the 1990's, the reverse was true.

But there is more to a perfume or cologne than just the scent. Another crucial ingredient is the fixative, a substance that makes the scent last longer by reducing its tendency to evaporate. Some of the best fixatives come from animals—and unusual animals at that. The fixative ambergris, for instance, is a black, waxy substance released from the intestines of sperm whales. Another fixative, civet, is a pungent secretion from the glands of civet cats, small mammals that live in Africa and Asia. Other fixatives include castor, a substance produced by beavers; and musk, secreted from the glands of the male musk deer, an animal native to Asia.

These substances, however, are much too scarce to meet the demands of today's manufacturers. So here again, chemical synthetics have largely replaced the natural ingredients.

Not quite as exotic as substances such as ambergris, but perhaps even more fascinating, are ingredients called pheromones. These odorless, invisible molecules are released by the bodies of many organisms and serve as airborne chemical messengers, affecting a variety of behaviors, including sexual activity.

Synthetic versions of human pheromones were being added to many fragrances in the 1990's. While some scientific studies have indicated that pheromones can cause certain people to be drawn to each other, experts say there is no conclusive evidence that fragrances containing artificial human pheromones have any particular power to attract the opposite sex.

The ingredient that takes up the bulk of the bottle in most fragrances is alcohol. Alcohol readily dissolves fats and oils, making it the perfect vehicle for mixing and matching a variety of fragrant oils to create an endless array of scents. It is precisely that mixing and matching that is at the heart of the perfumer's art.

The secret to a great fragrance, according to most perfume experts, lies in the way its ingredients, called notes, are arranged. Likening a perfume to a symphony, perfumers since the 1800's have worked on orchestrating fragrant compositions to release their scent in three basic phases: top note, middle note, and base note.

The top note, also called the head note, refers to how the fragrance smells during the first fleeting moments it is on the skin. At that time, the predominant scent is of the essence that is the most volatile—the one that evaporates the most quickly. Typically, these essences are delicate florals, citrus scents, and aldehydes.

The middle note, also called the heart note, is the phase in which ingredients in the fragrance interact with the chemistry of a person's skin. In this phase, floral fragrances release some of their richest scents.

The base note, also called the drydown, is not detectable until at least 15 minutes after the fragrance is applied. It is this note that appears after all the alcohol has evaporated and that gives the fragrance its staying power. Essences such as sandalwood, vanilla, and balsam, along with animal fixatives like musk, fall into this category.

Ideally, like the notes in a stirring musical composition, the three notes of a fragrance should complement each other, reflecting a common

**The nose knows**
Technicians at a perfume plant sample the latest fragrances on long strips of paper. Perfume makers can choose from about 2,000 ingredients to come up with new scents, but developing a new fragrance that is both different and pleasing takes skill.

theme and blending smoothly from one phase to the next. To get an idea of how a fragrance will smell in its long-lasting later stages, perfume experts recommend that consumers wait at least 10 minutes after applying a test fragrance before taking a whiff. And if you don't have enough time for that, experts advise, you should ask the salesperson to apply the fragrance to a tissue instead of your arm. That way you can smell the scent later when the middle and base notes have emerged from the fragrance.

Despite all the advances in modern chemistry, one aspect of fragrances is completely unpredictable: how the chemicals in a fragrance will interact with a person's body chemistry.

According to the Fragrance Foundation in New York City, fragrances are more intense on people with oily skin. Skin oils reduce the tendency of fragrances to evaporate. Dry skin does not retain a fragrance as well, forcing users to reapply the scent more often. Variations in diet can also alter skin chemistry and change the way fragrances smell and last on the skin. People who eat a lot of high-fat or spicy foods, for example, may find a scent to be stronger or sharper on their skin.

Perfumes and colognes can also interact with the body in undesirable ways. Fragrances are among the substances most likely to trigger allergic reactions in people, says John E. Bailey, Jr., director of the U.S. Food and Drug Administration's Office of Cosmetics and Colors. Among the problems that have been associated with perfumes and other fragrance products are headaches, dizziness, rashes, coughing, and vomiting. And the threat extends from the person wearing the fragrance to sensitive people who are nearby.

In addition to possibly causing allergic reactions in other individuals, fragrance users risk offending people who have a negative psychological reaction to strong scents. For those reasons, it is wise to heed the Fragrance Foundation's advice to use fragrances sparingly when going to work or attending social gatherings.

Even the most ardent users of perfumes and colognes say that other people should not realize you are wearing a fragrance until they enter your so-called scent circle, an area within an arm's length of your body. That space, however, belongs to you. And somewhere within that mysterious circle, the chemistry of perfume ends and its romance begins.

[Rebecca Kolberg]

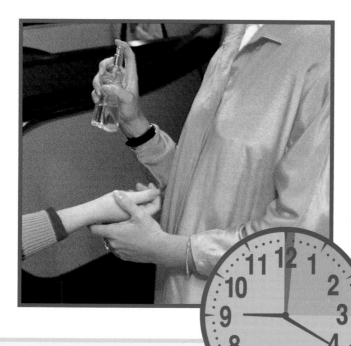

## How fragrances react with the skin

Colognes and perfumes are formulated to release their scents in three phases, called top, middle, and base notes. Each note reveals a different aspect of the fragrance.

| | |
|---|---|
| **Top note** | How the fragrance smells during the first minute—the scents of florals, citrus, and chemicals called aldehydes are noticed. |
| **Middle note** | An aroma created as the fragrance interacts with skin chemistry—rich florals may predominate. |
| **Base note** | How the fragrance smells after 15 minutes—sandalwood, vanilla, balsam, and musk may come to the fore. |

# Digital Cameras Gain in Popularity and Utility

During the 1990's, digital cameras played an increasingly important role in both personal and professional photography. Equipped with computer chips and sophisticated electronics, these cameras capture images in digital format—the patterns of 0's and 1's used by all computers.

Because the picture information is in a digital format, it can be adjusted, manipulated, or altered in many ways by computer software programs designed for such purposes. In addition, the picture can be stored or reproduced in a variety of formats. It can be printed on paper, saved on a magnetic disk, incorporated into documents, or posted on a World Wide Web site. Digital pictures can even be sent to friends and family as attachments to e-mail.

Although digital cameras were not yet as popular or widely used in 1999 as traditional cameras using film, their popularity was growing rapidly. They were seen by many photographic experts as representing the future of photography.

The first digital cameras debuted in the 1980's. They were quite expensive, costing thousands of dollars, and produced images that were far inferior to those made by even low-priced film cameras. By 1999, however, advances in electronic technology had brought about such great improvements in digital cameras that some of them approached the quality of conventional cameras.

As quality went up, prices came down. By 1999, a no-frills digital camera sold for as little as $300, though some models cost $1,000 or more. And as with any camera, the addition of special lenses and other equipment could add considerably to the price.

Digital cameras priced at the low end produce pictures of moderate quality that look fine when displayed on a computer screen or a Web site. Higher-priced digital cameras capture high-resolution images comparable to those produced by film cameras. They are designed for photographers who want the features of a digital camera but insist on images of the highest possible quality.

In some ways, digital cameras and film cameras are quite similar. Typical models of each type of camera are small enough to be held in the palm of a hand. Both have a lens that is used to focus the desired image. Either type of camera can be fitted with special lenses for telephoto, wide-angle, or close-up photography. And both digital and film cameras have a control button that activates a shutter, allowing light passing through the lens to enter the camera.

Once light has been admitted to the camera, though, the similarities end. Traditional cameras, which have been in use since the mid-1800's, capture images by way of chemical reactions. When the shutter is opened, light rays strike a piece of film coated with a gelatin, called an emulsion, containing light-sensitive chemicals. These chemicals respond to the varying levels of light reflected by the subjects of the photograph, creating a *latent* (invisible) negative image on the film.

When the film is removed from the camera and developed, the latent image becomes a visible negative—either black and white or color, depending on the kind of film that was used. The negative can then be used to make a positive image—the familiar photo print. With color *transparency* (slide) film, an additional processing step creates a positive image on the film itself. All of these processes must be carried out under special conditions, usually in a light-tight room called a darkroom. (Polaroid cameras, which produce prints that can be viewed just a few minutes after a picture is taken, combine the developer with the emulsion to allow the picture to be developed on the spot.)

Digital cameras grew in popularity in the 1990's as their image quality improved and their cost declined. In addition, many people preferred the high-tech utility of digital images, which can be used within computer-generated documents, sent with e-mail over computer networks, and posted on the World Wide Web.

Digital cameras use no chemicals at all. Nor is there any delay between the capture of an image and the photographer's ability to view or print it.

When a photographer opens the shutter of a digital camera, the focused image is captured with a *charge-coupled device* (CCD), a piece of electronic technology that converts light into electrical current. The CCD is composed of many light-sensitive photocells. When light strikes these cells, the properties of the light, such as brightness and color, are converted into electrical signals and broken down into *pixels* (picture elements). These tiny squares of color, each represented by a string of 0's and 1's (known as bits), are the smallest units into which the camera is able to digitally divide the visual information.

When a digital camera captures a black-and-white image, each pixel is given a brightness value appropriate to the level of black or white in the picture being created. For color images, pixels are divided into red, green, and blue, with corresponding brightness values for each. Combinations of these three colors result in an accurate reproduction of the images' actual colors.

A digital image is composed of hundreds of thousands, or even millions, of pixels. (A million pixels is called a megapixel.) The more pixels the image contains, the sharper the image and the higher its *resolution* (amount of detail). Typical digital cameras costing less than $1,000 can capture images at resolutions up to 1.5 megapixels. More sophisticated, and more expensive, digital cameras can achieve resolutions up to six megapixels.

Once the image has been captured by the CCD and converted into pixels, the data is transferred to the camera's storage device. This can be a removable magnetic disk, like those used in personal computers, or a built-in device that stores the information in electronic memory. In the storage device, the image is saved as a single, self-contained computer file, much as conventional photos are saved on individual frames of film. Because images composed of a megapixel or more of information require quite a large file of data, many

## Two different imaging processes

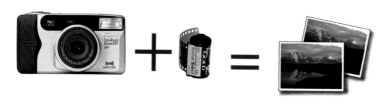

Conventional cameras capture images on light-sensitive film through a chemical process. When the film is developed, patterns of light are reproduced on special photographic paper. The result is the familiar photo print, which comes in several sizes.

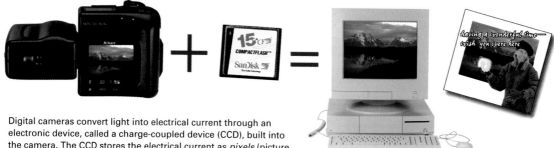

Digital cameras convert light into electrical current through an electronic device, called a charge-coupled device (CCD), built into the camera. The CCD stores the electrical current as *pixels* (picture elements) that are expressed as a string of 1's and 0's, the binary code used by all computer programs. The image can be transferred to a memory card, *above center,* a computer disk or directly to a computer monitor through a cable. Once on the computer, the image can be electronically retouched and manipulated, *right.*

## Film cameras versus digital cameras

|  | Film cameras | Digital cameras |
|---|---|---|
| Camera price | $10 to $500 | $300-$1,300 |
| Necessary extras | Film* | Computer, color printer, glossy photo paper or laser paper |
| End product | Prints or slides | Electronic images or prints |
| Pros | Large range of film types available | You can e-mail photos anywhere; photos can be retouched on computer and printed on color printer; no waiting for developing |
| Cons | Pictures cannot be viewed until film is developed | Requires extra equipment, costly, not as reliable as film cameras |
| Quality | High-quality prints and slides | Good-quality images and prints with a megapixel camera and a good color printer |

*Inexpensive, one-time-use cameras come already loaded with film.

digital cameras also contain software that compresses, or shrinks, the size of the image file for easier storage.

At the same time, other software in the camera adjusts the captured image to make it a more accurate representation of the subject. The edges of the pixels are digitally smoothed in order to make them flow together visually. Brightness and darkness factors are also adjusted by the software to achieve the best image quality.

If the camera is one that stores its images primarily on a disk, the disk can be removed from the camera and inserted into a computer, which can then be used to access and manipulate the digital images. Cameras that store digital photos electronically usually have a connector cable that plugs into a computer. This allows the visual data to be downloaded to the computer. The photographer can then view the images on the computer monitor.

The digital information in the photo files can then be used in a variety of ways. It can be kept in electronic form, or, as with a conventional negative, it can be used to make a paper print.

Initially, printing digital photos required the use of a special paper in order to achieve acceptable levels of clarity or resolution. Beginning in the 1990's, however, improvements in the quality of cameras, printers, and the software that runs both made it possible to make very high-quality prints on standard papers. This advance helped reduce the cost of printing digital photos. With the use of the least expensive papers, the price per photo could actually be less than that of a print made from a roll of film.

Prints can be made with either an inkjet or laser printer. Inkjet printers use tiny nozzles to spray quick-drying ink of various colors onto paper. Their inexpensive color capability made them a popular choice for individuals who take and print digital photos. Laser printers use a laser light to transfer electronic images to a light-sensitive drum and then to paper. They are generally more expensive than inkjet printers, but the quality of the resulting print is higher as well.

The output of the printer is the factor that most determines the quality of a digital print, including the clarity of the image and richness of the colors. Printer output is measured in dots per inch (dpi), the number of dots of ink that the printer can apply to 2.5 centimeters (1 inch) of paper. The higher a printer's dpi value, the closer its prints will resemble ones made from film. A typical computer monitor presents images at 72 dpi. Most laser and inkjet printers offer resolutions beginning at 300 dpi, while more expensive printers can achieve resolutions of more than 1,000 dpi. Digital photographs printed on special paper by the highest dpi printers are virtually indistinguishable from photos taken with film cameras.

The simplest approach to printing a photo is to just send the image file to the printer as is. But there are many other options available to the digital photographer. By using desktop-publishing software, for example, a student preparing a written report about a family vacation can supplement the text with digital pictures taken during the trip. The publishing software includes tools that adjust the photos to the appropriate size and align the columns of text around the images. The combined use of a digital camera, desktop-publishing software, and a color printer enables people to produce reports, newsletters, brochures, and other printed materials of near-professional quality.

For long-term storage, photographic prints are often stored in albums. Digital photographers can store their images on a compact disc (CD). Most CD's cannot be erased, so they serve as permanent digital photo albums. Because they can store hundreds of digital images, CD's are ideal for preserving a wide variety of photos for printing or for viewing on a computer monitor.

But you don't need a digital camera to make digital photographs. Photographs taken with film cameras can be converted into digital files using a device called a scanner. Once a conventional photograph is scanned and converted into pixels and saved as a file, it can be stored or reproduced in the same way as pictures made by digital cameras. This process allows the conversion of thousands of photos from family albums and public archives into digital resources ready to be shared with future generations.

Many people post their digital photographs on Web pages, adding images to the information they are presenting. Some small, private Web pages serve as photo albums that friends and relatives can call up on their computers. Commercial Web sites often use digital photographs to present catalog items for purchase or to dress up promotional information.

Digital images need not be stored or reproduced exactly as captured. Using computer software tools, the photographer can improve or alter a picture in a variety of ways. Colors can be adjusted and images enlarged or reduced. The photographer can also change the elements in a picture—for example, extracting one person from a group and inserting that individual in another setting. Special effects are easy to add, such as placing dinosaurs in the background of a photo of friends.

The ease with which digital images can be altered and manipulated has raised many questions about the trustworthiness of digital photos when used in newspapers or magazines. Traditional photography gave rise to a familiar saying: The camera never lies. Digital cameras, however, have made it very easy to convincingly falsify information in a photograph. Digital manipulation could, for instance, be used to show someone at a location where the person denies ever being. News and information organizations have been discussing ways to ensure the accuracy and reliability of the digital photos they use.

Although digital photography is a relatively new field, rapid technical progress is resulting in an almost constant stream of new products and product enhancements. Some of the latest digital cameras available in 1999 could transfer captured images directly to a printer, without going through a computer. Others incorporated digital sound recorders, enabling the photographer to add comments as pictures are taken or to record sounds at the scene, such as the crashing of waves at a seashore location. Many new digital cameras included a small monitor that let the photographer review pictures as they are taken.

Another new trend in the late 1990's was the arrival of digital video recorders. These devices operate in much the same way as digital cameras but record moving images and sound in digital files. Digital video cameras can also record still images. Once recorded, the digital video files can be edited and manipulated on a computer, just as any other digital images can be. Digital video was expected to play a large part in the future of motion-picture and television production.

As digital technology continues to advance and prices keep on dropping, more and more people are likely to switch to digital cameras and digital video recorders to capture and store images for the future. The utility of these devices is certain to increase in step with their popularity.

[Keith Ferrell]

# WORLD BOOK

## *Supplement*

Seven new or revised articles reprinted from the 1999
edition of *The World Book Encyclopedia*

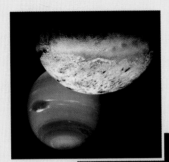

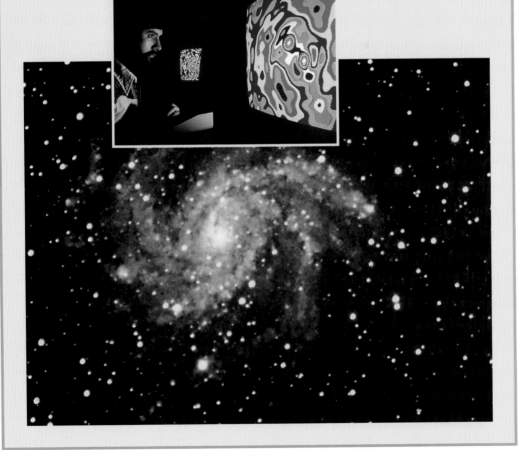

© Tony & Daphne Hallas, Astro Photo

**Stars fill the sky** in the large photograph, which shows what can be seen with the unaided eye. The familiar Big Dipper tilts downward toward the mountain at the left. The inset photo, taken by the powerful Hubble Space Telescope, shows how much more astronomers can see with telescopes. It reveals galaxies filling a tiny speck of sky near the Dipper's "handle."

NASA

# Astronomy

**Astronomy** is the study of the universe and the objects in it. Astronomers observe the sky with telescopes that gather not only visible light but also invisible forms of energy, such as radio waves. They investigate nearby bodies, such as the sun, planets, and comets, as well as distant galaxies and other faraway objects. They also study the structure of space and the past and future of the universe.

Astronomers seek answers to such questions as: How

did the universe begin? What processes release energy deep inside stars? How does one star "steal" matter from another? How do storms as big as Earth arise on Jupiter and last for hundreds of years?

To answer such questions, astronomers must study several subjects besides astronomy. Almost all astronomers are also astrophysicists because the use of physics is essential to most branches of astronomy. For example, some parts of *cosmology,* the study of the structure of the universe, require an understanding of the physics of elementary particles, such as the bits of matter called *quarks* that make up protons and neutrons. Astronomers use chemistry to analyze the dusty, gaseous matter between the stars. Specialists in the structure of planets use geology.

Astronomy is an ancient science. Like today's researchers, ancient scholars based their ideas of the uni-

*The contributor of this article is Jay M. Pasachoff, Field Memorial Professor of Astronomy and Director of the Hopkins Observatory at Williams College.*

verse on what they observed and measured and on their understanding of why objects move as they do. However, the ancients developed some incorrect ideas about the relationships between Earth and the objects they saw in the heavens. One reason for their errors was that they did not understand the laws of motion. For example, they did not know that a force—which we know as gravitation—controls the movements of the planets. Another reason was that their measurements did not reveal the movements of the planets in sufficient detail.

The ancients noted that the positions of the sun, moon, and planets change from night to night. We know that these movements are a result of the revolution of the moon about Earth and the revolution of Earth and the other planets about the sun. The ancients, however, concluded that the sun, moon, and planets orbit a motionless Earth. In many places, religious teachings supported this conclusion until the 1600's.

Although ancient people misinterpreted much of what they saw in the heavens, they put their knowledge of astronomy to practical use. Farmers in Egypt planted their crops each year when certain stars first became visible before dawn. Many civilizations used the stars as navigational aids. For example, the Polynesians used the positions of the stars to guide them as they sailed from island to island over thousands of miles or kilometers of the Pacific Ocean.

### Observing the sky

If you look at the night sky without a telescope or a pair of binoculars, you will see what the ancients saw. If the night is clear and you are far from city lights, you will see about 3,000 stars. Stretching across the sky will be a splotchy band of bright and dark areas called the Milky Way. In addition, a few fuzzy spots will be visible.

Ancient people in the Western world noticed that certain stars are arranged in patterns shaped somewhat like human beings, animals, or common objects. Many ancient civilizations associated such patterns, called *constellations,* with mythology. Many names of constellations have come to us from Greek myths. In one myth, Artemis, goddess of hunting, was greatly saddened by the death of a human hunter named Orion. In her sorrow, she placed Orion in the sky as a constellation.

**How the stars move.** If you map the location of several stars for a few hours, you will observe a regular motion. The stars move relative to Earth because of Earth's rotation about its own axis. All stars move in circles around a point in the sky known as a *celestial pole.* Stars rotate counterclockwise around the celestial north pole and clockwise around the celestial south pole. Stars that are far from the pole rise from below the horizon in the east, move upward, and then set in the west.

One bright star, Polaris, is so close to the celestial north pole that it moves very little. Because of its location, Polaris is also known as the North Star. There is no "South Star," but the constellation Octans (Octant) is close to the celestial south pole.

The sun, like the other stars, rises in the east and sets in the west. But the sun also moves eastward relative to the other stars about 1° each day. By noting which stars are visible above the horizon just before sunrise and just after sunset, people have mapped the path of the sun among the stars for thousands of years. This path is

known as the *ecliptic,* and the band of constellations near this path is called the *zodiac.*

**How the planets move.** With the unaided eye, you can see the planets Mercury, Venus, Mars, Jupiter, and Saturn. The planets move every night from east to west along the zodiac. In addition, their position in the zodiac changes from night to night. That is, when viewed at the same time on successive nights, the planets move relative to the background stars. Most of the year, they move from west to east.

The planets also have a motion that differs from that of any other celestial object. In their night-to-night movement relative to the stars, they slow down, stop, and then move westward in what is known as *retrograde motion.* They then slow down again, stop, and resume their eastward motion.

**How the moon moves.** The moon is the brightest and most easily seen object in the night sky. As a result, the most familiar observation is of the moon's *phases,* such as the full moon, half moon, and crescent. The moon moves from east to west as it rises and sets. From night to night, the moon moves eastward about 13° relative to the stars, rising almost an hour later each night.

**Earth-centered theories.** Ancient scholars produced elaborate schemes to account for the observed movements of the stars, sun, moon, and planets. In the 300's B.C., the Greek philosopher Aristotle developed a system of 56 spheres, all with the same center. The innermost sphere, which did not move, was Earth. Around Earth were 55 transparent, rotating spherical shells. The outermost shell carried the stars, believed to be merely points of light. Other shells carried the sun, moon, and planets. These shells rotated inside other shells that rotated within still other shells in ways that accounted for almost all the observed movements.

During the A.D. 100's, Ptolemy, a Greek astronomer who lived in Alexandria, Egypt, offered an explanation that better accounted for retrograde motion. Ptolemy said that the planets moved in small circles called *epicycles.* The epicycles moved in large circles called *deferents.* Earth was near the center of all the deferents.

**Sun-centered theories.** By the early 1500's, the Polish astronomer Nicolaus Copernicus had developed a theory in which the sun was at the center of the universe. This theory correctly explained retrograde motion as the changing view of the planets as seen from a moving Earth. The theory also correctly explained the east-to-west movement of the sun and stars across the sky. This movement is due to the west-to-east rotation of Earth about its own axis, rather than an actual motion of the sun and stars.

Several decades later, the Danish astronomer Tycho Brahe built gigantic instruments that he used to make precise measurements of planetary movements. German mathematician Johannes Kepler analyzed Tycho's measurements of the movement of Mars. He discovered that Mars moves in an *ellipse,* a type of squashed circle, with the sun at a key point inside the ellipse. Kepler also found a relationship between how far Mars is from the sun and how rapidly it moves. He concluded that all planets have elliptical orbits and that the relationship between distance from the sun and speed applies to all the planets. These findings, which became known as Kepler's first two laws of planetary motion, were published in 1609. In 1619, Kepler published his third law, which shows the relation between the sizes of the planets' orbits and the time they take to orbit the sun.

Thus, by the early 1600's, astronomers had used observations made with the unaided eye to determine the movement of the planets. But the Italian astronomer and physicist Galileo had already built and used the first of the great "aids to the eye," the optical telescope, ushering in the era of modern astronomy.

### Modern astronomy

Today's astronomers gather information in four major ways: (1) they use telescopes and other instruments to detect visible light and other forms of radiation that are *emitted* (sent out) by celestial objects; (2) they detect particles called neutrinos and cosmic rays from outer space; (3) they study chunks of matter that originated in outer space; and (4) they send spacecraft to land on other objects or to study them close-up.

Using these techniques, modern astronomers have made discoveries beyond the imagination of the ancients. They have discovered three planets outside the orbit of Saturn—Uranus, Neptune, and Pluto. They have found that more than 50,000 smaller bodies called *asteroids* revolve about the sun, most of them between the orbits of Mars and Jupiter. Astronomers have learned that the sun is merely one of hundreds of billions of stars in a vast, disk-shaped galaxy, the Milky Way. They now understand that many of the fuzzy spots visible with telescopes in the night sky are other galaxies.

In addition, astronomers have discovered exotic objects called *pulsars* and *quasars* that emit vast amounts of energy. They have found evidence of *black holes,* objects from which nothing—not even light—can escape.

**The stars and constellations of the Northern Hemisphere**

This map shows the sky as it appears from the North Pole with Polaris, the North Star, directly overhead. To use the map, face south and turn the map so that the current month appears at the bottom. The stars in about the bottom two-thirds of the map will be visible at some time of the night from most areas of the United States and southern Canada.

WORLD BOOK illustration by W. J. M. Tirion

They have studied exploding stars called *supernovae.*

**Units of distance.** Many distances involved in astronomy are so huge that they are measured in special units. One such unit is the *light-year,* the distance that light travels in a vacuum in a year. This distance equals about 5.88 trillion miles (9.46 trillion kilometers). The nearest star, Proxima Centauri, is about 4 light-years from Earth. The Milky Way is about 100,000 light-years across, and the sun is roughly 25,000 light-years from its center. The nearest large galaxy is the Andromeda Galaxy, which is about 2 million light-years away. Some galaxies are more than 10 billion light-years distant.

Astronomers measure distances in the solar system in *astronomical units* (AU). One AU is the average distance from Earth to the sun—about 93,000,000 miles (150,000,000 kilometers). This distance equals about 8 light-minutes. The average distance from the sun to Plu-

to, the farthest planet, is about 39.5 AU.

In their work with extremely long distances, astronomers use a unit called a *parsec,* rather than the light-year. The parsec is based on *parallax,* an angular measurement. One parsec equals about 3.26 light-years.

**Locating objects in space.** Astronomers still use two concepts developed by the ancients to specify locations of celestial objects: (1) an imaginary *celestial sphere* and (2) the constellations.

*The celestial coordinate system.* Astronomers specify locations in terms of the *celestial coordinate system,* a set of imaginary lines drawn on the *celestial sphere.* The celestial sphere is similar to the outermost shell in Aristotle's system—the shell that was thought to carry the stars. The lines are similar to the lines of longitude and latitude used by geographers. The poles of the celestial sphere are the *celestial north pole* and the *celestial*

**The stars and constellations of the Southern Hemisphere**

This map shows the sky as it appears from the South Pole. There is no "South Star," but the constellation Octans would be almost directly overhead if you were at the South Pole. To use the map, an observer in the Southern Hemisphere would face north and turn the map so that the current month appears at the bottom.

WORLD BOOK illustration by W. J. M. Tirion

*south pole,* which lie over Earth's north and south geographic poles. The sphere also has a *celestial equator* over the earthly equator.

Longitude in the sky, marked by half-circles going from the north celestial pole to the south celestial pole, is called *right ascension.* Latitude in the sky, marked by circles parallel to the celestial equator, is known as *declination.* Declination north of the celestial equator is positive, while declination south of the equator is negative.

*Using the constellations.* To locate and assign names to stars, astronomers have divided the sky into 88 parts, each associated with a constellation. Astronomers still use a system developed in the early 1600's to identify the brightest stars in the constellations. The brightest star of all in a constellation is usually designated by alpha, the first letter of the Greek alphabet; the second brightest by beta, the second letter; and so forth. The brightest star in the constellation Orion is thus Alpha Orionis. *Orionis* means *of Orion* in Latin.

Because the Greek alphabet has only 24 letters, this system is limited to 24 stars per constellation. Later astronomers developed naming systems in which numbers are assigned to fainter stars and Roman letters to *variable stars* (stars that vary in brightness).

**Electromagnetic radiation** is the most plentiful source of information about heavenly bodies. Its name comes from the fact that it consists of waves of electric and magnetic energy. Visible light is electromagnetic ra-

diation, and objects in space also emit many kinds of invisible electromagnetic radiation. Scientists can identify the various forms of this radiation by their wavelength, frequency, or energy.

*Wavelength* is the distance between successive crests of a wave. From the shortest wavelength to the longest, the forms of electromagnetic radiation are gamma rays, X rays, ultraviolet rays, visible light, infrared rays, and radio waves. Arranged in order of wavelength, the forms make up the *electromagnetic spectrum.*

*Frequency.* All electromagnetic radiation travels at the same speed. Therefore, a relatively short wave passes a given point more quickly than does a relatively long wave. Thus, more of the shorter waves pass that point each second. Scientists say that a shorter wave has a higher *frequency.* The unit used to measure frequency is the *hertz* (symbol Hz). One hertz represents the passing of one wave past a point in one second.

*Energy.* According to quantum theory, a cornerstone of modern physics, electromagnetic radiation can also be thought of as particles of energy called *photons.* The amount of energy of a given photon depends on the wavelength—or frequency—of the corresponding wave. Radiation that has a relatively short wavelength and therefore a relatively high frequency also has relatively high energy. Radiation with a relatively long wavelength has a relatively low frequency and relatively low energy.

**Optical astronomy** is the study of the heavens by de-

### Units of astronomical distance

| 1 astronomical unit | = | 150.0 million kilometers | = | 93.0 million miles | = | 0.0000158 light-years | = | 0.00000485 parsecs |
|---|---|---|---|---|---|---|---|---|
| 1 light-year | = | 9.46 trillion kilometers | = | 5.88 trillion miles | = | 63,200 astronomical units | = | 0.307 parsecs |
| 1 parsec | = | 30.9 trillion kilometers | = | 19.2 trillion miles | = | 206,000 astronomical units | = | 3.26 light-years |

## The size of the universe

The series of illustrations on this page and the next should enable you to appreciate the enormous size of the universe. Each succeeding diagram represents a cube in space 1,000 times larger on each side than the preceding cube. The 10's with small raised figures are a way of abbreviating numbers. For example, $10^6$ kilometers equals 1 followed by 6 zeroes, or 1,000,000 kilometers.

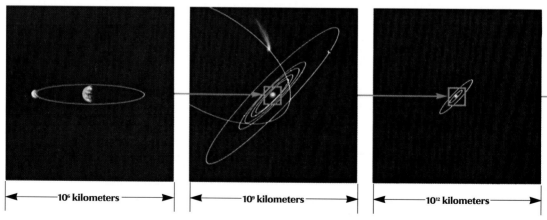

**$10^6$ kilometers**   **$10^9$ kilometers**   **$10^{12}$ kilometers**

The orbit of the moon around the earth would easily fit inside a cube whose length, width, and depth measure 1 million kilometers (620,000 miles).

This cube could contain the part of the solar system that lies within the orbit of Jupiter. The outermost planets would be billions of kilometers outside the cube.

The entire solar system would fill only a tiny portion of this cube. The rest would be empty space, which shows how distant even the nearest star is from the sun.

## The electromagnetic spectrum

| | Wavelength* | Frequency, hertz[†] | Sources |
|---|---|---|---|
| Radio waves | 1 millimeter and up | Up to $3.00 \times 10^{11\ddagger}$ | Pulsars; quasars; gas clouds orbiting the center of the Milky Way |
| Infrared rays | 700 nanometers[§] to 1 millimeter | $3.00 \times 10^{11}$ to $4.29 \times 10^{14}$ | Stars in the process of forming; relatively cool stars; planets |
| Visible light | 400 to 700 nanometers | $4.29 \times 10^{14}$ to $7.50 \times 10^{14}$ | Planets; stars; galaxies; asteroids; comets |
| Ultraviolet rays | 10 to 400 nanometers | $7.50 \times 10^{14}$ to $3.00 \times 10^{16}$ | Hydrogen gas between the stars; the sun |
| X rays | 0.1 to 10 nanometers | $3.00 \times 10^{16}$ to $3.00 \times 10^{18}$ | The sun's corona; disks of material around black holes; quasars |
| Gamma rays | Up to 0.1 nanometer | $3.00 \times 10^{18}$ and up | Collapsed stars; matter-antimatter annihilations |

* The ranges indicated are typically used by astronomers to distinguish between forms of electromagnetic radiation. Scientists working in other fields may use slightly different numbers.
† One hertz is one cycle of vibration per second.
‡ The small numeral next to the 10 indicates how many places the decimal point is moved to the right when the number is written out. For example, $3.00 \times 10^{11}$ equals 300,000,000,000.
§ One nanometer is 1 billionth of a meter.

tecting and analyzing visible light. Visible light of different wavelengths has different colors. The wavelengths range from about 400 *nanometers* for deep violet to 700 nanometers for deep red. One nanometer equals a billionth of a meter, or $\frac{1}{25,400,000}$ inch.

Modern observational astronomy began with Galileo's observations of the sky through an optical telescope. Today, astronomers often use observations made with visible light together with observations made in other parts of the electromagnetic spectrum. Astronomers use optical telescopes to observe objects ranging from nearby cool bodies, such as planets, to hotter objects, such as the sun and stars, and to extremely hot objects, such as supernovae.

An optical telescope gathers and focuses light with a lens or mirror. To study the brightest objects, such as the sun, a telescope with a lens or mirror only a few inches or centimeters across may do. For the faintest objects, a much larger light-collecting area is needed. The

largest all-purpose telescopes now in general use include the twin Keck Telescopes on Mauna Kea, an extinct volcano on the island of Hawaii. Each telescope has a mirror 33 feet (10 meters) in diameter.

Major optical telescopes are installed on mountains so that starlight does not have to travel far through the atmosphere. These locations minimize blurring due to the atmosphere. The atmosphere bends light due to a phenomenon known as *refraction,* and the atmosphere is constantly moving. As a result, starlight jiggles about and changes in brightness as it passes through the air. Thus, stars appear to twinkle. Twinkling blurs images.

Only a telescope operating in space can avoid blurring entirely. The largest orbiting telescope is the Hubble Space Telescope, which was launched in 1990, repaired in 1993, and upgraded in 1997. This telescope's main mirror measures 94 inches (2.4 meters) across. Because of the size of its mirror and its location above almost all of the atmosphere, the Hubble Space Tele-

WORLD BOOK illustrations by Ernest Norcia

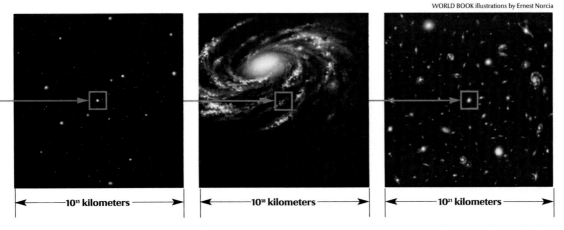

| ◄——— $10^{15}$ kilometers ———► | ◄——— $10^{18}$ kilometers ———► | ◄——— $10^{21}$ kilometers ———► |
|---|---|---|

This cube contains a few hundred nearby stars. It is about 100 *light-years* long on each side. One light-year equals about 9.46 trillion kilometers.

This region, about 100,000 light-years per side, would enclose the entire Milky Way. The solar system is about 25,000 light-years from the center of the Galaxy.

This cube would contain thousands of galaxies. It measures about 100 million light-years on each side. Most galaxies, if not all, are found in clusters.

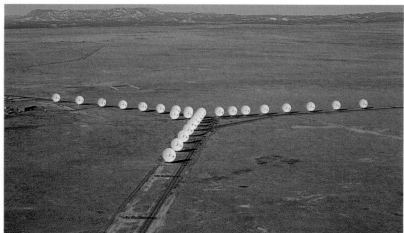

**The Very Large Array** radio observatory in Socorro, New Mexico, consists of 27 movable dish antennas, each 82 feet (25 meters) across. The arrangement of dishes can extend up to 22 miles (35 kilometers). Computers can combine signals from the antennas to produce images with the detail that would be provided by a single telescope 22 miles across.

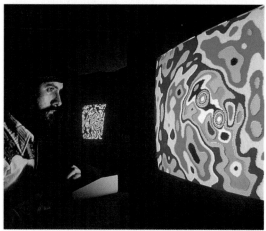

**An astronomer studies an image** produced by the Very Large Array of radio telescopes. The colors in the "painting" represent different amounts of radio energy sent out by objects in the sky.

scope can produce images that are approximately five times smaller in detail than images made from telescopes on Earth's surface.

Professional astronomers rarely look through telescopes. Instead, they study photographic images. Astronomers began to photograph images through telescopes in the 1850's. The use of long exposure times revealed faint objects that the eye could not see through a telescope. But film has largely been replaced by electronic devices that can detect and record even fainter light. The *charge-coupled device* (CCD), for example, is about 50 times or more sensitive to light than film is.

Optical astronomers have developed three special techniques that have also been used in other kinds of astronomy: (1) *spectroscopy,* (2) *interferometry,* and (3) *adaptive optics.*

*Spectroscopy* is the breaking down of the incoming radiation into a spectrum of its parts. The spectrum of visible light, for example, is a rainbowlike band of colors. At one end is red, which has the longest wave-

length. At the other end is violet, with the shortest wavelength. Spectroscopy is based on a discovery made in 1814 by German optician Joseph von Fraunhofer: The spectrum of the colors of sunlight contains dark lines where specific colors are absent. Later, scientists discovered that light from other stars also has such dark lines. When the object emitting the light is a hot gas, the spectrum has bright lines. Other kinds of electromagnetic radiation from celestial objects also have spectral lines.

Studies of spectral lines reveal the temperature, density, and chemical composition of the object emitting the radiation. The spectral lines arise in energy processes in atoms. An electron orbiting an atomic nucleus can have only certain definite levels of energy. These levels can be thought of as stairsteps. When light energy passes through a group of atoms, the electrons can absorb just the right amount of energy to jump from a lower step to a higher one. Because the energy is removed from the passing light, a kind of spectral line called an *absorption line* appears in the spectrum. The dark lines discovered by Fraunhofer were absorption lines. Each kind of atom has its own pattern of absorption lines for a given range of temperature.

Bright spectral lines called *emission lines* occur when electrons in atoms of hot objects "jump down the stairs" by emitting energy. In the early 1940's, an analysis of emission lines confirmed an earlier discovery that the sun's *corona,* the outer edge of its atmosphere, has a temperature of millions of degrees.

The entire band of radiation emitted by the surface of a star also contains information about the star's temperature. Relatively cool stars are red-hot and can appear slightly reddish to the eye. Hotter stars can become blue-white.

Spectroscopy can even reveal the distance to a star or galaxy. In the spectrum of any moving object, the spectral lines are shifted from where they would appear in the spectrum of a stationary object. Such shifts are interpreted in terms of the *Doppler effect.* One familiar example of this effect is a change in the pitch of sound waves emitted by a vehicle. For example, when a car approaches you, the pitch of the sound made by the engine is higher than it is when the car is going away. The shift of

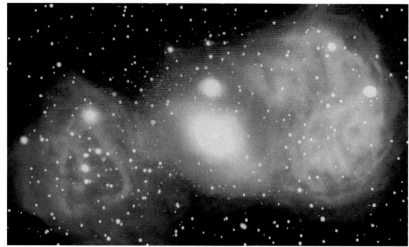

**A combined optical and radio image** shows radiation produced by what is probably a huge black hole in the center of a galaxy. Objects fall into the black hole, generating energy. The galaxy sends out the energy as visible light, shown in blue, and as radio waves, artificially colored red. The total distance from left to right is about 2 million light-years.

National Radio Astronomy Observatory

spectral lines also depends on whether the object is approaching or receding. If the lines shift toward the blue (shorter-wavelength) end of the spectrum, the object is approaching. If they shift toward the red (longer-wavelength) end, the object is moving away.

In 1929, the American astronomer Edwin Hubble discovered that the farther away a galaxy is, the faster it is retreating, and thus the greater is its red shift. Thus, astronomers can determine the distance of a remote galaxy by measuring its red shift.

*Interferometry* uses a phenomenon called *interference* in which rays of light combine to form a pattern. Astronomers use such patterns, often with the aid of computers, to produce extremely detailed images. In the simplest example in optical astronomy, a ray of light emitted by a star strikes the mirror of a telescope, and another ray emitted by the same star strikes the mirror of a nearby telescope. Optical and mechanical devices combine the rays so that a series of bright and dark bands of light called an *interference pattern* appears. For an explanation of why this pattern appears, see the article **Interference.** The pattern reveals any difference in the routes taken by the rays as they travel from the star

to the telescopes. For example, if the rays were emitted at opposite edges of the star, the pattern could reveal the diameter of the star. A computer helps astronomers analyze the interference pattern and use the results to produce an image.

*Adaptive optics* can make up for atmospheric blurring in ground-based telescopes. In this technique, light reflects from a telescope's main mirror to a special deformable mirror, then to a CCD. Pistons mounted on the underside of the deformable mirror can change its shape several hundred times a second to make up for atmospheric blurring. A special control system senses the amount of blurring and operates the pistons.

In one arrangement, the deformable mirror is in the main telescope and the control signals come from a smaller auxiliary telescope. A laser beam emitted by the auxiliary telescope reflects off atoms in the atmosphere and returns to the auxiliary telescope. As the beam travels, the atmosphere distorts it slightly. A computer analyzes the reflected beam, then operates the pistons in a way that would remove the distortion from the image of the beam. This operation removes much of the distortion in the image viewed by the main telescope.

WORLD BOOK diagram by Precision Graphics
University of Manchester/NASA

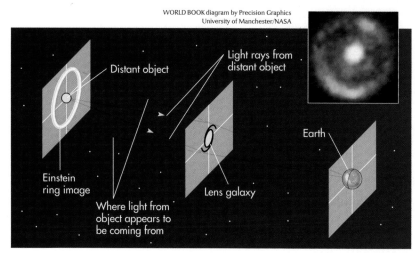

**Gravitational lensing** occurs when the gravitational force of a massive galaxy apparently acts as a lens, bending light rays sent out by an object on the other side of the galaxy. The light from the distant object appears to come from an arc or, as shown in the inset photo, a ring. Such rings are called Einstein rings—named after physicist Albert Einstein, who described how gravitation seems to bend light.

**Locating a star by declination and right ascension**

Astronomers locate objects in the sky by means of a *celestial coordinate system.* All the objects are considered to lie on an imaginary sphere surrounding Earth. Astronomers specify a location in terms of an angle measured from a horizontal circle and an angle measured from a vertical circle.

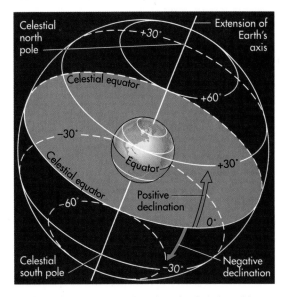

**Vertical angles** are measured northward and southward from the *celestial equator,* which lies above Earth's equator. A vertical angle is known as a *declination* and can be positive or negative.

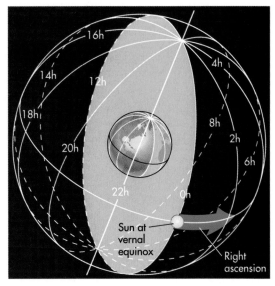

**Horizontal angles** are measured eastward from the *vernal equinox,* where the sun crosses the celestial equator. An angle, or *right ascension,* is given in hours, minutes, and seconds.

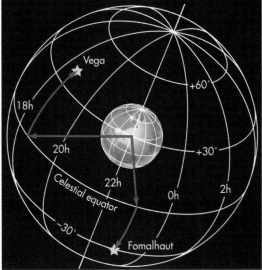

WORLD BOOK diagrams by Precision Graphics

**Celestial coordinates** locate the star Vega at right ascension (RA) 18 hours 36 minutes 56.3 seconds and at declination (Dec) + 38 degrees 47 minutes 1 second. Astronomers give the coordinates of the star Fomalhaut as RA 22$^h$57$^m$39.1$^s$, Dec −29°37'20".

**Infrared astronomy** deals with invisible electromagnetic waves whose wavelengths are longer than those of visible light. Objects that are bright in infrared wavelengths include relatively cool stars and stars in the process of forming. Planets and other objects that glow by reflecting sunlight or starlight are also best studied in the infrared spectrum.

The infrared spectrum covers a range from about 700 nanometers to 1 millimeter. Infrared astronomers commonly express wavelengths in *micrometers* (thousandths of a millimeter), specifying the overall range as about 0.7 to 1,000 micrometers.

A photon of infrared radiation has less energy than a photon of visible light. Most infrared photons do not have enough energy to cause the chemical reaction that produces images on film. Infrared astronomy therefore did not develop fully until about the 1960's, when electronic sensors that could produce infrared images were introduced.

One problem with infrared astronomy is that Earth's atmosphere absorbs rays of most wavelengths in the infrared spectrum. However, some of the shorter waves can be detected at mountaintop observatories. Notable infrared telescopes include the Infrared Telescope Facility of the National Aeronautics and Space Administration (NASA) of the United States and the United Kingdom Infrared Telescope. Both are on Mauna Kea. The telescopes are above most of the water vapor in Earth's atmosphere, one of the main absorbers of infrared rays.

In 1997, an infrared astronomy project called the Two Micron All Sky Survey (2MASS) began under the direction of the University of Massachusetts. Telescopes at Mount Hopkins, which is near Tucson, Arizona, and at Cerro Tololo, a mountain in Chile, are mapping the sky at a wavelength of 2 micrometers.

For 10 months in 1983, a multinational orbiting telescope called the Infrared Astronomical Satellite (IRAS) mapped infrared radiation across the entire sky. In 1995, the Infrared Space Observatory, a European spacecraft, took up where IRAS had left off, surviving until 1998.

An orbiting infrared telescope has a limited useful life because it must be cooled artificially. Its lifetime is limited by the amount of coolant it carries. Cooling is necessary to prevent the telescope's own infrared radiation from overwhelming the faint rays coming from outer space. The telescope must be cooled to the temperature of liquid helium—about 4 Celsius degrees above *absolute zero* (−459.67 °F or −273.15 °C). Absolute zero is the theoretical temperature at which atoms and molecules would have the least possible energy.

In 1989 and 1990, a satellite called the Cosmic Background Explorer (COBE) mapped the sky at the longest infrared wavelengths. In 1997, astronauts installed a device called the Near Infrared Camera/Multi-Object Spectrometer on the Hubble Space Telescope.

**Radio astronomy.** Astronomers use radio waves emitted by celestial objects to produce images of the objects and to study the objects spectroscopically. These are the same kinds of waves that radio and television broadcasters create to transmit programs—astronomers just use them differently.

The radio spectrum includes all electromagnetic waves longer than about 1 millimeter. Waves longer than about 1 millimeter and shorter than about 10 meters pass readily through Earth's atmosphere. Astronomers receive radio signals from a wide variety of objects, including particles swirling in the magnetic field of Jupiter, gas clouds orbiting the center of the Milky Way, *pulsars* (rapidly spinning collapsed stars from which regular bursts of radiation are received), distant galaxies, and *quasars.* A quasar is a distant object that produces an enormous amount of radiation. Almost all astronomers believe that a quasar is powered by an enormous black hole at the center of a galaxy.

Karl G. Jansky, an American engineer, discovered radio waves from outer space in 1931. The science of radio astronomy was not established until after World War II (1939-1945), however.

A radio telescope is essentially a large dish antenna. Unlike the reflecting surfaces of other kinds of telescopes, the dish surface does not have to be extremely smooth. The smoothness required of a reflecting surface depends on the length of the waves to be reflected. The shorter the wavelength is, the smoother the surface must be. Because radio waves are long, some dish surfaces are even made of metal mesh.

The largest radio telescopes that can be steered to point anywhere in the sky are 328 feet (100 meters) in diameter. One is in Effelsberg, Germany, near Bonn. The other is in Green Bank, West Virginia. The largest radio telescope of all is a nonsteerable telescope near Arecibo, Puerto Rico. Its reflecting mesh covers a natural bowl in the ground 1,000 feet (305 meters) in diameter.

Radio astronomy is the field in which interferometry is most useful. To use separate telescopes as an interferometer, the distance between them must be controlled to a fraction of the wavelength of the radiation to be detected. This requirement severely limits the use of interferometry in the other branches of observational astronomy, where wavelengths are shorter. Because radio waves are so long, however, the dishes of radio interferometers can be ten, hundreds, or even thousands of miles or kilometers apart.

**The Cassini probe,** launched in 1997, is due to arrive at Saturn in 2004. Cassini will study Saturn, its rings, and its moons. It will also drop a small probe into the atmosphere of the moon Titan.

The Very Large Array (VLA) in Socorro, New Mexico, links 27 movable dishes, each 82 feet (25 meters) in diameter. The telescopes move along railroad tracks built in the shape of a Y. Individual dishes can be up to 22 miles (35 kilometers) apart. The Very Long Baseline Array (VLBA) consists of 10 telescopes, each 82 feet across, spread across one side of Earth. Their locations range from the Virgin Islands north to New Hampshire and west to Hawaii. As an interferometer, the VLBA is equivalent to a single telescope with a diameter roughly equal to the diameter of Earth.

Radio astronomers have learned much by studying red and blue shifts in spectral lines. Because the radio spectrum is not visible, astronomers see these "lines" as low and high points in a graph of wavelength. The low points represent wavelengths of radiation absorbed by celestial objects. The high points represent wavelengths of radiation emitted by strong radio sources.

Red shift and blue shift work with radio waves exactly as they work with waves of visible light: If the lines in the spectrum of radiation emitted by an object shift toward the shorter-wavelength end, the object is approaching. The object is said to be blue-shifted. If the lines shift toward the longer-wavelength end, the object is moving away and is said to be red-shifted.

Astronomers have analyzed red shifts and blue shifts in radio radiation emitted by gas clouds in the Milky Way. Their analysis showed how rapidly the galaxy is rotating and how the speed of the revolution of its stars changes with the stars' distance from the galactic center. The astronomers then applied a formula relating the

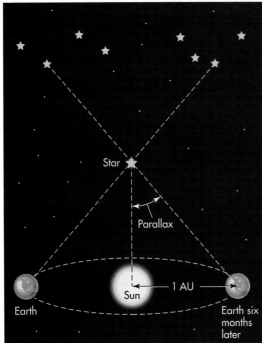

WORLD BOOK diagram by Precision Graphics

**Parallax** is the angle by which a star's location relative to background stars differs when measured from two points 1 astronomical unit (AU) apart and on a line perpendicular to the line from the star. One AU is the distance from Earth to the sun—about 93 million miles (150 million kilometers). Parallax is thus half the difference seen from opposite sides of Earth's orbit.

speed of the stars' revolution to the stars' masses. They discovered that the amount of mass in the galaxy is about 1 trillion times the mass of the sun.

Both radio astronomers and optical astronomers have studied a phenomenon known as *gravitational lensing.* This phenomenon occurs, for example, where radiation emitted by a small, distant galaxy passes near a massive galaxy that is between the object and Earth. The gravitational force of the galaxy apparently bends the radiation much as an ordinary optical lens bends light rays that pass through it. Gravitational lensing can produce an image of the small galaxy in the shape of an arc or even a

ring. Astronomers can study the radiation in the arc or ring to learn about the small galaxy.

**Ultraviolet astronomy.** The ultraviolet part of the electromagnetic spectrum has wavelengths shorter than those of visible light. Wavelengths of ultraviolet light range from near the limit of the shortest waves that the eye can see, about 400 nanometers, down to about 10 nanometers. Radiation from 400 to 300 nanometers (near ultraviolet) passes through Earth's atmosphere and therefore can be detected on the ground. But ultraviolet astronomers get much of their information from shorter wavelengths—in the *far ultraviolet,* from 300 to about 100 nanometers; and in the part of the *extreme ultraviolet* from 100 to 10 nanometers.

Studies in the far and extreme ranges must be carried out by satellites. From 1978 to 1997, the International Ultraviolet Explorer studied wavelengths from 320 down to 115 nanometers. NASA sent the Extreme Ultraviolet Explorer aloft in 1992 to study even shorter wavelengths.

The largest and most sensitive ultraviolet telescope in orbit is the Hubble Space Telescope. Astronomers have used this telescope to study hydrogen gas, whose strongest spectral lines are in the ultraviolet spectrum. The researchers have studied both normal hydrogen and a heavy form known as *deuterium* in the space between the sun and nearby stars. All this deuterium formed in the first 1,000 seconds after the *big bang,* the explosion that began the universe. Astronomers understand how the amount of deuterium that formed is related to present amounts of other kinds of matter. Knowing the amount of deuterium in space therefore can help them determine the density of the universe.

Another major space telescope was the Solar and Heliospheric Observatory (SOHO), which orbits between Earth and the sun. The European Space Agency launched SOHO in 1995 to monitor the sun with visible-light and ultraviolet cameras and spectrographs. NASA provided some of this equipment. Highly detailed images from SOHO show the sun in, for example, the ultraviolet light of helium gas at 60,000 °C or iron gas at 1,500,000 °C. Ground controllers lost control of SOHO in 1998. NASA's Transition Region and Coronal Explorer (TRACE) is sending images that are even more detailed.

**X-ray astronomy.** X rays have wavelengths from about 10 nanometers down to about 0.1 nanometer. The hottest regions in space produce X rays. These regions include the sun's corona and disks of material around

**Detecting red shift and blue shift**

Astronomers can learn about a star's motion by comparing the spectral lines in its visible light with those in a spectrum produced in a laboratory. If the star's spectral lines are shifted toward the red end of the rainbowlike spectrum of colors, the star is moving away from Earth. If the lines are shifted toward the blue end, the star is approaching. Astronomers use a similar technique to analyze shifts in the spectral "lines" of invisible forms of radiation, such as infrared and ultraviolet rays.

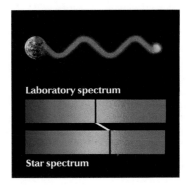

Laboratory spectrum

Star spectrum

**Red shift**

Laboratory spectrum

Star spectrum

**Blue shift**

black holes. The material in these disks heats up due to friction as it spirals into the black holes. As the material heats up, it emits X rays. The hot gas at the center of clusters of galaxies also emits X rays. Quasars are another source of X rays.

Celestial X rays do not penetrate Earth's atmosphere and therefore can be studied only from spacecraft. X rays would pass through ordinary telescope mirrors and lenses, so one kind of X-ray telescope has specially designed mirrors. X rays strike these mirrors at low angles, then skip away like stones skipping off water. Rays from all the mirrors meet at a single focal point.

Other X-ray telescopes do not have mirrors. The rays enter the telescope through openings between lead or iron slats, then strike special detectors.

In the 1970's and 1980's, a series of High-Energy Astronomy Observatories mapped the sky in X rays and studied certain objects in detail. Since 1990, a German-U.S.-British spacecraft named Rosat has been surveying the sky. A smaller Japanese satellite, Yohkoh, sends back X-ray images of the sun. These images show the corona and *solar flares,* explosive events that reach millions of degrees. NASA's Rossi X-ray Timing Explorer, launched in 1995, captures X rays with the largest collector ever sent aloft.

NASA also planned to launch the Advanced X-ray Astrophysics Facility (AXAF). AXAF would produce much more detailed images than any other X-ray telescope. It would be to X-ray astronomy what the Hubble Space Telescope is to ultraviolet and optical astronomy.

**Gamma-ray astronomy.** Electromagnetic waves that have the shortest wavelengths—about 0.1 nanometer and shorter—are known as gamma rays. Gamma-ray photons have the highest energy in the electromagnetic spectrum. Thus, they form in the regions of the highest energy in the universe.

Gamma-ray sources include places where matter and *antimatter* are annihilating each other. For every type of ordinary subatomic particle, there also exists an antiparticle. An antiparticle has the same amount of *mass* (total matter) as its corresponding particle, but it carries an opposite electric charge. For example, the antiparticle of an electron is a *positron.* An electron carries a negative charge, while a positron carries a positive charge. If a particle and its antiparticle collide, they annihilate each other, releasing gamma rays and other energy.

Astronomers have observed that matter-antimatter annihilation occasionally occurs near the center of the Milky Way. Other gamma-ray sources include the Crab Nebula, in the constellation Taurus, and a nearby collapsed star known as Geminga. The Crab Nebula consists of matter that was thrown out into space during a supernova observed in A.D. 1054.

The Compton Gamma Ray Observatory, launched in 1991, has several gamma-ray instruments. One detector followed up on a previous discovery of gamma-ray bursts that apparently come at random intervals from random places in the sky. The instrument has detected one such burst approximately every day.

Locating the sources of the bursts was difficult because gamma-ray detectors do not indicate clearly the positions from which the rays strike them. An Italian-Dutch satellite named BeppoSAX provided a breakthrough in 1997. After detecting a burst, it could turn quickly enough to enable another instrument to detect X rays coming from the same burst. Spectroscopy of the optical spectrum of one source located in this way showed that a distant galaxy was between Earth and the burster. Gamma-ray bursters must therefore be extremely far away. Furthermore, to give off gamma rays as strong as those that have been detected, they must be among the most powerful emitters in the universe.

**Neutrino astronomy.** Another type of particle that arrives from outer space is the neutrino. Neutrinos interact so rarely with particles on Earth that they have been difficult to detect. But several detectors now routinely monitor neutrinos that arrive from the sun. One detector, called Super-Kamiokande, is deep underground in a zinc mine in Japan. Its main part is a cylindrical tank of water that measures 131 feet (40 meters) deep and 131 feet in diameter and has more than 13,000 electronic detectors. The detectors can sense flashes of light produced when a neutrino collides with an an atomic nucleus or an electron in the water. Super-Kamiokande began operating in 1996.

Beginning in the 1960's, scientists working with an older kind of detector measured only about one-third of the expected number of neutrinos coming from the sun. This discovery led to a crisis in astronomy. Scientists wondered which was incorrect—the physics associated with the prediction or the measurements. If there really were fewer solar neutrinos than predicted, for example, the center of the sun might be cooler than astronomers had thought. Astronomers would then have to revise their ideas about the source of energy that fuels the sun and other stars. Careful testing and newer types of measurements have confirmed that there is indeed a difference between theory and results. Astronomers believe the difference may be explained if neutrinos have a small amount of mass. In 1998, Super-Kamiokande provided strong evidence that neutrinos have mass.

**Cosmic-ray astronomy.** Cosmic rays are electrically charged, high-energy particles. There are two kinds of cosmic rays: (1) *primary cosmic rays,* often called *primaries,* which originate in outer space; and (2) *secondary cosmic rays,* or *secondaries,* which form in Earth's atmosphere. Secondaries originate when primaries collide with atoms at the top of the atmosphere.

Most primaries are protons or other nuclei of atoms. They do not usually penetrate the atmosphere. Astronomers therefore use instruments aboard high-flying airplanes or satellites, such as the Compton Gamma Ray Observatory, to detect them. Secondaries can reach low altitudes. A small fraction of them even strike Earth's surface, where special sensors can detect them.

Some primary cosmic rays come from the sun, but most of them are *galactic cosmic rays,* which originate outside the solar system. Many galactic cosmic rays have tremendous energy. Astronomers do not know how they acquire this energy—though it may come from supernova explosions.

**Gravitational-wave astronomy.** Astronomers are building equipment to detect another type of radiation: gravitational waves. These waves are predicted by the general theory of relativity announced in 1915 by German-born physicist Albert Einstein. No gravitational waves have ever been directly detected. Researchers have found indirect evidence for them, however—certain variations in the orbits of two dense stars that revolve about each other.

**Direct sampling** is the examination of pieces of material from celestial objects. In their examinations, scientists often use techniques of geology, including chemical analysis. The most common samples are *meteorites,* rocks that fell through the atmosphere from farther out in our solar system. Thousands of meteorites have been found. Most come from asteroids, and only a handful are known to have come from the moon or Mars. The best place to find meteorites is Antarctica. They can be seen on the polar ice much more easily than they can be detected among ordinary rocks at other locations.

Scientists have also studied hundreds of kilograms of moon rocks brought to Earth by astronauts from 1969 to 1972. In 1970, the Soviet Union's Luna 16 spacecraft, a remotely controlled vehicle, returned small samples of soil from the moon. In addition, researchers have analyzed bits of space dust collected by devices on high-altitude aircraft. They have determined that some of the dust came from beyond our solar system.

### Computer modeling

Astronomers use computers to build *scientific models* (sets of mathematical equations) that represent certain processes, such as the formation of a star. After entering the equations into a computer, an astronomer inserts numbers into the equations, and the computer then *simulates* (represents) how the process would develop. In some cases, the computer produces a moving picture on a computer screen. The picture can run much faster than the actual process. This kind of model can help astronomers because many important processes occur much too slowly for astronomers to observe. Other important processes that can be simulated occur in inaccessible places, such as the interiors of stars.

### Learning about astronomy

It is easy for students and the general public to take part in astronomy. Amateur astronomers range from individuals who make casual nighttime or solar observations to people for whom astronomy is a serious pursuit.

Amateur astronomers have their own associations and local and regional clubs. Some of the larger clubs hold star parties at which members set up their own telescopes and observe the skies.

Amateur astronomers make major contributions to various branches of astronomy. For example, the American Association of Variable Star Observers, whose headquarters are in Cambridge, Massachusetts, collects observations from amateurs throughout the world and puts them together. The association shares the data with professional observers, including astronomers who want to know what certain stars are doing before they conduct observations. Other amateur astronomers search for new comets or supernovae.

Much technical information is available to amateur astronomers. Books on observing the heavens contain tables that indicate where stars and planets can be found each month. Many cities have planetariums, which can present shows demonstrating the movements of celestial objects. Some planetariums sponsor lectures by professional astronomers. Computer programs available on CD-ROM simulate sky conditions for any location and any date and time. Many of the programs contain photographs and motion pictures of astronomical objects. Such images are also available on the Internet.

In addition, certain computer programs enable users to combine or enhance existing astronomy photographs. For example, individual images taken at total eclipses can be combined to show the structure of the solar corona in more detail.

### History

The roots of astronomy extend back to the dawn of civilization. More than 4,000 years ago, in what is now England, several generations of people built Stonehenge, an "observatory" consisting of huge, cut stones arranged in circles. Certain stones and alignments of stones appear to mark locations of astronomical importance, such as the point at which the sun rose on the longest day of the year. Stonehenge was apparently also a place of worship.

Some of today's constellations have their roots in pat-

**At a star party,** amateur astronomers set up their telescopes, chat with friends, and wait for sunset and a night of viewing the heavens. Astronomy clubs throughout the world sponsor star parties.

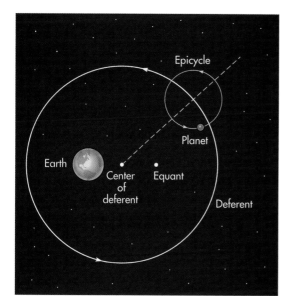

WORLD BOOK diagram by Precision Graphics

**The old theory** of Ptolemy put Earth at the center of the universe. Each of the other planets revolved in a circle called an *epicycle,* which revolved along a *deferent.* The epicycle's center moved at a constant speed about a point called an *equant.* This complex movement, *right,* explained why the planets sometimes appear to change direction relative to the stars.

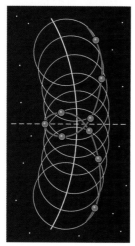

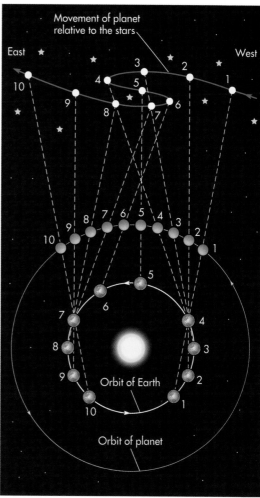

WORLD BOOK diagram by Precision Graphics

**The modern explanation** is that the planets appear to change direction because of differences in orbital speed as one planet passes another. A planet close to the sun orbits the sun more rapidly than does a more distant planet. In this diagram, each set of three numbers—for example, the three 5's—represents the same moment of time.

terns in the sky noted by the Sumerians in perhaps 2000 B.C. Chinese constellation patterns, which are largely different from those used in Europe, may also date from that time. Babylonian tablets show that astronomers there were noting the positions of the moon and planets by 700 B.C. The Babylonians also noted eclipses.

The ancient Greeks carried forward ideas from the Babylonians and invented some of their own. Aristotle's system of physics and astronomy, developed in the 300's B.C., survived for almost 2,000 years. In Aristotle's system of astronomy, Earth was the center of the universe. During the A.D. 100's, Ptolemy modified Aristotle's system to account for the retrograde motion of the planets. Ptolemy also maintained that Earth was the center of the universe, however.

**Developing the modern view.** By the early 1500's, Nicolaus Copernicus had developed a theory in which Earth and the other planets revolved about the sun. In the early 1600's, Johannes Kepler analyzed precise measurements of planetary positions that had been made by

Tycho Brahe. Kepler then developed three laws that correctly describe the shapes of the orbits of planets, indicate how rapidly a planet moves at various times of its year, and account for the length of the planet's year.

While Kepler based his laws of planetary motion on observations, the English astronomer and mathematician Isaac Newton proved them mathematically. In 1687, Newton completed a book usually called *Principia Mathematica* setting forth not only laws of motion but also the law of gravity that is still in general use.

By Newton's time, the field of optical astronomy had already begun. In 1609, Galileo heard that an optical device had been built that made distant objects appear closer. He soon built his own telescope. The discoveries Galileo made with this instrument backed the Copernican theory over the theories of Aristotle and Ptolemy. In 1616, however, the Roman Catholic Church warned

Galileo not to teach that Earth revolves about the sun. A book of Galileo's published in 1632 was interpreted as a violation of the ban, and Galileo was put under house arrest. Only in 1992 did the Catholic Church confirm that Galileo should not have been tried or convicted.

**Finding new planets.** The British astronomer William Herschel discovered a new object in the sky in 1781. At first, he thought the object was a comet. It turned out to be a planet—later named Uranus. This was the first planet discovered since ancient times.

In 1845, astronomers John C. Adams of Britain and Urbain Leverrier of France declared that another planet must lie a certain distance beyond Uranus. They based their statement on their calculations of differences between the observed orbit of Uranus and the orbit predicted by one of Kepler's laws. They said that this difference was due to the gravitational attraction of the then-unknown planet. Using one of their predictions the next year, the German astronomer Johann G. Galle found the planet, Neptune.

The American astronomer Clyde W. Tombaugh found the last of the known planets in the solar system in 1930. He discovered Pluto on photos he had taken for the purpose. He used a wide-angle telescope at Lowell Observatory in Flagstaff, Arizona.

**Discovering other galaxies.** In the early days of optical astronomy, the fuzzy regions of the sky became known as *nebulae*—Latin for *clouds*. When viewed through the telescopes then available, the nebulae resembled comets. Someone who was trying to discover comets could easily mistake a nebula for a comet. To prevent such errors, the French astronomer Charles Messier made a list from the 1750's to 1784 of the most prominent nebulae. This *Messier Catalog* now contains 110 objects, known by their *Messier numbers.*

In Messier's time, no one knew what the nebulae were. But in the mid-1800's, Lord Rosse of Ireland built a telescope whose superior light-gathering power enabled him to discover that many nebulae have spiral shapes. The telescope's mirror measured about 6 feet (1.8 meters) across—gigantic for that time.

It took decades to discover what the spiral nebulae were. The answer came only in 1924 with the discovery by Edwin Hubble that the spiral nebulae are so far away that they must be beyond the Milky Way. Astronomers concluded that they are independent galaxies. The "nebula" with Messier number M31, for example, is actually the Andromeda Galaxy. Astronomers now use *nebula* to mean a cloud of dust and gas. The remaining objects in the Messier catalog are *star clusters,* groups of closely placed stars.

**Advances in astrophysics.** By the end of the 1930's, the German-born physicist Hans Bethe had suggested how nuclear fusion powers the stars. For example, a process known as the *proton-proton chain* powers the sun. In this process, six protons come together in several steps to produce a helium nucleus and two protons. The final products contain slightly less mass than the original ingredients. The missing matter is converted to energy according to Einstein's formula $E = mc^2$, where $E$ is energy, $m$ is the missing mass, and $c^2$ is the speed of light multiplied by itself.

British astronomers Geoffrey Burbidge, Margaret Burbidge, and Fred Hoyle and American physicist William

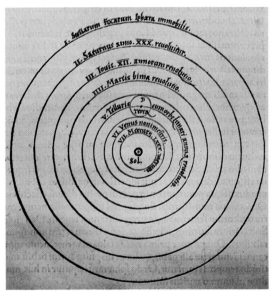

© Jay M. Pasachoff

**The sun-centered theory** proposed by Nicolaus Copernicus of Poland in 1543 revolutionized astronomy. This diagram is from his *Concerning the Revolutions of the Celestial Spheres.*

Ann Ronan Picture Library

**Tycho Brahe's observatory** is pictured in this drawing, which dates from 1598. Brahe, a Dane, is shown seated behind his giant *quadrant,* an instrument used to measure the altitude of stars and planets. Brahe's observations of the planets were far more accurate than any previous ones.

A. Fowler showed in the 1950's how nuclear reactions could have built up all but the lightest chemical elements. Astronomers now know that the lighter elements formed minutes after the big bang. Moderately heavy elements formed inside stars; the heaviest elements, in supernova explosions.

In 1965, American physicists Arno Penzias and Robert Wilson discovered faint radio radiation coming equally from all directions in space. Scientists showed that the radiation was emitted about 300,000 years after the big bang and has been cooling ever since. Its temperature is now about 3 Celsius degrees above absolute zero.

For four years starting in 1989, instruments aboard the Cosmic Background Explorer satellite measured more precisely the temperature of the radiation detected by Penzias and Wilson. Another instrument aboard the satellite found small variations in the temperature from one location in the sky to another. These so-called ripples in space may be the "seeds" from which the galaxies and clusters of galaxies grew long ago.

**Finding quasars and pulsars.** In 1963, the Dutch-born astronomer Maarten Schmidt identified the starlike objects now known as quasars. Schmidt showed that the spectra of quasars have huge red shifts, indicating that the spectra are produced by powerful sources of energy in distant galaxies. Most of them are billions of light-years away.

In 1967, the British astronomer Jocelyn Bell Burnell identified a new type of object in radio observations she was making as part of her Ph.D. thesis. These objects emit radio waves that arrive at Earth in regular pulses about 1 second apart. The objects came to be known as pulsars. Later work showed that pulsars are rapidly spinning *neutron stars* (stars made mostly of neutrons). With every spin, a narrow beam of radio waves sweeps over Earth, producing a pulse. Astronomers have found pulsars that pulse as often as 600 times per second.

**Space probes.** The United States began to send space probes to other planets and to the moon in the late 1900's. From 1979 to 1981, Voyager 1 and Voyager 2 flew near Jupiter and Saturn. Voyager 2 flew past Uranus in 1986 and Neptune in 1989. Both craft sent back data and photographs that greatly enriched scientists' knowledge of those planets.

The United States has put a number of scientific satellites into orbit around celestial objects. From 1990 to 1994, the orbiter Magellan mapped the surface of Venus with radar. The Galileo spacecraft entered an orbit around Jupiter in 1995 on a mission to explore that planet and several of its satellites. In 1997, the Mars Global Surveyor went into orbit to study Mars's magnetic fields, monitor its weather, and measure visible light and other radiation emitted by the planet. Also in 1997, the United States launched the Cassini mission to study Saturn, its rings, and its satellites. The Cassini craft is due to reach Saturn in 2004. It will drop a probe into the atmosphere of Saturn's moon Titan. Lunar Prospector entered an orbit over the moon's poles in 1998. The spacecraft's mission was to map the mineral composition of the moon.

## Careers

**What astronomers do.** Most professional astronomers work at observatories or research institutes or teach and conduct research at colleges and universi-

J. R. Eyerman, *Life* magazine, © 1950 Time Inc.
**Edwin Hubble,** an American astronomer, demonstrated that the universe is expanding. He is shown in the observer's cage of the Hale telescope at the Palomar Observatory near San Diego.

ties. Planetariums employ astronomers to lecture and conduct classes for the public. A few astronomers work for companies that build equipment for scientific satellites and space probes. Others work for firms that do such work as monitoring the environment from space.

**Becoming an astronomer.** The most important characteristic for a person who wishes to become an astronomer is a powerful spirit of inquiry. The person should also have a strong ability to learn mathematics.

High-school students who are interested in becoming astronomers should take as many math courses as they can to prepare for college mathematics and physics. A high-school physics course is also useful. Some branches of astronomy deal more with chemistry or geology than physics, so courses in those subjects can also help. Visits to planetariums and science museums as well as participation in an amateur astronomy club can help prepare a student for a career in astronomy. Other useful skills include keyboarding and an ability to work with basic scientific and mathematical software.

To conduct research and teach astronomy at the college level requires a Ph.D. degree. Students usually take about six years to obtain this degree after receiving their bachelor's degree. During most of this time, Ph.D. students perform research, and they are almost always supported through research grants or teaching salaries. After obtaining their degree, most astronomers take postdoctoral positions for two or more years before searching for permanent jobs.

**Astronomy associations.** Astronomers from throughout the world gather every three years at the General Assembly of the International Astronomical Union. Professionals in the United States and Canada belong to the American Astronomical Society. Both professional and amateur astronomers may join the Astronomical Society of the Pacific. Many countries also have organizations devoted to astronomy, such as the Astronomical Society of India, the Royal Astronomical Society of Canada, and the United Kingdom's Royal Astronomical Society.    Jay M. Pasachoff

**Related articles** in *World Book*. See **Star** and its list of *Related articles*. See also the following articles :

### Astronomers

| | | |
|---|---|---|
| Aristarchus | Galileo | Laplace, Marquis |
| Banneker, Benjamin | Hale, George E. | de |
| Barnard, Edward E. | Halley, Edmond | Leavitt, Henrietta |
| Bessel, Friedrich | Herschel, Caroline | S. |
| W. | L. | Lovell, Sir Bernard |
| Bowditch, | Herschel, Sir John | Lowell, Percival |
| Nathaniel | F. W. | Messier, Charles |
| Bradley, James | Herschel, Sir | Mitchell, Maria |
| Brahe, Tycho | William | Newton, Sir Isaac |
| Burnell, Jocelyn B. | Hipparchus | Omar Khayyam |
| Cannon, Annie J. | Hogg, Helen S. | Payne-Gaposchkin, |
| Copernicus, Nicolaus | Hubble, Edwin P. | Cecilia H. |
| laus | Jansky, Karl G. | Ptolemy |
| De Sitter, Willem | Kepler, Johannes | Russell, Henry N. |
| Eddington, Sir | Kuiper, Gerard P. | Sagan, Carl E. |
| Arthur S. | Langley, Samuel P. | Shapley, Harlow |
| | | Struve, Otto |

### Instruments

| | |
|---|---|
| Astrolabe | Interferometer |
| Bolometer | Spectrometer |
| Hubble Space Telescope | Telescope |

### Solar system

| | | |
|---|---|---|
| Asteroid | Leonids | Pluto (planet) |
| Ceres (asteroid) | Mars (planet) | Satellite |
| Comet | Mercury (planet) | Saturn (planet) |
| Earth | Meteor | Solar system |
| Eclipse | Moon | Sun |
| Fireball | Neptune (planet) | Uranus (planet) |
| Jupiter (planet) | Planet | Venus (planet) |

### Terms

| | | |
|---|---|---|
| Azimuth | Magnitude | Perihelion |
| Baily's beads | Nadir | Red shift |
| Corona | Opposition | Sunspot |
| Evening star | Orbit | Transit |
| Horizon | Parallax | Zenith |

### Time and astronomy

| | | |
|---|---|---|
| Day | Leap year | Solstice |
| Equinox | Midnight sun | Sundial |
| Hour | Month | Time |
| International Date | Season | Twilight |
| Line | Sidereal time | Year |

### Additional resources

**Level I**
Asimov, Isaac, and Reddy, Francis. *Astronomy in Ancient Times*. Rev. ed. Gareth Stevens, 1995.
Dickinson, Terence. *The Universe…and Beyond*. Rev. ed. Camden Hse., 1992.
Hathaway, Nancy. *The Friendly Guide to the Universe*. Viking, 1994.
Schaaf, Fred. *The Amateur Astronomer: Explorations and Investigations*. Watts, 1994.

**Level II**
Barbree, Jay, and Caidin, Martin. *A Journey Through Time: Exploring the Universe with the Hubble Space Telescope*. Penguin, 1995.
Hoskin, Michael, ed. *The Cambridge Illustrated History of Astronomy*. Cambridge, 1997.
Kolb, Rocky. *Blind Watchers of the Sky: The People and Ideas That Shaped Our View of the Universe*. Addison-Wesley, 1996.

**Universe** consists of all matter and all light and other forms of radiation and energy. It consists of everything that exists anywhere in space and time.

The universe includes the earth, everything on the earth and within it, and everything in the solar system. The solar system contains nine major planets along with thousands of comets and minor planets called *asteroids*. It also contains the sun, the star around which the planets revolve.

All stars, including the sun, are part of the universe. Some other stars also have planetary systems. In addition to planets and stars, the universe contains gas, dust, magnetic fields, and high-energy particles called *cosmic rays*.

Stars are grouped into *galaxies*. The sun is one of more than 100 billion stars in a giant spiral galaxy called the Milky Way. This galaxy is about 100,000 *light-years* across. A light-year is the distance that light travels in a vacuum in a year—about 5.88 trillion miles (9.46 trillion kilometers).

Galaxies tend to be grouped into *clusters*. Some clusters appear to be grouped into *superclusters*. The Milky Way is part of a cluster known as the Local Group. This cluster is about 3 million light-years in diameter. Also in the cluster are two giant spirals known as the Andromeda Galaxy and M33 and about 30 small galaxies, also known as *dwarf galaxies*. The Local Group is part of the Local Supercluster, which has a diameter of about 100 million light-years.

### Size of the universe

No one knows whether the universe is finite or infinite in size. Studies of the sky indicate that there are at least 100 billion galaxies in the observable universe. Measurements show that the most distant galaxies observed to date are about 12 billion to 16 billion light-years from the earth. They are observed in every direction across the sky.

Astronomers determine the distance to a faraway object by measuring the object's *red shift*. This phenomenon occurs when an object that is moving away from an observer *emits* (gives off) light. Red light has the longest *wavelength* of any visible light—that is, the longest distance between successive wave crests. The observer sees light from a receding object at wavelengths longer than those that would be seen by an observer who was moving with the object. The observed change in wavelength is the object's red shift. The amount of red shift depends upon the speed at which the object recedes from the observer. The larger the red shift is, the more rapidly the object is moving away.

Among the most distant objects ever observed are tremendously bright objects called *quasars*. Individual quasars are as much as 1,000 times brighter than the entire Milky Way. No one knows the structure of quasars, because a quasar appears pointlike in a photographic image. However, a quasar seems likely to contain a giant *black hole* in its center. A black hole is an object whose gravitational field is so strong that nothing—not even light—can escape from it. Matter is apparently falling into the massive black hole in the center of the quasar, radiating energy before being swallowed up.

Astronomers interpret the red shifts that have been measured as evidence that the universe is expanding.

That is, every part of the universe is moving away from every other part. The matter within a particular object does not expand, however. For example, the stars in a galaxy do not move away from one another because gravity holds the galaxy together. But the galaxies are moving away from one another. The expansion of the universe is a basic observation that any successful theory of the universe must explain.

### Changing views of the universe

In ancient times, people of many cultures thought that the universe consisted of only their own locality, distant places of which they had heard, and the sun, moon, planets, and stars. Many people thought that the heavenly bodies were gods and spirits. But the Polish astronomer Nicolaus Copernicus suggested in 1543 that the earth is like the other planets and that the planets revolve around the sun. Later astronomers showed that the sun is a typical star.

The development of the telescope, the photographic plate, and the *spectroscope* (an instrument that analyzes light) led to a great increase in knowledge. Astronomers discovered that the sun is moving within a large system of stars, the Milky Way. In about 1920, astronomers realized that not all of the fuzzy patches of light seen in the night sky are part of the Milky Way. Rather, many of these objects, called *nebulae,* are other galaxies. The discovery of the red shift of distant galaxies led to the theory of the expanding universe.

The big bang theory provides the best explanation of the basic observations. According to this theory, the universe began with an explosion—called the big bang—10 billion to 20 billion years ago. Immediately after the explosion, the universe consisted chiefly of intense radiation and particles of matter. This radiation and matter formed a rapidly expanding region called the primordial fireball. After thousands of years, the fireball consisted mostly of matter, largely in the form of hydrogen gas. Today, faint radio waves coming from all directions of space are all that remain of the radiation from the fireball.

In time, the matter broke apart into huge clumps. The clumps became galaxies, many of them grouped into clusters and superclusters. Smaller clumps within the galaxies formed stars. Part of one of these clumps became the sun and other objects of the solar system.

The best available evidence indicates that the galaxies will move apart forever. This evidence and the universe's current rate of expansion indicate that the present age of the universe is about 15 billion to 20 billion years. This estimate agrees with observations of the oldest stars in groups known as globular star clusters.

Astronomers do not rule out the possibility, however, that all the galaxies will come together again in about 70 billion years. This will happen if the universe contains enough matter to slow, halt, and reverse the expansion. If that were to happen, the universe would eventually collapse into a "big crunch," perhaps to explode once again in another big bang. The universe would then enter a new phase, possibly resembling the present one.

Kenneth Brecher

**Related articles** in *World Book* include:

**Planet** is a large, round heavenly body that orbits a star and shines with light reflected from the star. We know of nine planets that orbit the sun in our solar system. In the 1990's, astronomers also discovered many planets orbiting distant stars.

All but two of the planets in our solar system have smaller objects revolving around them called *satellites* or *moons.* Our solar system also contains thousands of smaller bodies known as *asteroids.* The asteroids are often called minor planets, and the term *major planet* is used to distinguish the nine planets from the asteroids. The remainder of this article uses *planet* to mean *major planet.*

The usual order of the planets in our solar system, outward from the sun, is Mercury, Venus, Earth, Mars, Jupiter, Saturn, Uranus, Neptune, and Pluto. To help remember the order, some people use the phrase *My Very Educated Mother Just Sent Us Nine Pizzas* as a memory aid. The initial letters of the words in that phrase match the initial letters of the planet names.

Pluto is not always the farthest planet from the sun, however. Its orbit is such a long oval that Pluto moves inside the path of Neptune for about 20 years every 248 years. One such 20-year period lasted from Jan. 23, 1979, to Feb. 11, 1999. After February 11, however, Pluto will again be the most distant planet for hundreds of years.

The planets of our solar system can be divided into two groups. The innermost four planets—Mercury, Venus, Earth, and Mars—are small, rocky worlds. They are called the *terrestrial* (earthlike) planets, from the Latin word for Earth, *terra.* Earth is the largest terrestrial planet. The other earthlike planets have from 38 percent to 95 percent of Earth's diameter and from 5.6 percent to 81 percent of Earth's *mass* (total quantity of matter).

The next four planets—Jupiter, Saturn, Uranus, and Neptune—are called *gas giants* or *Jovian* (Jupiterlike) *planets.* They have gaseous atmospheres and no solid surfaces. All four Jovian planets consist mainly of hydrogen and helium. Smaller amounts of other materials also occur, including traces of ammonia and methane in their atmospheres. They range from 3.9 times to 11.2 times Earth's diameter and from 15 times to 318 times Earth's mass. Jupiter, Saturn, and Neptune give off more energy than they receive from the sun. Most of this extra energy takes the form of *infrared* radiation, which is felt as heat, instead of visible light. Scientists think the source of some of the energy is probably the slow compression of the planets by their own gravity.

The ninth planet, Pluto, is only 18 percent the diameter of Earth and $\frac{1}{500}$ of its mass. As a small, rocky planet with a larger orbit than the gas giants, it does not fit in either group. Some astronomers think that Pluto may not be a major planet at all.

### Observing the planets

People have known the inner six planets of our solar system for thousands of years because they are visible

*Jay M. Pasachoff, the contributor of this article, is Field Memorial Professor of Astronomy and Director of the Hopkins Observatory at Williams College.*

from Earth without a telescope. The outermost three planets—Uranus, Neptune, and Pluto—were discovered by astronomers, beginning in the 1780's. All three can be seen from Earth with a telescope.

To the unaided eye, the planets look much like the background stars in the night sky. However, the planets move slightly from night to night in relation to the stars. The name *planet* comes from a Greek word meaning *to wander.* The planets and the moon almost always follow the same apparent path through the sky. This path, known as the *zodiac,* is about 16° wide. At its center is the *ecliptic,* the apparent path of the sun. If you see a bright object near the ecliptic at night or near sunrise or sunset, it is most likely a planet. You can even see the brightest planets in the daytime, if you know where to look.

Planets and stars also differ in the steadiness of their light when viewed from Earth's surface. Planets shine with a steady light, but stars seem to twinkle.

The twinkling is due to the moving layers of air that surround Earth. Stars are so far away that they are mere points of light in the sky, even when viewed through a telescope. The atmosphere bends the starlight passing through it. As small regions of the atmosphere move about, the points of light seem to dance and change in brightness.

Planets, which are much closer, look like tiny disks through a telescope. The atmosphere scatters light from different points on a planet's disk. However, enough light always arrives from a sufficient number of points to provide a steady appearance.

### How planets move

Planets move in two main ways. They travel around their parent star in paths called *orbits.* As each planet orbits its star, it also rotates on its *axis,* an imaginary line through its center.

**Orbits.** Viewed from Earth's surface, the planets of the solar system and the stars appear to move around Earth. They rise in the east and set in the west each night. Most of the time, the planets move westward across the sky slightly more slowly than the stars do. As a result, the planets seem to drift eastward relative to the background stars. This motion is called *prograde.* For a while each year, however, the planets seem to reverse their direction. This backward motion is called *retrograde.*

In ancient times, most scientists thought that the moon, sun, planets, and stars actually moved around Earth. One puzzle that ancient scientists struggled to explain was the annual retrograde motion of the planets. In about A.D. 150, the Greek astronomer Ptolemy developed a theory that the planets orbited in small circles, which in turn orbited Earth in larger circles. Ptolemy

**Inner planets**

Mercury
Venus
Earth
Mars

Venus
Sun
Mercury
Earth
Mars

**Outer planets**

Jupiter
Saturn
Uranus
Neptune
Pluto

Uranus
Jupiter
Mars
Pluto
Saturn
Neptune

**The orbits of the planets** around the sun are shown here. Two diagrams are needed because the orbits of the outer planets would extend off the page if they were drawn to the same scale as those of the inner planets.

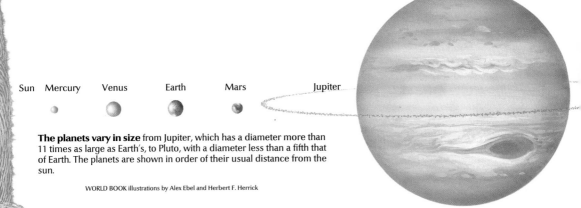

Sun　Mercury　Venus　Earth　Mars　Jupiter

**The planets vary in size** from Jupiter, which has a diameter more than 11 times as large as Earth's, to Pluto, with a diameter less than a fifth that of Earth. The planets are shown in order of their usual distance from the sun.

WORLD BOOK illustrations by Alex Ebel and Herbert F. Herrick

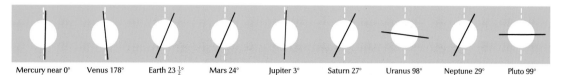

| Mercury near 0° | Venus 178° | Earth 23 ½° | Mars 24° | Jupiter 3° | Saturn 27° | Uranus 98° | Neptune 29° | Pluto 99° |

**The axes of the planets,** represented by the solid lines, are imaginary lines around which the planets rotate. A planet's axis is not perpendicular to the path of the planet's orbit around the sun. It tilts at an angle from the perpendicular position indicated by the broken line.

thought that retrograde motion was caused by a planet moving on its small circle in an opposite direction from the motion of the small circle around the big circle.

In 1543, the Polish astronomer Nicolaus Copernicus showed that the sun is the center of the orbits of the planets. Our term *solar system* is based on Copernicus's discovery. Copernicus realized that retrograde motion occurs because Earth moves faster in its orbit than the planets that are farther from the sun. The planets that are closer to the sun move faster in their orbits than Earth travels in its orbit. Retrograde motion occurs whenever Earth passes an outer planet traveling around the sun or an inner planet passes Earth.

In the 1600's, the German astronomer Johannes Kepler used observations of Mars by the Danish astronomer Tycho Brahe to figure out three laws of planetary motion. Although Kepler developed his laws for the planets of our solar system, astronomers have since realized that Kepler's laws are valid for all heavenly bodies that orbit other bodies.

*Kepler's first law* says that planets move in *elliptical* (oval-shaped) orbits around their parent star—in our solar system, the sun. An *ellipse* is a closed curve formed around two fixed points called *foci*. The ellipse is formed by the path of a point moving so that the sum of its distances from the two foci remains the same. The orbital paths of the planets form such curves, with the parent star at one focus of the ellipse. Before Kepler, scientists had assumed that the planets moved in circular orbits.

*Kepler's second law* says that an imaginary line joining the parent star to its planet sweeps across equal areas of space in equal amounts of time. When a planet is close to its star, it moves relatively rapidly in its orbit.

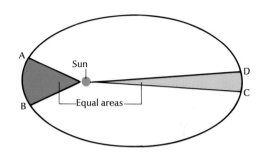

**Kepler's second law** shows how a planet covers equal areas of its orbit in equal lengths of time. The planet travels at a higher speed near the sun, from *A* to *B*, than far from the sun, *C* to *D*.

The line therefore sweeps out a short, fat, trianglelike figure. When the planet is farther from its star, it moves relatively slowly. In this case, the line sweeps out a long, thin figure that resembles a triangle. But the two figures have equal areas.

*Kepler's third law* says that a planet's *period* (the time it takes to complete an orbit around its star) depends on its average distance from the star. The law says that the square of the planet's period—that is, the period multiplied by itself—is proportional to the cube of the planet's average distance from its star—the distance multiplied by itself twice—for all planets in a solar system.

In 1687, the English scientist, astronomer, and mathematician Isaac Newton completed his theory of gravity and explained why Kepler's laws work. Newton showed how his expanded version of Kepler's third law could be used to find the mass of the sun or of any other object around which things orbit. Using Newton's explanation, astronomers can determine the mass of a planet by

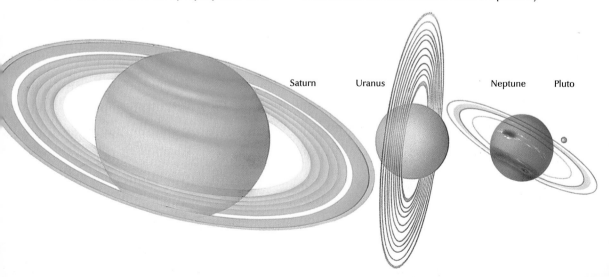

| Saturn | Uranus | Neptune | Pluto |

NASA/JPL/MSSS

**A canyon on Mars** is Earthlike in appearance. However, numerous craters produced by the impact of meteorites give the surface near the edge of the canyon a pitted, unearthly look.

NASA/JPL

**The Great Red Spot of Jupiter** is one of the most spectacular features in the solar system. This swirling mass of gas, which resembles a hurricane, is about three times as wide as Earth.

studying the period of its moon or moons and their distance from the planet.

**Rotation.** Planets rotate at different rates. One day is defined as how long it takes Earth to rotate once. Jupiter and Saturn spin much faster, in only about 10 hours. Venus rotates much slower, in about 243 earth-days.

Most planets rotate in the same direction in which they revolve around the sun, with their axis of rotation standing upright from their orbital path. A law of physics holds that such rotation does not change by itself. So astronomers think that the solar system formed out of a cloud of gas and dust that was already spinning.

Uranus and Pluto are tipped on their sides, however, so that their axes lie nearly level with their paths around the sun. Venus is tipped all the way over. Its axis is almost completely upright, but the planet rotates in the direction opposite from the direction of its revolution around the sun. Most astronomers think that some other objects in the solar system must have collided with Uranus, Pluto, and Venus and tipped them.

### The planets of our solar system

Astronomers measure distances within the solar system in *astronomical units* (AU). One astronomical unit is the average distance between Earth and the sun, which is about 93 million miles (150 million kilometers). The inner planets have orbits whose diameters are 0.4, 0.7, 1.0, and 1.5 AU, respectively. The orbits of the gas giants are

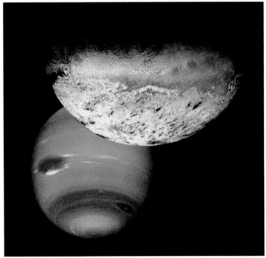

NASA/JPL

**The blue clouds of Neptune** are mostly frozen methane, the main chemical in natural gas—a fuel used for heating and cooking on Earth. The other object shown is Neptune's moon Triton.

NASA/JPL

**A river of lava on Venus** split in two as it flowed from left to right, producing a delta like those made by rivers of water on Earth. Venus's surface has many long channels of hardened lava.

much larger: 5, 10, 20, and 30 AU, respectively. Because of their different distances from the sun, the temperature, surface features, and other conditions on the planets vary widely.

**Mercury,** the innermost planet, has no moon and almost no atmosphere. It orbits so close to the sun that temperatures on its surface can climb as high as 800 °F (430 °C ). But some regions near the planet's poles may be always in shadow, and astronomers speculate that water or ice may remain there. No spacecraft has visited Mercury since the 1970's, when Mariner 10 photographed about half the planet's surface at close range.

**Venus** is known as Earth's twin because it resembles Earth in size and mass, though it has no moon. Venus has a dense atmosphere that consists primarily of carbon dioxide. The pressure of the atmosphere on Venus's surface is 90 times that of Earth's atmosphere. Venus's thick atmosphere traps energy from the sun, raising the surface temperature on Venus to about 860 °F (460 °C), hot enough to melt lead. This trapping of heat is known as the *greenhouse effect.* Scientists have warned that a similar process on Earth is causing permanent global warming. Several spacecraft have orbited or landed on Venus. In the 1990's, the Magellan spacecraft used radar—radio waves bounced off the planet—to map Venus in detail.

**Earth,** our home planet, has an atmosphere that is mostly nitrogen with some oxygen. Earth has oceans of liquid water and continents that rise above sea level. Many measuring devices on the surface and in space monitor conditions on our planet. In 1998, the National Aeronautics and Space Administration (NASA) launched the first of a series of satellites called the Earth Observing System (EOS). The EOS satellites will carry remote-sensing instruments to measure climate changes and other conditions on Earth's surface.

**Mars** is known as the red planet because of its reddish-brown appearance, caused by rusty dust on the Martian surface. Mars is a cold, dry world with a thin atmosphere. The *atmospheric pressure* (pressure exerted by the weight of the gases in the atmosphere) on the Martian surface is less than 1 percent the atmospheric pressure on Earth. This low surface pressure has enabled most of the water that Mars may once have had to escape into space.

The surface of Mars has giant volcanoes, a huge system of canyons, and stream beds that look as if water flowed through them in the past. Mars has two tiny moons, Phobos and Deimos. Many spacecraft have landed on or orbited Mars.

**Jupiter,** the largest planet in our solar system, has more mass than the other planets combined. Like the other Jovian planets, it has gaseous outer layers and may have a rocky core. A huge storm system called the Great Red Spot in Jupiter's atmosphere is larger than Earth and has raged for hundreds of years.

Jupiter's four largest moons—Io, Europa, Ganymede, and Callisto—are larger than Pluto, and Ganymede is also bigger than Mercury. Circling Jupiter's equator are three thin rings, consisting mostly of dust particles. A pair of Voyager spacecraft flew by Jupiter in 1979 and sent back close-up pictures. In 1995, the Galileo spacecraft dropped a probe into Jupiter's atmosphere and went into orbit around the planet and its moons.

NASA/JPL

**The rings of Saturn** consist of billions of pieces of ice, ranging in size from particles the size of dust grains to chunks over 10 feet (3 meters) wide. The rings are shown here in false color.

**Saturn,** another giant planet, has a magnificent set of gleaming rings. Its gaseous atmosphere is not as colorful as Jupiter's, however. One reason Saturn is relatively drab is that its hazy upper atmosphere makes the cloud patterns below difficult to see. Another reason is that Saturn is farther than Jupiter from the sun. Because of the difference in distance, Saturn is colder than Jupiter. Due to the temperature difference, the kinds of chemical reactions that color Jupiter's atmosphere occur too slowly to do the same on Saturn.

Saturn's moon Titan is larger than Pluto and Mercury. Titan has a thick atmosphere of nitrogen and methane. In 1980 and 1981, the Voyager 2 spacecraft sent back close-up views of Saturn and its rings and moons.

The Cassini spacecraft will orbit Saturn in 2004. It will also drop a small probe into Titan's atmosphere.

**Uranus** was the first planet discovered with a telescope. German-born English astronomer William Herschel found it in 1781. He at first thought he had discovered a comet. Almost 200 years later, scientists detected 10 narrow rings around Uranus when the planet moved in front of a star and the rings became visible. Voyager 2 studied Uranus and its rings and moons close-up in 1986.

**Neptune** was first observed in 1846 by German astronomer Johann G. Galle after other astronomers predicted its position by studying how it affected Uranus's orbit. In 1989, Voyager 2 found that Neptune had a storm system called the Great Dark Spot, similar to Jupiter's Great Red Spot. But five years later, in 1994, the Hubble Space Telescope found that the Great Dark Spot had vanished. Neptune has four narrow rings, one of which has clumps of matter. Neptune's moon Triton is one of the largest in the solar system and has volcanoes that emit plumes of frozen nitrogen.

**Pluto.** Tiny, distant Pluto has been difficult to study because it is so far from Earth. The American astronomer Clyde W. Tombaugh discovered Pluto in 1930. Only in 1978, when astronomers discovered a moon orbiting

Pluto, could they determine the planet's mass. They found that Pluto was much less massive than expected. The Hubble Space Telescope found a dozen areas of contrasting light and dark on Pluto. NASA plans to launch a mission called the Pluto-Kuiper Express in the early 2000's. The mission will send a probe to explore Pluto and the *Kuiper* (pronounced *KY pur) belt,* a band of small rocky objects orbiting beyond Neptune. Astronomers hope the probe reaches Pluto before the planet has traveled so far from the sun that its thin atmosphere freezes and snows.

### Planets in other solar systems

**How planets are detected.** Even with the most advanced telescopes, astronomers cannot see planets orbiting other stars directly. The planets shine only by reflected light and are hidden by the brilliance of their parent stars. The planets and their stars are also much farther away than our sun. The nearest star is 4.2 light-years away, compared to 8 light-minutes for the sun. One *light-year* is the distance that light travels in one year—about 5.88 trillion miles (9.46 trillion kilometers). Thus, it takes light 4.2 years to reach Earth from the nearest star beyond the sun and only 8 minutes to reach Earth from the sun.

Although astronomers cannot see planets around distant stars, they can detect the planets from tiny changes in the stars' movement. These changes are caused by the slight pull of the planet's gravity on its parent star. To find new planets, astronomers use a technique called *spectroscopy,* which breaks down the light from stars into its component rainbow of colors. The scientists look for places in the rainbow where colors are missing. At these places, dark lines known as *spectral lines* cross the rainbow. The spectral lines change their location in

the rainbow slightly as a star is pulled by the gravity of an orbiting planet toward and away from Earth. These apparent changes in a star's light as the star moves are due to a phenomenon known as the *Doppler effect.* The changes not only show that a planet is present but also indicate how much mass it has.

**The first discoveries.** Astronomers announced the discovery of the first planets around a star other than our sun in 1992. The star is a pulsar named PSR B1257+12 in the constellation Virgo. *Pulsars* are dead stars that have collapsed until they are only about 12 miles (20 kilometers) across. They spin rapidly on their axes, sending out radio waves that arrive on Earth as pulses of radio energy. Some pulsars spin hundreds of times each second. If a pulsar has a planet, the planet pulls the star to and fro slightly as it orbits. These pulls cause slight variations in the radio pulses. From measurements of these variations, the Polish-born American astronomer Alexander Wolszczan and American Dale A. Frail discovered three planets in orbit around PSR B1257+12. The star emits such strong X rays, however, that no life could survive on its planets.

Astronomers soon began to find planets around stars more like the sun. In 1995, Swiss astronomers Michel Mayor and Didier Queloz found the first planet orbiting a sunlike star, 51 Pegasi, in the constellation Pegasus. American astronomers Geoffrey W. Marcy and R. Paul Butler confirmed the discovery and found planets of their own around other stars. By the late 1990's, the stars known to have planets included 70 Virginis in the constellation Virgo, 47 Ursae Majoris in the Big Dipper (Ursa Major), and Rho¹ Cancri in the constellation Cancer.

Some stars have planets orbiting them at a distance that could support life. Most scientists consider liquid

## The planets at a glance*

| | Mercury ☿ | Venus ♀ | Earth ⊕ | Mars ♂ |
|---|---|---|---|---|
| Average distance from the sun | 35,980,000 mi. (57,900,000 km) | 67,230,000 mi. (108,200,000 km) | 92,960,000 mi. (149,600,000 km) | 141,000,000 mi. (227,900,000 km) |
| Closest approach to Earth | 57,000,000 mi. (91,700,000 km) | 25,700,000 mi. (41,400,000 km) | ———————— ———————— | 34,600,000 mi. (55,700,000 km) |
| Length of year (earthdays) Average orbital speed | 87.97 29.76 mi. per sec. (47.89 km per sec.) | 224.7 21.77 mi. per sec. (35.03 km per sec.) | 365.26 18.51 mi. per sec. (29.79 km per sec.) | 686.98 14.99 mi. per sec. (24.13 km per sec.) |
| Diameter at equator | 3,031 mi. (4,878 km) | 7,521 mi. (12,104 km) | 7,926 mi. (12,756 km) | 4,223 mi. (6,796 km) |
| Rotation period Tilt of axis (degrees) | 59 earthdays about 0 | 243 earthdays 178 | 23 hrs. 56 min. 23.44 | 24 hrs. 37 min. 23.98 |
| Temperature | −280 to +800 °F (−170 to +430 °C) | +860 °F (+460 °C) | −130 to +140 °F (−90 to +60 °C) | −220 to +60 °F (−140 to +20 °C) |
| Mass (Earth = 1) Density (g/cm³) Gravity (Earth = 1) | 0.056 5.42 0.386 | 0.815 5.25 0.879 | 1 5.52 1 | 0.107 3.94 0.38 |
| Number of known satellites | 0 | 0 | 1 | 2 |

*Many of these figures are approximations or obtained by scientific calculations.

water essential for life. For a planet to support living things, it must orbit its star at the right distance so that it is neither too hot nor too cold to have liquid water. Astronomers call this region the *habitable zone*. The planets around 70 Virginis and 47 Ursae Majoris have orbits in the habitable zone. However, all the planets found so far are probably gaseous with no solid surface.

Astronomers were surprised to find that other solar systems have huge, gaseous planets in close orbits. In our own solar system, the inner planets are rocky and small, and only the outer planets, except for Pluto, are huge and gassy. But several newly discovered planets have at least as much mass as Jupiter, the largest planet in our solar system. Unlike Jupiter, however, these massive planets race around their stars in only a few weeks. Kepler's third law says that for a planet to complete its orbit so quickly, it must be close to its parent star. Several of these giant planets, therefore, must travel around their stars even closer than our innermost planet, Mercury, orbits our sun. Such close orbits would make their surfaces too hot to support life as we know it.

Some newly discovered planets follow unusual orbits. Most planets travel around their stars on nearly circular paths, like those of the planets in our solar system. But a planet around the star 16 Cygni B follows an extremely elliptical orbit. It travels farther from its star than the planet Mars does from our sun, and then draws closer to the star than Venus does to our sun. If a planet in our solar system traveled in such an extreme oval, its gravity would disrupt the orbits of the other planets and toss them out of their paths.

### How the planets formed

Astronomers have developed a theory about how our solar system formed that explains why it has small, rocky planets close to the sun and big, gaseous ones farther away. Astronomers believe our solar system formed about 4.6 billion years ago from a giant, rotating cloud of gas and dust called the *solar nebula*. Gravity pulled together a portion of gas and dust at the center of the nebula that was denser than the rest. The material accumulated into a dense, spinning clump that formed our sun.

The remaining gas and dust flattened into a disk called a *protoplanetary disk* swirling around the sun. Protoplanetary disks around distant stars were first observed through telescopes in 1983. Rocky particles within the disk collided and stuck together, forming bodies called *planetesimals*. Planetesimals later combined to form the planets. At the distances of the outer planets, gases froze into ice, creating huge balls of frozen gas that formed the Jovian planets.

Hot gases and electrically charged particles flow from our sun constantly, forming a stream called the *solar wind*. The solar wind was stronger at first than it is today. The early solar wind drove the light elements—hydrogen and helium—away from the inner planets like Earth. But the stronger gravity of the giant outer planets held on to more of the planets' hydrogen and helium, and the solar wind was weaker there. So these outer planets kept most of their light elements and wound up with much more mass than Earth.

Astronomers developed these theories when they thought that rocky planets always orbited close to the parent star and giant planets farther out. But the "rule" was based only on our own solar system. Now that astronomers have learned something about other solar systems, they have devised new theories. Some scientists have suggested that the giant planets in other solar systems may have formed far from their parent stars and later moved in closer.     Jay M. Pasachoff

| Jupiter ♃ | Saturn ♄ | Uranus ♅ | Neptune ♆ | Pluto ♇ |
|---|---|---|---|---|
| 483,600,000 mi. (778,300,000 km) | 888,200,000 mi. (1,429,400,000 km) | 1,786,400,000 mi. (2,875,000,000 km) | 2,798,800,000 mi. (4,504,300,000 km) | 3,666,200,000 mi. (5,900,100,000 km) |
| 390,700,000 mi. (628,760,000 km) | 762,700,000 mi. (1,277,400,000 km) | 1,607,000,000 mi. (2,587,000,000 km) | 2,680,000,000 mi. (4,310,000,000 km) | 2,670,000,000 mi. (4,290,000,000 km) |
| 4,332.7 8.12 mi. per sec. (13.06 km per sec.) | 10,759 5.99 mi. per sec. (9.64 km per sec.) | 30,685 4.23 mi. per sec. (6.81 km per sec.) | 60,190 3.37 mi. per sec. (5.43 km per sec.) | 90,800 2.95 mi. per sec. (4.74 km per sec.) |
| 88,846 mi. (142,984 km) | 74,898 mi. (120,536 km) | 31,763 mi. (51,118 km) | 30,800 mi. (49,500 km) | 1,430 mi. (2,300 km) |
| 9 hrs. 55 min. 3.08 | 10 hrs. 39 min. 26.73 | 17 hrs. 14 min. 97.92 | 16 hrs. 7 min. 28.80 | 6 earthdays 98.8 |
| −220 °F (−140 °C) | −290 °F (−180 °C) | −360 °F (−220 °C) | −350 °F (−210 °C) | −390 to −370 °F (−230 to −220 °C) |
| 317.892 1.33 2.53 | 95.184 0.69 1.07 | 14.54 1.27 0.91 | 17.15 1.64 1.14 | 0.0022 2.0 0.07 |
| 16 | 18 | 17 | 8 | 1 |

**Galaxy** is a system of stars, dust, and gas held together by gravity. Galaxies are scattered throughout the universe. They range in diameter from a few thousand to half a million *light-years.* A light-year is the distance light travels in a year—about 5.88 trillion miles (9.46 trillion kilometers). Large galaxies have more than a trillion stars. Small galaxies have fewer than a billion.

Astronomers have photographed millions of galaxies through telescopes. They estimate that there are about 100 billion galaxies in the universe.

The solar system is in a galaxy called the Milky Way. The solar system lies halfway to the edge of this galaxy. Only three galaxies outside the Milky Way are visible from the earth without a telescope. These three galaxies appear as small, hazy patches of light. People in the Northern Hemisphere can see the Andromeda Galaxy, which is about 2 million light-years away. People in the Southern Hemisphere can see the Large and the Small Magellanic Clouds, which are about 160,000 and 180,000 light-years away.

Galaxies are distributed unevenly in space. Some galaxies are found alone in space, but most are grouped in formations called *clusters.* Clusters of galaxies range in size from a few dozen members to several thousand.

**Kinds of galaxies.** There are three types of galaxies. These types are called *spiral galaxies, elliptical galaxies,* and *irregular galaxies.* A spiral galaxy is shaped like a disk with a bulge in the center. The disk resembles a pinwheel, with bright spiral arms that coil out from the central bulge. The Milky Way is a spiral galaxy. Elliptical galaxies range in shape from almost perfect spheres to flattened globes. The light from an elliptical galaxy is brightest in the center and gradually becomes fainter toward its outer region. Irregular galaxies have no particular shape.

All spiral galaxies rotate. Some elliptical galaxies may also rotate, but more slowly than do spirals. Observations show that new stars are constantly forming out of the gas and dust in spirals. Ellipticals have much less dust and gas than do spirals, so no new stars are forming in them.

**The study of galaxies.** Galaxies give off many kinds of radiation, including the chief kinds of *electromagnetic waves.* These are, in order of decreasing *wavelength* (distance between successive wave crests): radio waves, infrared waves, visible light, ultraviolet rays, X rays, and gamma rays. Astronomers study this radiation with optical and radio telescopes, and other instruments. They estimate the distance and motion of a galaxy by measuring its *red shift.* Red shift is an apparent lengthening of electromagnetic waves radiated by an object moving away from the earth. A red shift may be seen when light from a galaxy is broken up into a band of colors called a *spectrum.* Lines of certain colors will be shifted toward the red end of the spectrum if the galaxy is receding. See **Red shift.**

Almost all galaxies appear to be moving away from the earth. The galaxies farthest from the earth seem to be moving away the fastest. Scientists interpret these observations as evidence that the universe is expanding.

There are various theories about the origin of galaxies. In the *big bang theory,* it is thought that masses of gas condensed soon after the universe began to expand billions of years ago. Gravity slowly compressed these masses into galaxies. No new galaxies—or very few—have formed since then. Another theory, the *steady state theory,* suggests that new galaxies constantly form as old ones move apart. Kenneth Brecher

**Related articles** in *World Book* include:
Astronomy (picture: Great Galaxy)
Cosmology

**Solar energy** is a term that usually means the direct use of sunlight to produce heat or electric power. The sun's energy is plentiful, but it is thinly distributed over a large area and must be collected and concentrated to produce usable power. As a result, solar energy is a more expensive power source than fossil fuels for most applications. Solar technology is improving rapidly, though. Someday, it may provide a clean and abundant source of power.

There are two chief ways that sunlight may be converted into electric power: (1) directly, in a process called *photovoltaic conversion,* or (2) by *solar thermal conversion,* which converts light to heat and then to electric power. Most solar thermal devices heat water to produce steam, which drives a steam turbine.

Nearly all the energy that we use is actually solar energy—energy from the sun. For example, solar energy stored in plants millions of years ago makes up such *fossil fuels* as coal, petroleum, and natural gas. Hydroelectric power plants harness the energy of moving water, and there would be no moving water without the sun. The sun's heat evaporates moisture so that it falls back to earth as rain and other forms of precipitation. The sun also powers the air currents that cause the wind to blow. This article, however, discusses only solar electric and solar heating technologies. For information on other forms of solar energy, see the articles on the fossil fuels, such as **Coal** and **Petroleum,** and the articles on other forms of power, such as **Water power** and **Wind power.**

### Photovoltaic conversion

Devices called *photovoltaic cells* or *solar cells* produce electric current directly from sunlight. This ability results from the *photovoltaic effect,* a phenomenon in which the energy in sunlight causes electric charges to flow through layers of a conductive material to produce a useful electric current.

**The development of photovoltaic cells.** The French physicist Alexandre Edmond Becquerel discovered the photovoltaic effect in 1839. He immersed two metal plates in a solution and observed a small voltage when one plate was exposed to sunlight. The first photovoltaic cells were made of a semimetallic element called selenium. Selenium cells could convert only 1 percent of sunlight to electric power, so they remained just a curiosity for many years.

In 1954, scientists at Bell Telephone Laboratories (now part of Lucent Technologies) invented the first photovoltaic cell that could produce a useful amount of electric power. The Bell scientists, chemist Calvin S. Fuller and physicists Daryl M. Chapin and Gerald L. Pearson, developed a solar cell with an efficiency of 6 percent, six times better than the best selenium cells. A solar cell's efficiency measures the percentage of sunlight striking the cell that it turns into electric power. The Bell scientists made their cell from purified silicon, the material used to make computer chips. Silicon is a *semiconductor*—that is, a material that conducts electric current better than an insulator but not as well as a conductor.

**How a photovoltaic cell is made.** The most common type of photovoltaic cells are *crystalline silicon cells,* so named because every atom in the cell is part of a single crystal structure. To make a crystalline silicon cell, a manufacturer begins with a thin wafer of silicon that has been *doped* (treated) with an impurity. The impurity, usually the element boron, is called a *p-type dopant,* the *p* standing for *positive.* The addition of the dopant causes local *deficits* (shortages) of electrons, called *holes,* to appear in the material. The doped material is called a *p-type material.* A hole can pass from one atom to another and migrate around the crystal.

An *n-type dopant (n* for *negative),* such as phosphorous, is then *diffused* (spread out) part way into the p-type material. The addition of this dopant produces localexcesses of electrons. This layer is then called the *n-type material.* A *potential* (difference in electrical charge) is thus set up between the two layers. When sunlight strikes the photovoltaic cell, the light's energy forces the negative and positive charges in the semiconductor to separate and to accumulate at electrodes joined to each of the two layers. An electric current will then flow through a wire connecting the two electrodes.

**Using photovoltaic cells.** For most applications, engineers wire together many cells in a grouping called a *module* to produce a desired voltage. Multiple modules may be connected in an arrangement called an *array* to produce the required current for the application. Silicon cell efficiencies have reached 25 percent without special equipment to concentrate sunlight. Compound photovoltaic cells combined with solar concentrating systems have achieved 35 percent efficiency. Researchers have also made solar cells from a number of other semiconductor materials.

Today, solar cells provide power for spacecraft and artificial satellites, handheld calculators, and wristwatches. Solar cells are also used for electric power generation in remote areas, where extending power lines would be difficult or costly. Most photovoltaic systems require a storage facility, which normally consists of batteries. Excess energy is stored in the batteries during the day and extracted as needed during the night.

**Other solar cells.** Researchers are also studying solar cells called *photoelectrochemical cells.* The simplest such cell is similar to the device made by Becquerel in 1839. However, today's cells are much more stable and efficient. In addition to producing electric current, photoelectrochemical cells can be used to chemically split water directly into oxygen and hydrogen gas. The hydrogen can then be burned as fuel.

### Solar thermal conversion

Solar thermal conversion systems, also called *solar concentrators,* use one or more reflectors to concentrate solar energy to extremely high levels. There are three major kinds of systems: (1) parabolic trough systems, (2) parabolic dish systems, and (3) central receivers.

**Parabolic trough systems** are the simplest of the solar thermal systems. A *parabola* is a type of curve. In such a system, a *parabolic* (curved) trough covered with

### How a photovoltaic cell produces energy

Artwork adapted courtesy of Solar Energy Research Institute

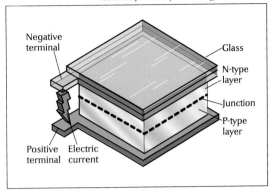

A photovoltaic cell produces electric current when exposed to sunlight. The most common such cell consists of a thin wafer of silicon treated with impurities to create a *p-type* layer with local deficits of electrons and an *n-type* layer with local excesses of electrons. An electrode is joined to each of the layers. Current will flow through a wire connecting the two electrodes.

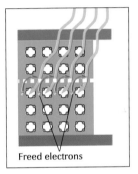

The absorption of light energy in the p-type layer near the *junction,* where the p- and n-type layers join, frees electrons in the p-type layer. The electrons jump across the junction to collect at the negative electrode.

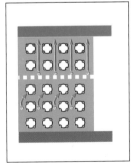

The loss of electrons from the p-type layer produces local deficits, called *holes,* in the layer. Other electrons fill the holes, creating holes in the atoms from which they have come. In this way, the holes migrate from atom to atom.

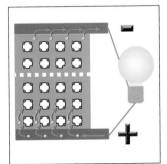

When a wire connects the two terminals, electrons flow from the negative terminal to the positive one, creating electric current. The electrons will then fill some holes in the p-type layer. Electrons will migrate through the cell as long as the cell absorbs light.

rows of reflectors moves to track the sun. The reflectors focus sunlight to a line that strikes a fluid-filled pipe at the center of the trough. The fluid in the pipe is a heat-absorbing substance that may reach temperatures above 750 °F (400 °C). The heat can be used to make steam to generate electric power or to merely heat water. Trough systems can concentrate solar radiation to 100 times the intensity of normal sunlight. Power plants using this technology in southern California provide about 350 megawatts of electric power per year—enough for a community of about 350,000 people.

**Parabolic dish systems** resemble parabolic trough systems except that they focus light to a point instead of a line. Parabolic dish systems have reflectors arrayed along the contour of a bowl-shaped structure called a *dish.* Such systems can achieve concentrations up to 10,000 times the intensity of normal sunlight.

**Central receiver systems** use an array of sun-tracking mirrors, called *heliostats,* that reflect light onto a single central tower called a *receiver.* A central receiver system called Solar Two operates in the Mojave Desert near Barstow, California. It has 1,926 heliostats that focus sunlight on a single receiver. The receiver is filled with a mixture of molten sodium and potassium salts, which

holds heat longer than other fluids. After the molten salt is heated, the system pumps it into insulated tanks. When power is needed, the molten salt is pumped into a device called a *heat exchanger,* where it produces steam that turns a turbine. Solar Two's storage capability enables it to operate even after dark or when the sun is covered by clouds. The plant supplies 10 megawatts of electrical power. Other central receiver test facilities operate in Almería, Spain, and Rehovot, Israel.

### Solar heating

Solar heating requires an efficient absorber to collect sunlight and convert it to heat. The absorber may be as simple as a coating of black paint, or it may be a textured, heat-absorbing ceramic. A good absorber collects 95 percent or more of the solar radiation while emitting 20 percent or less of the heat energy an ordinary hot surface would.

There are several methods of solar heating. One common method uses windows as solar collectors, as in a greenhouse. The windows trap the sun's heat, and the heat passes back through the windows, roof, and walls slowly.

The simplest solar collectors are *flat-plate collectors.*

### How solar energy heats a house

WORLD BOOK illustrations by Oxford Illustrators Limited

**A solar-heated home** has large south-facing windows that let in heat from the sun. The walls and floor absorb the heat during the day and release it at night. A wood-burning stove provides heat on cloudy days. Overhangs shade the windows in summer when the sun is high. Sunlight also heats collectors on the roof. Liquid inside the collectors flows to a heat exchanger in the basement, where water is heated for household use.

**A flat-plate collector** has a black plate that absorbs heat from sunlight. When the plate gets hot, it heats a liquid that flows in channels inside the collector. Glass or plastic sheets and insulation prevent heat loss.

The plates are fixed, and the sun shines on them at various angles as it moves. The sun heats fluid inside the plates to a temperature of up to 212 °F (100 °C). The hot fluid flows to a heat exchanger, a device like an automobile radiator through which water circulates, and transfers its heat to the water. The hot water is used to warm buildings in a conventional hot-water heating system or to heat their hot-water supplies.

Another kind of collector, designed specifically for heating air, is the *transpired solar collector*. Such collectors consist of flat or ridged plates pierced by an array of small holes. Air is drawn through the holes and is heated by the sun-warmed plates. As much as 80 percent of the solar energy collected by the plates is transferred to the air stream.

A *solar furnace* is a type of solar collector that concentrates sunlight to produce temperatures high enough for use in industrial processes. Scientists use solar furnaces to process steel, ceramics, and other materials; to *pump* (provide energy for) solid-state lasers; and to destroy hazardous wastes.                    J. Roland Pitts

**Solar system** is a group of heavenly bodies consisting of a star and the planets and other objects orbiting around it. We are most familiar with our own solar system, which includes Earth, eight other major planets, and the sun. Our solar system also includes many smaller objects that revolve around the sun, such as asteroids, meteoroids, and comets; and a thin cloud of gas and dust known as the *interplanetary medium*. More than 60 moons, also called satellites, orbit the planets.

Besides the sun, Earth, and Earth's moon, many objects in our solar system are visible to the unaided eye. These objects include the planets Mercury, Venus, Mars, Jupiter, and Saturn; the brightest asteroids; and occasional comets and meteors. Many more objects in the solar system can be seen with telescopes.

In the 1990's, astronomers discovered many planets orbiting distant stars, though the planets could not be seen directly. By studying the masses and orbits of these planets, astronomers hope to learn more about solar systems in general. For example, our own solar system contains four small, rocky planets near the sun—Mercury, Venus, Earth, and Mars—and four giant, gaseous planets farther out—Jupiter, Saturn, Uranus, and Neptune. Astronomers were surprised to find that other stars have giant, gaseous planets in close orbits. For example, a planet nearly the size of Jupiter orbits the star 51 Pegasi closer than Mercury orbits our own Sun.

### Our solar system

**The sun** is the largest and most important object in our solar system. It contains 99.8 percent of the solar system's *mass* (quantity of matter). The sun provides most of the heat, light, and other energy that makes life possible.

The sun's outer layers are hot and stormy. The hot gases and electrically charged particles in those layers continually stream into space and often burst out in solar eruptions. This flow of gases and particles forms the *solar wind*, which bathes everything in the solar system.

**Planets** orbit the sun in oval-shaped paths called *ellipses*, according to a law of planetary motion discovered by German astronomer Johannes Kepler in the early 1600's. The sun is slightly off to the side of the center of each ellipse at a point called a *focus*. The focus is actually a point inside the sun—but off its center—called the *barycenter* of the solar system.

The inner four planets consist chiefly of iron and rock. They are known as the *terrestrial* (earthlike) planets because they are somewhat similar in size and composition. The outer planets, except for Pluto, are giant worlds with thick, gaseous outer layers. Almost all their mass consists of hydrogen and helium, giving them compositions more like that of the sun than that of Earth. Beneath their outer layers, the giant planets have no solid surfaces. The pressure of their thick atmospheres turns their insides liquid, though they may have rocky cores.

A distant object called Pluto has been referred to as the ninth planet since its discovery in the 1930's. But Pluto has so many unusual features that some astronomers think it may not be a planet at all. For example, it travels around the sun in an elongated oval path much different from the nearly circular orbits of the other planets. Unlike the other outer planets, Pluto is small and solid. But Pluto contains only $\frac{1}{500}$ the mass of Earth.

During the 1990's, astronomers discovered dozens of small rocky objects orbiting the sun beyond Neptune and Pluto. Astronomers had long suspected that the outer solar system had such a band of rocky material, called the *Kuiper* (pronounced *KY pur) belt*. The belt is named for the Dutch-born American astronomer Gerard P. Kuiper, who first predicted its existence. Pluto may merely be the largest of the objects in the Kuiper belt.

**Moons** orbit all the planets except Mercury and Venus. The inner planets have few moons. Earth has one, and Mars has two tiny satellites. The giant outer planets, however, resemble small solar systems, with many moons orbiting each planet. Jupiter has 16 moons. The largest 4 are known as the Galilean satellites because the Italian astronomer Galileo discovered them in 1610 with one of the first telescopes. The largest Galilean satellite—and the largest satellite in the solar system—is Ganymede, which is even bigger than Mercury and Pluto. Saturn has 18 moons. The largest of Saturn's moons, Titan, has an atmosphere thicker than Earth's and a diameter larger than that of Mercury or Pluto. Uranus has 17 moons, and Neptune has 8. The smallest of these moons are less than 20 miles (32 kilometers) across, and the giant planets probably have more small moons not yet discovered. Pluto has one moon.

**Rings** of dust, rock, and ice chunks encircle all the giant planets. Saturn's rings are the most familiar, but thin rings also surround Jupiter, Uranus, and Neptune.

**Comets** are snowballs composed mainly of ice and rock. When a comet approaches the sun, some of the ice in its *nucleus* (center) turns into gas. The gas shoots out of the sunlit side of the comet. The solar wind then carries the gas outward, forming it into a long tail.

Astronomers divide comets into two main types, *long-period comets*, which take 200 years or more to orbit the sun, and *short-period comets*, which complete their orbits in fewer than 200 years. The two types come from two regions at the edges of the solar system. Long-period comets originate in the *Oort* (pronounced *oort* or *ohrt) cloud*, a cluster of comets far beyond the orbit of Pluto. The Oort cloud was named after the Dutch astronomer Jan H. Oort, who first suggested its existence.

WORLD BOOK illustration by Rob Wood

**The solar system includes many different objects** that travel around the sun. These objects vary from planets much larger than the earth to tiny meteoroids and dust particles.

Short-period comets come from the Kuiper belt, the band of rocky objects orbiting the sun just beyond Pluto. Many of the objects in the Oort cloud and the Kuiper belt may be rocky chunks known as *planetesimals* left over from the formation of the solar system.

**Asteroids** are minor planets. Some have elliptical orbits that pass inside the orbit of Earth or even that of Mercury. Others travel on a circular path among the outer planets. Astronomers estimate that more than 50,000 asteroids exist, but they have accurately determined the orbits of fewer than 10,000 of them. Most asteroids circle the sun in a region called the *asteroid belt,* between the orbits of Mars and Jupiter. Dozens of asteroids measure more than 125 miles (200 kilometers) across. Astronomers have even found several large asteroids with smaller asteroids orbiting them.

**Meteoroids** are chunks of metal or rock smaller than asteroids. When meteoroids plunge into Earth's atmosphere, they form bright streaks of light called *meteors as* they disintegrate. Some meteoroids reach the ground, and then they become known as *meteorites.* Most meteoroids are broken chunks of asteroids that resulted from collisions in the asteroid belt. During the 1990's, astronomers discovered a number of meteoroids that came from Mars and from the moon. Many tiny meteoroids are dust from the tails of comets.

**Heliosphere** is a vast, teardrop-shaped region of space containing electrically charged particles given off by the sun. Scientists do not know the exact distance to the *heliopause,* the limit of the heliosphere. Many astronomers think that the heliopause is about 9 billion miles (15 billion kilometers) from the sun at the blunt end of the "teardrop."

### Formation of our solar system

Many scientists believe that our solar system formed from a giant, rotating cloud of gas and dust known as the *solar nebula.* According to this theory, the solar nebula began to collapse because of its own gravity. Some astronomers speculate that a nearby *supernova* (explod-

ing star) triggered the collapse. As the nebula contracted, it spun faster and flattened into a disk.

The nebular theory indicates that particles within the flattened disk then collided and stuck together to form asteroid-sized objects called planetesimals. Some of these planetesimals combined to become the nine large planets. Other planetesimals formed moons, asteroids, and comets. The planets and asteroids all revolve around the sun in the same direction, and in more or less the same plane, because they originally formed from this flattened disk.

Most of the material in the solar nebula, however, was pulled toward the center and formed the sun. According to the theory, the pressure at the center became great enough to trigger the nuclear reactions that power the sun. Eventually, solar eruptions occurred, producing a solar wind. In the inner solar system, the wind was so powerful that it swept away most of the lighter elements—hydrogen and helium. In the outer regions of the solar system, however, the solar wind was much weaker. As a result, much more hydrogen and helium remained on the outer planets. This process explains why the inner planets are small, rocky worlds and the outer planets, except for Pluto, are giant balls composed almost entirely of hydrogen and helium.

### Other solar systems

Several other stars have disk-shaped clouds around them that seem to be solar systems in formation. In 1983, an infrared telescope in space photographed such a disk around Vega, the brightest star in the constellation Lyra. This discovery was the first direct evidence of such material around any star except the sun. In 1984, astronomers photographed a similar disk around Beta Pictoris, a star in the southern constellation Pictor.

By the 1990's, astronomers had discovered many planets orbiting stars like our sun. Stars known to have solar systems include 51 Pegasi in the constellation Pegasus, 70 Virginis in the constellation Virgo, and 47 Ursae Majoris in the Big Dipper (Ursa Major).          Jay M. Pasachoff

**Carbon** is one of the most familiar and important chemical elements. All living things are based on carbon, and industry uses it in a wide variety of products. Yet carbon makes up only 0.032 percent of the earth's crust. Carbon is the main component of such fuels as coal, petroleum, and natural gas. Carbon is also found in most plastics.

### Chemical properties

Carbon has the chemical symbol C. Pure carbon does not react readily with other chemicals at room temperature. Most naturally occurring forms of carbon, such as diamond and graphite, do not dissolve in acid or any other common solvent. Carbon solids are stable up to very high temperatures in the absence of oxygen. At reduced pressures, some forms of carbon *sublime* (change from a solid to a vapor without melting).

Carbon's *atomic number* (number of protons) is 6. The most abundant *isotope* of carbon is carbon 12. The isotopes of an element have the same number of protons but different numbers of neutrons. Carbon 12 is the international standard for atomic weight. By agreement, C-12 has an atomic weight of exactly 12 of the units known as *atomic mass units*. The average atomic weight of carbon's natural isotopes is 12.0107.

Carbon atoms are unusual because they can form strong chemical bonds with two, three, or four other atoms. These atoms can be carbon atoms or atoms of other chemical elements. Carbon atoms can link together to form long chains, rings, or combinations of chains and rings. This unique linking ability enables carbon to form the complex molecules that make up living things. Carbon atoms also combine to form balls and tubes.

### Carbon compounds

Much of the carbon on earth exists in combination with other elements. There are more than 1 million known carbon compounds, the largest number of compounds formed by any element except hydrogen. The most abundant carbon compounds are the gas carbon dioxide, which is part of the atmosphere; the carbonate minerals, such as limestone (also known as calcium carbonate) and marble; and the *hydrocarbons*, compounds of carbon and hydrogen that are the chief ingredients of the fuels petroleum and natural gas.

Carbon compounds make up the living tissues of all plants and animals. Organic chemistry—the study of trial compounds made by and derived from living things—is primarily the study of carbon compounds. Most organic compounds consist mainly of carbon combined with hydrogen, nitrogen, and oxygen in various proportions.

### Forms of carbon

Pure carbon occurs in four forms: (1) diamond, (2) graphite, (3) amorphous carbons, and (4) fullerenes. The four forms have different *crystalline structures*—that is, their atoms are arranged differently. The various forms of carbon differ greatly in hardness and other properties, depending on how their atoms are arranged.

**Diamond** is the hardest naturally occurring substance and one of the most valuable. Natural diamonds form in the rock beneath the earth's crust, where high temperature and pressure cause carbon atoms to make strong bonds with four other carbon atoms each and to crystallize. Volcanic activity then forces the diamonds to the surface. Manufacturers produce artificial diamonds by heating and compressing pure carbon, usually graphite. Scientists grow synthetic diamond coatings by placing the object to be coated in a special chamber where a carbon-rich gas separates chemically and deposits a carbon film on the surface of the object.

The atoms in a diamond are arranged in a pyramid-shaped pattern called a *tetrahedron* that makes the structure extremely rigid. As a result, diamonds are the hardest known substance. The density of diamond is 3.5 grams per cubic centimeter.

Only a small percentage of natural diamonds are pure and perfect enough to become gemstones. Most diamonds, whether natural or synthetic, are used for industrial purposes. Because of diamonds' hardness, manufacturers use them to shape, cut, grind, and polish hard materials. Diamonds have another unique pair of properties—they are good conductors of heat but do not conduct electrical current. Diamond films are thus used in high-power electronic devices to remove excess heat without affecting the device's electrical characteristics.

The crystalline structure of diamond is the same as that of silicon, the chief material used in transistors. As a result, transistors can also be made from diamond. Diamond transistors can be safely used under much harsher conditions, such as extremely high temperatures, than ordinary silicon transistors can.

**Graphite** is a soft, black mineral that feels slick to the touch. Like diamond, natural graphite forms beneath the surface of the earth. Perfect graphite crystals are rare and hard to find, but low-grade graphite is plentiful. Industry produces synthetic graphite by heating *coke,* a solid fuel that contains about 90 percent carbon.

Graphite consists of carbon atoms arranged in flat, parallel layers. The layers slide easily over one another, making the graphite soft and slippery. Graphite is much less dense than diamond, with a density of only 2.2 grams per cubic centimeter.

Because graphite is slick and soft, it is used in powdered lubricants and for the "lead" in some pencils. Unlike diamond, graphite is a good electrical conductor. As a result, it is used to make the contacts in electric motors and other machinery. Because graphite fibers are strong, they are used to reinforce plastic. Graphite and plastic form a strong, lightweight composite material that is used to make dish antennas, tennis rackets, fishing rods, bicycle frames, and spacecraft parts.

**Amorphous carbons,** also called *glassy carbons,* are made of tiny, irregularly arranged particles of graphite with no regular crystalline structure. Familiar amorphous carbons include the fuels charcoal and coke.

Amorphous carbons form, along with ash, when carbon-rich substances are heated or burned in an airtight furnace without enough oxygen to convert all the carbon to carbon dioxide. Charcoal, for example, is obtained by burning wood in the absence of air. A powdery soot called *carbon black* forms when natural gas or a petroleum-based fuel, such as kerosene, is burned in the same way. Carbon black is used as the black pigment in automobile tires and printing inks. A similar process using coal or petroleum produces coke and a tarry residue called *pitch.* Coke is an essential raw material in converting iron to steel.

**Forms of carbon**

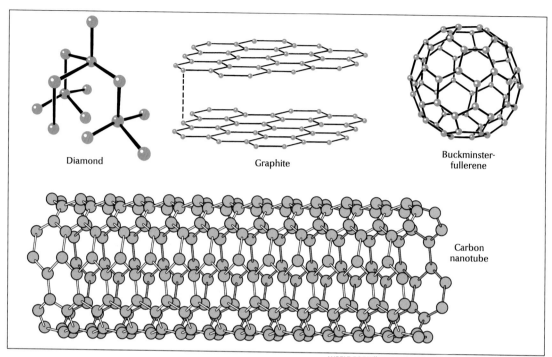

Diamond

Graphite

Buckminster-
fullerene

Carbon
nanotube

**Pure, solid carbon** occurs in three crystalline forms—rigid, pyramid-shaped diamond; flat layers of
graphite; and large, hollow fullerenes. Two kinds of fullerenes are shown here: buckminsterfullerene, also known as a *buckyball;* and a carbon nanotube, sometimes called a *buckytube.*

Amorphous carbons have a wide range of properties.
They have low densities and are quite porous. *Carbon
aerogels,* also called "frozen smoke," are among the
world's lightest solids, with densities as low as 0.04
gram per cubic centimeter. The plentiful pores in charcoal trap many substances effectively, so charcoal is
used to filter impurities from liquids and the air. The
pores also enable oxygen to penetrate rapidly inside the
charcoal, making it a good fuel. Amorphous carbons are
also hard, resistant to high temperatures, and chemically
*inert*—that is, they do not react with most other chemicals. Because of their heat resistance, they are used for
shields to protect missiles and spacecraft from getting
too hot when they reenter the earth's atmosphere.

**Fullerenes** are hollow molecules made up of a large,
even number of carbon atoms, 32 or more. The best
known of these molecules are *buckminsterfullerenes,*
also known as $C_{60}$'s or *buckyballs.* Each buckminsterfullerene consists of 60 carbon atoms bonded together
in the shape of a soccer ball. Small amounts of fullerenes occur naturally in rock and in sooty flames, such as
those of candles, but scientists make almost all fullerenes in the laboratory.

A fullerene with 70 atoms is shaped somewhat like a
rugby ball. Fullerenes with more than 70 atoms can be
ball-shaped or tubular. The tubes can have open or
closed ends. Tubular fullerenes are sometimes called
*buckytubes* or *carbon nanotubes.*

Fullerenes were first produced in 1985 by chemists
Harry W. Kroto of the United Kingdom, Richard E. Small-
ey and Robert F. Curl of the United States. They won the
1996 Nobel Prize in chemistry for their major contributions to the discovery. The scientists vaporized graphite
with a laser, producing clusters of 60 and 70 carbon
atoms each. They named the $C_{60}$ molecule *buckminsterfullerene* because its structure resembles a *geodesic
dome,* a type of structure designed by American engineer R. Buckminster Fuller. They named the entire
group of hollow carbon molecules *fullerenes.*

It became much easier to study fullerenes in 1990. In
that year, physicists Donald R. Huffman of the United
States and Wolfgang Krätschmer of Germany devised a
simple method for large-scale production of the molecules. As Kroto and his colleagues had done, Huffman
and Krätschmer generated fullerenes by vaporizing
graphite. They did it, however, by setting up two
graphite rods with their ends almost touching, then
sending an electric current across the gap. This process
generated a sooty material containing about 10 percent
fullerenes, which the scientists extracted and purified.

Buckytubes and ball-shaped fullerenes have a number
of properties that may prove to be of commercial value.
Filled with metal atoms, for example, buckytubes form
the smallest wire imaginable. The buckyball ($C_{60}$) can be
chemically modified to block a key step in the reproduction of the human immunodeficiency virus (HIV), which
causes AIDS. Fullerenes can also be made into *super-
conductors,* substances that conduct electric current
with no resistance at extremely low temperatures.

John E. Fischer

## INDEX

### How to use the index

This index covers the contents of the 1998, 1999, and 2000 editions.

---

Amphorae (vessels), **00:** 12
**Amplification, 99:** 209 (il.), **98:** 254
**Analog signals, 00:** 101
**Angiosperms, 00:** 243 (il.), 244
**Angiostatin** (drug), **99:** 254
**Angular momentum, 00:** 35
**Animals**
    brain diseases, **98:** 101-113
    consciousness, **00:** 169
    early multicellular, **00:** 242-243
    Galapagos, **98:** 60-71
    habitat destruction, **98:** 150-151, 158
    land-dwelling, **99:** 238-239
    soft-bodied, fossils of, **98:** 230
    zoo, **00:** 84-97
    see also **Biology; Cloning; Conservation;**
       **Ecosystems; Endangered species;**
       **Extinction; Fossil studies**
**Anomalies** (oceanography), **00:** 16, 20
**Antarctica, 99:** 190-191, 199 (il.), 251 (il.),
    266, **98:** 178
**ANTHROPOLOGY, 00:** 180-184, **99:** 180-
    182, **98:** 168-171
    books, **98:** 194
    forensic science, **99:** 162
**Antibiotics, 99:** 256-257, **98:** 159
**Antidepressants, 99:** 273-274, **98:** 259
**Antimatter, 00:** 271-272
**Antiparticles, 00:** 271-272
**Antiperspirants, 99:** 297-300
**Ants, 99:** 216, 225 (il.)
**Apligraf** (surgery), **00:** 75, 78 (il.)
**Apocrine glands, 99:** 297-298
**Apollo program, 00:** 30-32, 34
**Apoptosis, 99:** 246-247
**Aquariums, 00:** 89, 286-289
**Arabs, 00:** 116, 117
**Aragonite, 00:** 253-254
**Arava** (drug), **00:** 228
*Archaea* (organisms), **00:** 206-207, **98:** 116,
    123, 126, 128 (il.)
*Archaeopteryx* (bird), **00:** 58-69, 242, **99:**
    242, **98:** 228, 229
**Arizona-Sonora Desert Museum, 00:** 91
**Arson, 99:** 167, 172
**Arthritis, 00:** 228
**Artificial intelligence, 00:** 133 (il.), **98:** 204
    (il.)
**Asthma, 98:** 212
**Astronauts.** See **Space technology**
**ASTRONOMY, 00:** 191-199, **99:** 189-196, **98:**
    177-186
    books, **00:** 210, **99:** 206
    history of science, **00:** 115, 116, 118-120,
    128
    *WBE*, **00:** 306
    see also **Space technology; Universe**
**Atherosclerosis, 99:** 237
**Atlantis** (space shuttle), **99:** 280, **98:** 266

---

Each entry gives the last two digits of the edition year, followed by a colon and the page number or numbers. For example, this entry indicates that information on analog signals may be found on page 101 of the 2000 edition.

When there are many references to a topic, they are grouped alphabetically by clue words under the main topic. For example, the clue words under **Animals** group the references to that topic under several subtopics.

The "see" and "see also" cross-references indicate that references to a topic are listed under another entry in the index.

An entry in all capital letters indicates that there is a Science News Update article with that name in at least one of the three volumes covered by this index. References to the topic in other articles may also be listed in the entry.

An entry that only begins with a capital letter indicates that there are no Science News Update articles with that title but that information on this topic may be found in the editions and on the pages listed.

The indication (il.) after a page number means that the reference is to an illustration only. For example, this entry refers to the picture illustrating artificial intelligence on page 133 of the 2000 edition.

An entry followed by *WBE* refers to a new or revised *World Book Encyclopedia* article in the supplement section. This entry means that there is a *World Book Encyclopedia* article on astronomy beginning on page 306 of the 2000 edition.

# Index

# Index

# Index

# Index

**MEDICAL RESEARCH, 00:** 255-258, **99:** 253-257, **98:** 240-246
  Benford interview, **00:** 144-146
  books, **00:** 211, **99:** 207
  brain diseases, **98:** 100-113
  funding priorities, **00:** 278
  history of science, **00:** 116, 122-123, 127 (il.), 128-129
  Nobel Prizes, **00:** 259, **99:** 258, **98:** 248
  see also **Genetics; Transplants**
**Medicine, Forensic, 99:** 158, 161
**Mediterranean Sea, 00:** 230 (il.)
**Megafauna extinctions, 00:** 224
**Melatonin** (hormone), **98:** 278-281
**Melissa** (computer virus), **00:** 220
*Melkarth* (shipwreck), **00:** 13-15
**Menageries, 00:** 85-86
**Mendeleev, Dmitri, 00:** 122
**Mental illness. See Psychology**
**Mentalism, 00:** 167-168
**Mercury** (metal), **00:** 209, **98:** 283
**Mercury** (planet), **98:** 18, 19 (il.)
**Meridia** (drug). See **Sibutramine**
**Mesenchymal stem cells, 00:** 80, 81, 255
**Mesh** (network), **00:** 151
**Mesons** (particles), **00:** 271-272
**Metazoans** (organisms), **98:** 229
**Meteorites, 00:** 174, 175, 252 (il.), **99:** 34-35, 189-191, **98:** 178-179
**Meteoroids, 00:** 36, 41, 192
**Meteorology. See Atmospheric science**
**Methane, 98:** 24, 25
**Mexican gray wolves, 00:** 222
**Meyer, Julius Lothar, 00:** 122
**Miami River Circle** (site), **00:** 185-187
**Mice, 00:** 246, 250 (il.), **99:** 146 (il.), 150 (il.), **98:** 260 (il.)
**Michelangelo, 99:** 59, 60, 63
**Microbursts, 00:** 44
**Microchips, 00:** 212 (il.), 218, **99:** 211, 215
**Microgravity, 99:** 132, 138
**Microlock, 00:** 236-238
**Micromotors. See Nanotechnology**
**Microorganisms**
  deep-sea, **98:** 114-116
  disease-causing, **98:** 157-159
  extreme environments, **00:** 206-207, **99:** 204-205
  see also **Bacteria; Viruses**
**Microsaurs** (animals), **99:** 238-239
**Microscopes, 99:** 144-154
  art conservation, **99:** 60, 65 (il.), 69 (il.)
  forensic science, **99:** 162 (il.), 164, 172
  nanotechnology, **99:** 112, **98:** 255
  see also **Electron microscopes**
**Microsoft Corp., 00:** 220-221, **99:** 212, 214, 215, **98:** 202
**Microtubules, 00:** 169
**Microwave radiation, 00:** 109-110, 162, 264-265, **99:** 212, **98:** 74, 75, 87
**Midocean ridge, 00:** 254 (il.), **98:** 117, 118, 126
**Mighty Whale** (power plant), **00:** 234 (il.)
**Migrations, 98:** 160, 172, 209
**Milk, 99:** 261, **98:** 249 (il.)
**Milky Way Galaxy, 00:** 128, 161-162, 197, 198, **98:** 73, 182-184
**Millennium. See Year 2000**
**Millennium Dome** (England), **00:** 235
**Mind-body problem, 00:** 167

**Minerals, Ocean-floor, 98:** 126-127
**Minnesota Zoo, 00:** 92, 94, 95
**Mir** (space station), **00:** 280, **99:** 136, 139, 280, 281, **98:** 266-267
**Mirapex** (drug). See **Pramipexole**
**Mitochondria, 00:** 170, 172 (il.), **99:** 81
**Mitochondrial DNA, 99:** 20 (il.), 22-25
**Molecules, 99:** 192-193, **98:** 196-197
**Mollusks, 99:** 240
**Mongooses, 98:** 193
*Monitor*, U.S.S. (ship), **00:** 24
**Monkeys, 00:** 95, 168, **99:** 201 (il.), 205, **98:** 190
**Monod, Jacques, 00:** 165, 166
**Monte Verde** (site), **98:** 171-172
**Moon, 00:** 126 (il.), 155, **99:** 189, 283, 284, **98:** 13, 18-19
  origins, **00:** 28-41
  see also **Satellites, Planetary**
**Morokweng Structure, 99:** 251-252
**Mosquitoes, 99:** 179, **98:** 98 (il.), 158, 159, 262
**Moths, 98:** 191-192
**Motion, Laws of, 00:** 119
**Motion pictures, 98:** 42-57, 200, 276-277
**Motor oil, 98:** 283, 284
**Motorola Corp., 00:** 104
**Motors, Electric, 99:** 228
**Mounds, Earthen, 99:** 186-188
**Mountains, Volcanic. See Volcanoes**
**MP3 format, 00:** 216-217
**MRI. See Magnetic resonance imaging**
**mRNA, 00:** 165
**mtDNA. See Mitochondrial DNA**
**Mudpuppies, 99:** 46
**Multiphoton excitation microscopes, 99:** 150 (ils.)
**Mummy, 00:** 186 (il.)
**Muons** (particles), **00:** 268
**Murad, Ferid, 00:** 259
**Murder. See Crime**
**Museums, Science, 99:** 116-129
**Music, on Internet, 00:** 216-217
**Mutations, 00:** 171, 172, 241, 251, **99:** 22-25, 52, 302-304, **98:** 106-109, 229-230, 241-242
**Mycorrhizal fungi, 00:** 232

# N

**N-body method, 00:** 37
**Nanobes** (organisms), **00:** 206
**Nanorobots** (machines), **00:** 213
**Nanotechnology, 00:** 146-147, 213-214, **99:** 103-115, 267 (il.), **98:** 255
**Nanotubes, 00:** 213-214, **99:** 151 (il.), **98:** 255
**NASA. See National Aeronautics and Space Administration**
**NASA Scatterometer** (satellite), **00:** 264
**National Aeronautics and Space Administration, 00:** 280-282, **99:** 280-284, **98:** 268-269
  art restoration, **99:** 70
  communication satellites, **00:** 103
  Mars exploration, **99:** 28, 31, 34, 37
  moon exploration, **00:** 30, 37
  ocean studies, **00:** 264
  radar maps, **99:** 183
  solar-powered aircraft, **99:** 228
  space stations, **99:** 132-140, 143
  see also **Astronomy; Hubble Space Telescope; Space probes**
**National Bioethics Advisory Commission, 00:** 277-278, **98:** 263-264

**National Institute of Mental Health, 00:** 278
**National Institutes of Health, U.S., 00:** 277
**National Oceanographic Partnership Program, 99:** 263-266
**National Tallgrass Prairie Preserve, 98:** 207
**National Zoo** (Washington, D.C.), **00:** 86, 89, 95
**Native Americans, 00:** 182-183, 236, 240 (il.), 278, **99:** 182, 186-188, 278, **98:** 171-172
  see also **Archaeology; Inca; Maya**
**Natural gas, 98:** 156
**Natural resources, 98:** 154-156
**Natural selection. See Evolution**
**Nature, Forces of, 98:** 82
**Nature, Laws of, 98:** 80, 86
**Navigation, 99:** 286-289
**Neanderthals, 00:** 184, **99:** 13, 16 (il.), 18-25, 180, **98:** 174
**NEAR** (spacecraft), **00:** 282, **99:** 281-282
**Near-frictionless carbon, 99:** 229
**Near Infrared Mapping Spectrometer, 99:** 191-192
**Neomorphogenesis** (biology), **00:** 76-77
**Neoproterozoic Era, 00:** 252-253
**Neptune, 98:** 24-25
**Netscape Communications Corp., 00:** 220, 221
**Network Associates** (company), **99:** 279
**Neural networks** (software), **00:** 292
**Neurons, 00:** 258, **99:** 272
**Neutrinos** (particles), **00:** 268-271
**Neutron autoradiography, 99:** 61
**Neutron stars, 00:** 195-197, **99:** 196
**Neutrons, 00:** 127-128, 196, **99:** 189, 196
*New Carissa* (ship), **00:** 239 (il.)
**New molecular entities** (drugs), **00:** 228
**Newton, Isaac, 00:** 119-120, 158, 159
**Newts, 99:** 45, 49 (il.)
**NEWTSUIT** (oceanography), **00:** 15 (il.)
**NEXRAD system, 00:** 51
**NeXT Software, Inc., 98:** 202
**Nicaragua, 00:** 153-154
**Nichols, Terry, 99:** 168 (il.)
**Nicotine, 98:** 262
**Niemann-Pick type C disease, 99:** 248-249
**Night Stalker murders, 99:** 170
**Night Vision** (imaging system), **00:** 237 (il.)
**NiMH battery, 00:** 232
**Nitric acid, 98:** 152, 226
**Nitrocellulose lacquer, 00:** 296
**Nitrogen, 99:** 226, 237, **98:** 23-25, 186-187, 225-226
**Nitrogen dioxide, 00:** 241
**Nitrous oxide, 98:** 187, 226
**NOBEL PRIZES, 00:** 259, **99:** 258, **98:** 246-248
**Nolvadex** (drug). See **Tamoxifen citrate**
**Nomad** (vehicle), **99:** 229-232
**Norepinephrine** (chemical), **99:** 223
**North Carolina Zoological Park, 00:** 91 (il.), 92 (il.), 96 (il.)
**Norvir** (drug), **98:** 244
**Notes** (fragrance), **00:** 299-300
**Nozomi** (space probe), **00:** 282, **99:** 38
**NR-1** (submersible), **00:** 17

# Index

# Index

# Index

Water pressure, in ocean, **98**: 118, 127
Watson, James, **00**: 125 (il.), 128, 165
Watson Brake (site), **99**: 186-188
Weak force (physics), **98**: 82, 85 (il.)
Weapons. See **Firearms; Nuclear weapons**
Weather, *WBE*, **99**: 314
  see also **Atmospheric science; Climate**
Web. See **World Wide Web**
WebTV Plus, **99**: 213-214
Webzter (computer), **00**: 215
Welding, Underwater, **98**: 223
Werner syndrome, **00**: 171
Wet cleaning, **99**: 290-293
Wetlands, **99**: 48 (il.), 50-51, 53-54, **98**: 150, 209
Whales, **99**: 92, 95-101, 218, 266, **98**: 250-251
Wheat, **00**: 178, **98**: 166-167
White dwarf stars, **00**: 162
Whole-genome shotgun sequencing, **00**: 249
Wieschaus, Eric, **00**: 165-166
Wild-2 (comet), **00**: 282
Wildfires, **98**: 188
Wildlife conservation. See **Conservation**
Wildlife Conservation Society, **00**: 95
Wilmut, Ian, **99**: 77, 80, 82 (il.), 84, **98**: 230
Wind, **00**: 263, **99**: 264, 265, **98**: 188
Wind energy, **00**: 232-233, **98**: 218 (il.)
Wind shear, **00**: 43-44, 47-48, 53

Wind speed, **99**: 197-198
Windows, Switchable, **99**: 228 (il.)
Windows 95 (software), **00**: 220-221, **99**: 212, **98**: 202, 203
Windows 98 (software), **99**: 212
Wizards Beach Man (fossil), **00**: 182
Wolves, **00**: 222, **99**: 218-219, **98**: 30
Women, **99**: 260, 273, **98**: 163
  see also **Breast cancer; Cervical cancer; Pregnancy**
Wood, **00**: 294-296
Wool, **00**: 179 (il.)
World Wide Web, **00**: 293, **99**: 129, 173, 279, **98**: 205, 217
  digital photos, **00**: 301, 304
  growth, **00**: 217, **99**: 212-213
  music, **00**: 216-217
  WebTV, **99**: 213-214
  see also **Internet**
WorldCom, Inc., **99**: 214
Worms, **00**: 242-244, **99**: 204-205, 263 (il.), **98**: 230
Writing, Early, **00**: 189-191
  see also **Handwriting**

## X

X-ray fluorescence spectroscopy, **99**: 63
X-ray spectrometry, **99**: 172
X rays, **00**: 195-198, **99**: 60, 63, 271 (il.), **98**: 241, 254 (il.)
Xenical (drug), **00**: 229
Xenotransplantation (surgery), **99**: 277

## Y

Y2K Bug. See **Year 2000**
Yakutian horses, **00**: 231-232
Yanagimachi, Ryuzo, **00**: 246
Year 2000
  computer compatibility, **00**: 218-219, **98**: 203-204
  midcentury predictions, **00**: 130-141
Yellow fever, **98**: 159
Yellowstone National Park, **00**: 206, 207
Yohkoh (satellite), **00**: 193 (il.), **98**: 181
*Yorktown* (ship), **00**: 25
Yucatan Peninsula, **00**: 175, **98**: 239, 251

## Z

Z-Trim (fat substitute), **98**: 199
Zafirlukast (drug), **98**: 212
ZAG (chemical), **99**: 299
Zarya (space module), **00**: 280, 283 (il.)
Zidovudine (drug). See **AZT**
Zileuton (drug), **98**: 212
Zinc, **98**: 248
Zinkernagel, Rolf M., **98**: 248
Zoll, Paul, **00**: 227
Zoos, **00**: 84-97
Zweig, George, **00**: 128
Zyflo (drug), **98**: 212-213

# ACKNOWLEDGMENTS

The publishers gratefully acknowledge the courtesy of the following artists, photographers, publishers, institutions, agencies, and corporations for the illustrations in this volume. Credits are listed from top to bottom, and left to right, on their respective pages. All entries marked with an asterisk (*) denote illustrations created exclusively for this yearbook. All maps, charts, and diagrams were staff-prepared unless otherwise noted.

| | |
|---|---|
| **2** | AP/Wide World; © David Ball, Tony Stone Images; NASA |
| **3** | NASA/JPL/CALTECH; Henry Sobel, University of California, Irvine |
| **4** | © Joe Tucciarone; Davis Meltzer, National Geographic Society Image Collection |
| **5** | AP/Wide World; © SIU from Photo Researchers |
| **10** | North Carolina Zoo; Zoo Atlanta |
| **11** | © Wayne R. Bilenduke, Tony Stone Images; Jonathan Blair, National Geographic Society Image Collection |
| **12** | © The Quest Group Ltd, from Woods Hole Oceanographic Institution |
| **15** | Priit Veslind, National Geographic Society Image Collection |
| **16** | © Jason Foundation for Education and the City of Hamilton; Paul Perreault* |
| **18** | Paul Perreault* |
| **19** | Jonathan Blair, National Geographic Society Image Collection; Jason Foundation for Education and the City of Hamilton; © Polaris Imaging Inc. |
| **22** | Thomas Abercrombie, National Geographic Society Image Collection; © Odyssey Marine Exploration |
| **23** | Emory Kristof © Hamilton-Scourge Foundation |
| **25** | © RMS Titanic, Inc. |
| **26** | Black Sea Trade Project of the University of PA Museum of Archeology and Anthropology |
| **28** | © Franz Kohlhauf |
| **33** | WORLD BOOK photos by Herbert Herrick |
| **35** | NASA |
| **37** | Alastair Cameron, Harvard University |
| **38** | © Joe Tucciarone |
| **39** | William K. Hartmann |
| **42** | Corbis/Digital Stock |
| **51** | National Weather Service; National Center for Atmospheric Research |
| **52** | Rockwell Collins |
| **53** | NOAA Forecast Systems Laboratory; National Center for Atmospheric Research |
| **55** | Federal Aviation Administration |
| **56** | AP/Wide World |
| **59** | Tim Hayward, Bernard Thornton Artists* |
| **60** | WORLD BOOK illustration by Alex Ebel |
| **62-65** | Tim Hayward, Bernard Thornton Artists* |
| **66** | O. Lewis Mazzatenta, National Geographic Society Image Collection |
| **67** | O. Lewis Mazzatenta, National Geographic Society Image Collection; Jon Blumb* |
| **68** | Tim Hayward, Bernard Thornton Artists*; © Mark Hallett. |
| **70** | © Chuck Carlton, Index Stock |
| **77** | © Hulton Getty from Liaison Agency; Reuters/Archive Photos; © Art Stein, Photo Researchers |
| **78** | Organogenesis Inc.; AP/Wide World |
| **79** | Joseph P. Vacanti, M.D.; © Brooks Kraft, Sygma |
| **80** | © Roberto Osti |
| **81** | David J. Mooney, University of Michigan |
| **82** | © Breck P. Kent, Animals Animals |
| **84** | North Carolina Zoo; © Disney Enterprises, Inc. |
| **86** | Minnesota Zoo |
| **87** | Corbis; Corbis; Zoological Society of London |
| **88** | Zoo Atlanta |
| **89-92** | North Carolina Zoo |
| **93** | © Michael Nichols, Magnum; Brookfield Zoo/Chicago Zoological Society; Brookfield Zoo/Chicago Zoological Society |
| **94** | Minnesota Zoo; North Carolina Zoo |
| **95** | © Karen Kasmauski, Woodfin Camp, Inc. |
| **96** | Arthur W. Clark |
| **97** | North Carolina Zoo |
| **98** | Iridium L.L.C. |
| **99** | Teledesic L.L.C. |
| **102** | Intelsat; NASA |
| **103** | Iridium L.L.C. |
| **110** | © Wayne R. Bilenduke, Tony Stone Images |
| **112** | *Milady's Boudoir* by Edmund Emshwiller © Carol Emshwiller (Collection of Phyllis and Alex Eisenstein); Culver |
| **113** | Granger Collection; Seth Joel*; SST/NASA/R. Williams & the Hubble Deep Field Team |
| **114** | Culver |
| **115** | From Wen Wu magazine 1975 |
| **118-119** | Granger Collection |
| **120** | *Portrait of Antoine Lavoisier and Wife* (1788) oil on canvas by Jacques Louis David; Rockefeller Institute for Medical Research; Granger Collection; Granger Collection |
| **121** | Granger Collection |
| **122** | Corbis/Bettmann |
| **123** | Corbis/Hulton |
| **124** | Granger Collection; Culver |
| **125** | © Bar-Brown, Camera Press Ltd. |
| **126** | NASA; AP/Wide World |
| **127** | © Jonathan Levine, Liaison Agency |
| **130** | *Dome World* by Edmund Emshwiller © Carol Emshwiller (From the Collection of Phyllis and Alex Eisenstein) |
| **132** | David Meltzer, National Geographic Society Image Collection |
| **133** | © Warner Bros. |
| **134** | Jackson-Zender* |
| **135** | *Hostile Reception on Aldebaran IV* by Edmund Emshwiller © Carol Emshwiller (From the Collection of Phyllis and Alex Eisenstein) |
| **136** | Syd Mead (From the Collection of the U.S. Steel Corporation); George Jones* |
| **138** | Photofest |
| **139** | *Milady's Boudoir* by Edmund Emshwiller © Carol Emshwiller (From the Collection of Phyllis and Alex Eisenstein) |
| **140** | *Powder Keg* by Edmund Emshwiller © Carol Emshwiller (From the Collection of Phyllis and Alex Eisenstein); Paul Alexander* |
| **141** | Paul Alexander* |
| **142-147** | Seth Joel* |
| **148-149** | From the Collection of Gregory Benford |
| **151-153** | Seth Joel* |
| **156** | © Michel Tcherevkoff; © Lennart Nilsson/Albert Bonniers Forlag AB, *Behold Man*, Little, Brown & Co.; Sauabh Jha, Peter Garnavich, Peter Challis & Robert P. Kirshner, Harvard University |
| **157** | © Ellen Dooley, Liaison Agency; © Mehau Kulyk/SPL from Photo Researchers |
| **158** | © Michel Tcherevkoff |
| **160** | © Peter Gonzales |
| **161** | SST/NASA/R. Williams & the Hubble Deep Field Team |
| **162** | Barbara Cousins* |
| **163** | Sauabh Jha, Peter Garnavich, Peter Challis & Robert P. Kirshner, Harvard University |
| **164** | © Lennart Nilsson/Albert Bonniers Forlag AB, *Behold Man*, Little, Brown & Co. |
| **165** | Barbara Cousins* |
| **167** | © Mehau Kulyk/SPL from Photo Researchers |
| **168** | Barbara Cousins* |
| **169** | Stuart Hameroff and Dave Cantrell, University of Arizona |
| **170** | © Bobby Neel Adams |
| **172** | Barbara Cousins* |
| **173** | © Justin Lane, Liaison Agency |
| **175** | J. D. Griggs, U.S. Geological Survey; © Ellen Dooley, Liaison Agency; Mark Gabbana* |
| **176** | Linda Degenstein and Chuck Wellek; Reuters/Archive Photos |
| **177** | Charles Piskoti; © Anton Pauw; Reuters/Archive Photos |
| **179** | CSIRO Australia |
| **180** | © *New Scientist*; Kenneth Jarrett, National Geographic Society Image Collection |
| **181** | Reuters/Archive Photos; AP/Wide World |
| **182** | Reconstruction by Sharon Long, David Hunt & Doug Owsley, Nevada State Museum/NMNH/Smithsonian Institution |

World Book Encyclopedia, Inc., provides high-quality educational and reference products for the family and school. They include THE WORLD BOOK MEDICAL ENCYCLOPEDIA, a 1,040-page fully illustrated family health reference; THE WORLD BOOK OF MATH POWER, a two-volume set that helps students and adults build math skills; and THE WORLD BOOK OF WORD POWER, a two-volume set that is designed to help your entire family write and speak more successfully. For further information, write WORLD BOOK ENCYCLOPEDIA, INC., 525 W. Monroe St., Chicago IL 60661.